METHODS IN
MICROBIOLOGY

ADVISORY BOARD

METHODS IN MICROBIOLOGY

Volume 18

Edited by

GERHARD GOTTSCHALK

Institute for Microbiology
University of Göttingen
Göttingen, Federal Republic of Germany

1985

ACADEMIC PRESS

(Harcourt Brace Jovanovich, Publishers)
London   Orlando   San Diego   New York
Toronto   Montreal   Sydney   Tokyo

ACADEMIC PRESS INC. (LONDON) LTD.
24–28 Oval Road
LONDON NW1 7DX

United States Edition published by
ACADEMIC PRESS, INC.
Orlando, Florida 32887

LIBRARY OF CONGRESS CATALOG CARD NUMBER: 68–57745

ISBN 0–12–521518–5

PRINTED IN THE UNITED STATES OF AMERICA

85 86 87 88   9 8 7 6 5 4 3 2 1

CONTENTS

CONTRIBUTORS

Arthur G. Andrewes Chemistry Department, Saginaw Valley State College, University Center, Michigan 48710, U.S.A.

August Böck Institut für Genetik und Mikrobiologie, Universität München, 8000 Munich 19, Federal Republic of Germany

M. D. Collins Department of Microbiology, National Institute for Research in Dairying, Shinfield, Reading RG2 9AT, England

G. E. Fox Department of Biochemical and Biophysical Sciences, University of Houston, University Park, Houston, Texas 77004, U.S.A.

John L. Johnson Department of Anaerobic Microbiology, Virginia Polytechnic Institute and State University, Blacksburg, Virginia 24061, U.S.A.

C. W. Jones Department of Biochemistry, University of Leicester, Leicester LE1 7RH, England

Synnøve Liaaen-Jensen Organic Chemistry Laboratories, Norwegian Institute of Technology, University of Trondheim, N-7034 Trondheim-NTH, Norway

W. Ludwig Institut Botanik und Mikrobiologie, Technische Universität München, D-8000 Munich, Federal Republic of Germany

H. Mayer Max-Planck-Institut für Immunbiologie, Freiburg i. Br., Federal Republic of Germany

J. Oelze Institut für Biologie II, Mikrobiologie, der Universität, Freiburg i. Br., D-7800 Freiburg, Federal Republic of Germany

R. K. Poole Department of Microbiology, Queen Elizabeth College, University of London, London W8 7AH, England

Karl Heinz Schleifer Technische Universität München, D-8000 Munich 2, Federal Republic of Germany

E. Stackebrandt[1] Institut für Botanik und Mikrobiologie, Technische Universität München, D-8000 Munich, Federal Republic of Germany

R. N. Tharanathan[2] Discipline of Biochemistry, C. F. T. R. I., Mysore 570013, India

Thomas G. Tornabene School of Applied Biology, Georgia Institute of Technology, Atlanta, Georgia 30332, U.S.A.

J. Weckesser Institut für Biologie II, Mikrobiologie, der Universität, Freiburg i. Br., D-7800 Freiburg, Federal Republic of Germany

[1] Present address: Institut für Allgemeine Mikrobiologie, Universität Kiel, D-23 Kiel, Federal Republic of Germany.

[2] Present address: Max-Planck-Institut für Immunbiologie, Freiburg i. Br., Federal Republic of Germany.

PREFACE

Volume 18 of "Methods in Microbiology" is concerned with components of bacterial cells that are of taxonomic interest. The techniques used for the analysis of some of these components are already widely applied and belong to the repertoire of many microbiological laboratories. This holds for the determination of the G + C content of bacterial DNA and of DNA–RNA homologies. Nevertheless, an up-to-date presentation of the appropriate methods seemed necessary. It is included in this volume together with procedures for the analysis of cellular components that are applied mainly in specialist laboratories. These methods deal with the analysis of 16 S ribosomal RNA, murein, lipopolysaccharides, ribosomal proteins, lipids, cytochromes, quinones, carotenoids, and bacteriochlorophylls. It is hoped that this volume will contribute to a wider application of these procedures and to a wider consideration of the taxonomic relevance of the components analyzed.

The editor wishes to express his gratitude first of all to the authors, who did an excellent job and who delivered their manuscripts, more or less, on time. Thanks are also due to the staff of Academic Press Inc. in London and to the Advisory Board Chairman, John R. Norris, who did everything to assure the rapid publication of this volume.

Gerhard Gottschalk

1

Determination of DNA Base Composition

JOHN L. JOHNSON

*Department of Anaerobic Microbiology, Virginia Polytechnic Institute and
State University, Blacksburg, Virginia, USA*

I. Introduction

The deoxyribonucleic acid (DNA) of most organisms contains the purine bases adenine (A) and guanine (G) and the pyrimidine bases thymine (T) and cytosine (C). Occasionally an organism may contain modified bases such as 5-methylcytosine or hydroxymethylcytosine. In most cases the modified base substitutes for only a small fraction of the parent base. The DNA in prokaryotic and eukaryotic organisms and in many of the viruses consists of double strands with specific base pairs (A + T) or (G + C) located at each position along the complementary strands. Since the recognition that the ratio of A + T to G + C base pairs may differ from one organism to the next and that the ratio in a given organism, usually expressed as mole percentage G + C (mol% G + C), is relatively con-stant, the mol% G + C values have been used for the comparative charac-terisation of organisms.

The chemical and physical properties of DNA are such that the base composition can be measured in several ways. Initially DNA preparations were hydrolysed and the bases separated by paper chromatography and then eluted from the paper and measured spectrophotometrically (Chargaff, 1955). Since then several physical and spectrophotometric properties of the DNA or DNA components have been correlated with the chromatographic data. This has resulted in several indirect methods for

1

METHODS IN MICROBIOLOGY
VOLUME 18

estimating mol% G + C values. It is the purpose of this chapter to review the methods presently available and provide protocols that incorporate recent advances in nucleic acid chemistry.

II. Methodology

The determination of the mol% G + C values of DNA can be separated into two parts, the isolation of the DNA and the determination of base composition.

A. DNA isolation

The isolation of DNA is the most time-consuming part of the mol% G + C analysis, and is the part most fraught with technical problems. Some of the procedures, such as CsCl buoyant density centrifugation, work best with high molecular weight DNA, and others, for example, base or nucleotide chromatography and spectrophotometric analysis, require DNA free from ribonucleic acid (RNA). In many instances, the mol% G + C analyses are done in conjunction with DNA homology studies (Chapter 2); therefore, most of the DNA isolation procedures will be covered in that chapter. Several procedures have been described for the rapid isolation of DNA from cells from medium-size culture volumes (Britten *et al.,* 1970; Cashion *et al.,* 1977; Gibson and Ogden, 1979; and Zadrazil *et al.,* 1973). Presented in this chapter is a rapid isolation procedure for the isolation of small amounts of RNA-free DNA (50–100 μg) suitable for an analysis.

1. Buffers and reagents

> *Sodium chloride–EDTA, pH 8.0:* 0.15 *M* NaCl, 0.01 *M* ethylenediaminetetraacetic acid (EDTA).
> *Sodium dodecylsulphate (SDS):* 20% (w/v) solution.
> *Sodium perchlorate:* 5 *M* solution.
> *Phenol–chloroform:* 1 : 1 (v/v) mixture of chromatography-grade liquid phenol equilibrated against the NaCl–EDTA salt solution and chloroform–isopentanol (24 : 1 v/v) to which is added 0.1% 8-hydroxyquinoline.
> *Tris–EDTA (TE) buffer, pH 7.2:* 10 m*M* tris(hydroxymethyl)aminomethane (Tris), 1 m*M* EDTA.
> *TE-0.5 M NaCl:* 0.5 *M* NaCl in TE buffer.

TE-1.0 M NaCl: 1.0 *M* NaCl in TE buffer.

TE-2.0 M NaCl: 2.0 *M* NaCl in TE buffer.

80% ethanol: 95% ethanol–water (24 : 1 v/v).

Proteinase K: prepare a 5-mg ml^{-1} solution in TE buffer just prior to use.

Ribonuclease (RNase): bovine pancreatic RNase, 0.5 mg ml^{-1} in 0.15 *M* NaCl, pH 5.0. Heat the solution at 80°C for 10 min to inactivate any traces of deoxyribonucleases (DNase). Store at −20°C. Just prior to use add T$_1$ RNase to the pancreatic RNase preparation to a concentration of 500 units ml^{-1}.

NACS-52: This is a gravity flow chromatography matrix which fractionates primarily by ion-exchange mechanisms. Bound nucleic acids are eluted from columns predominantly in order of increasing molecular weight. Prepare for use according the manufacturer's manual (Bethesda Research Laboratories, Inc.).

2. Protocol

1. Grow the cells in 20–40 ml of a suitable medium to the late log or early stationary phase of growth. Centrifuge the cells and resuspend them in 5.0 ml NaCl–EDTA salt solution. Using a French pressure cell, disrupt the cells into a flask containing another 5.0 ml of the salt solution, 0.5 ml of 20% SDS, and 2.5 ml of 5 *M* perchlorate. Add 0.15 ml of proteinase K solution and incubate for 1 h at 60°C.

2. Add 5.0 ml of phenol–chloroform and shake the flask for 20 min on a wrist-action shaker. Centrifuge the emulsion at 17,000 *g* for 10 min. Remove the aqueous supernatant to a second tube and add 2 volumes of cold 95% ethanol, mix, store in a −20°C freezer for 1 h, and again centrifuge as above. Pour off the ethanol, allow the tube to drain inverted on a paper towel for a few minutes, and then carefully add 10 ml of 80% ethanol (−20°C) so as not to disrupt the nucleic acid pellet. Rotate the tube so that the 80% ethanol will make contact with all of the surfaces of the tube. Place the tube in the −20°C freezer for 0.5–1 h and recentrifuge with the tube being placed in the same orientation as during the first centrifugation so that the pellet will not be dislodged. Pour off the ethanol and allow the tube to drain well and partially air dry.

3. Dissolve the nucleic acids in 2.0 ml of TE buffer and add 0.3 ml of the RNase mixture, mix and incubate at 37°C for 0.5 h. Add 0.5 ml of chloroform–isopentanol. Briefly mix two or three times on a vortex mixer and centrifuge at 17,000 *g* for 100 min.

4. Transfer the aqueous supernatant to another tube, add 0.2 ml of 3

M sodium acetate, mix, add 2 volumes of 95% ethanol, and mix again. Place the tube in the $-20°C$ freezer for 1 h, centrifuge at 17,000 g for 10 min, and wash the pellet with 80% ethanol as outlined above. Air dry the pellet; a 37°C incubator works well.

5. Dissolve the pellet in 1.0 ml of TE. After the DNA is in solution, add 1.0 ml of TE-1.0 M NaCl.

6. Prepare a NACS-52 column, 0.5–0.75 ml bed volume in a Pasteur pipette or other small column, as recommended by the manufacturer (Bethesda Research Laboratories, Inc.). The NACS-52 used to pack the column must first be hydrated with TE-2.0 M NaCl buffer. Wash the column with several column equivalents of TE-0.5 M NaCl. Load the sample by gravity flow and pass it through a couple of times to ensure maximum binding of the DNA. The RNA fragments should not bind and will be eluted by running an additional 5.0–6.0 ml of the TE-0.5 M NaCl buffer through the column. The last of the buffer is removed from the column by gentle air pressure (a small rubber bulb works well for this) and the DNA eluted by sequentially adding 3 0.2-ml volumes of TE-1.0 M NaCl buffer. Force the buffer out between each addition with gentle air pressure.

7. Add 2 volumes of 95% ethanol to the eluate, mix, then place in the $-20°C$ freezer and proceed with the centrifugation and 80% ethanol wash as outlined above. Dissolve the pellet in 0.5 ml of distilled water, and 50 μl of 3.0 M sodium acetate, mix, and add 1.0 ml of 95% ethanol for a final precipitation and pellet wash. After air drying the pellet, dissolve the DNA in 0.5 ml of distilled water. DNA isolated in this manner is ready for use in the optical and chromatographic methods of mol% G + C analysis.

3. Comments

1. This procedure works well for bacteria that are readily disrupted by passage through a French pressure cell; it may not work as well for Gram-positive cocci or small rods. Because of the small fragment size of DNA prepared in this manner, it will not work very well for buoyant density centrifugation.

2. The NACS-52 column step probably would not be needed for the T_m method.

3. Many of the procedures described below require DNA preparation free of RNA. The presence of RNA contamination can be demonstrated and an estimation of the amount obtained by electrophoresing DNA preparations in polyacrylamide gels (partially degraded RNA will migrate near the dye front), and then staining the gel with a silver stain (Beidler *et al.*, 1982; Igloi, 1983).

B. Methods for measuring DNA base compositions

Unique physical, chemical, and optical properties of DNA enable one to determine mol% G + C values in several different ways. Initially the bases were released from the DNA by acid hydrolysis, separated by paper chromatography, quantitatively eluted from the paper, and measured spectrophotometrically (Chargaff, 1955; Wyatt, 1955; Werner *et al.,* 1966). Although seldom, if ever, used at the present time, it is interesting to note that most other procedures for estimating mol% G + C values are based upon correlations with results obtained by quantitative paper chromatography.

1. Buoyant density centrifugation

The density of DNA in CsCl increases linearly with the mol% G + C. Schildkraut *et al.* (1962) were the first to publish an equation correlating the density of DNA preparations with mol% G + C values determined chemically. Subsequently, other investigators have carried out similar investigations (De Ley, 1970; Gasser and Mandel, 1968). Shown in Fig. 1 are the regression lines from the various studies. The slopes of the lines

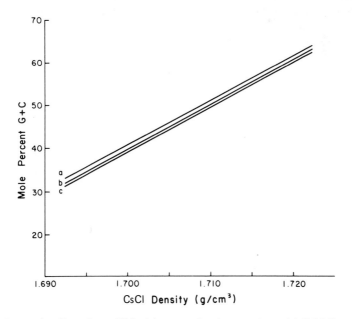

Fig. 1. Regression lines for published buoyant density equations: (a) Schildkraut *et al.* (1962); (b) De Ley (1970); and (c) Gasser and Mandel (1968).

are essentially the same although the constants do vary a little, such that the differences in mol% G + C using the different equations range from 1 to 2%.

The buoyant densities are best measured using high-molecular-weight DNA, since the width of the bands is a function of the fragment length. Advantages of the procedure are that only small amounts (a few micrograms) of DNA are required and it is not overly sensitive to RNA and protein contamination, although tenacious binding of pigments to DNA has caused density alterations (Enquist and Bradley, 1971). Disadvantages of the procedure are that an analytical ultracentrifuge is required and the typical centrifugation run is 44 h.

For a detailed description of the procedure the reader is referred to Mandel *et al.* (1968).

2. Thermal melting profile (T_m)

When a preparation of DNA is heated to high temperature, the bonding between the base pairs collapses and the two DNA strands separate (denature). This denaturation phenomenon can be measured readily with a UV spectrophotometer. As the DNA goes from the native state to the denatured, the absorbance at 260 nm will increase by about 40%. Although the melting curves appear as smooth transitions when the rate of temperature increase is in the range of 0.5–1.0°C min^{-1}, when high-resolution melting curves (temperature increase in the range of 0.05°C min^{-1}) are analysed as the derivative curves, the melting profile appears as a collection of subtransitions (thermalites) which have an average length of about 900 bp (Ansevin *et al.*, 1976; Vizard and Ansevin, 1976). This chapter will be limited to melting profiles obtained by rates of temperature increase in the 0.5–1.0°C range.

The midpoint temperatures of the thermal melting profiles (T_m) of DNA increase with increases in the mol% G + C. These values were first correlated with the chromatographically determined base composition of DNA by Marmur and Doty (1962). Although the A + T and G + C polymers did not fit on the linear regression line, there was a linear correlation for DNA preparations ranging from about 25 to 75 mol% G + C. De Ley (1970) and Mandel *et al.* (1970) have re-examined this correlation and have also correlated the T_m values with the CsCl buoyant density values. Regression lines for equations are shown in Fig. 2. Also included in Fig. 2 is the regression line for the equation by Owen *et al.* (1969) which includes the Na$^+$ molarity of the salt solution. The equations of De Ley (1970) and Marmur and Doty (1962) are nearly identical as is the slope of

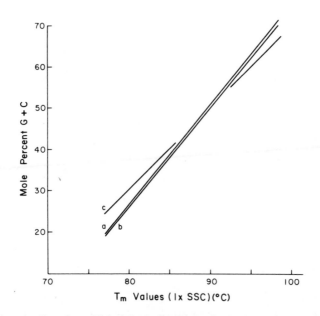

Fig. 2. Regression lines for published thermal melting point (T_m) equations: (a) Marmur and Doty (1962) and De Ley (1970); (b) Owen *et al.* (1969); and (c) Mandel *et al.* (1970).

the Owen *et al.* (1969) equation. The Mandel *et al.* (1970) equation has less of a slope, and as a result, when using this equation, mol% G + C values will be higher than with the other equations for mol% G + C DNA preparations that are less than 51 and lower for DNA preparations greater than 51 (when using *Escherichia coli* DNA as the standard).

The midpoint of the hyperchromic shift has been established by several different methods, including a graphic determination (Marmur and Doty, 1962), and the use of normal probability paper (Knittle *et al.*, 1968). Ferragut and Leclerc (1976) compared four methods and concluded that the graphic determination of T_m was the method of choice for routine analysis. This method is illustrated in Fig. 3.

The melting profile curve is generated by gradually heating a DNA sample in the cuvette chamber of an ultraviolet spectrophotometer and continuously measuring the absorbance of the sample(s) and the temperature within the cuvettes or cuvette chamber. Historically, the temperature of one of the cuvettes was measured using a mercury thermometer and the temperature was increased stepwise using an external temperature bath. Many of the newer spectrophotometers have electronically heated cuvette holders with thermistors embedded within them and are either microprocessor or computer controlled. When performing melting profiles, the two major variables about which an investigator must be concerned

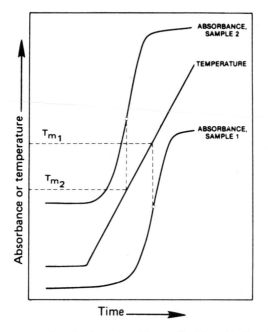

Fig. 3. Graphic representation of a thermal melting profile. The midpoint of the profile (T_m) is used for comparison with other T_m values. From Johnson (1981).

are the ionic strength of the buffer and the accuracy of the temperature. The ionic strength of the buffer has a large effect on the T_m of a DNA preparation (Schildkraut and Lifson, 1965). The ionic strength of the buffer can be standardised for the experiment by dialysing all of the DNA samples against the same batch of buffer, and the temperature is best standardised by including a DNA standard in each spectrophotometer run (*E. coli* b or K-12 is usually used).

(*a*) *Buffers and materials.* 1. *0.5× standard saline citrate (SSC):* Standard saline citrate is 0.15 *M* NaCl, 0.015 *M* sodium citrate, pH 7.0, and other concentrations of it are indicated by a number such as 20× (20-fold concentration) or 0.1× (one-tenth the concentration). For most DNA preparations, the T_m values in 0.5× SSC are low enough to ensure that the hyperchromic shift is complete at a temperature of less than 100°C.

2. Wash pieces of dialysis tubing ($\frac{1}{2}$ × 51 in.) by placing them into a 1–2% solution of Na_2CO_3 and boiling for about 5 min. Rinse the tubing extensively with tap water and then place into a beaker of distilled water.

(*b*) *Protocol.* 1. Prepare 2–4 litres of 0.5× SSC, saving some of it for diluting the DNA preparations and for rinsing out cuvettes just prior to putting the DNA samples into them. Divide the rest so the buffer can be changed once during dialysis.

2. Prepare 2- to 5-ml samples (depending on the size of the cuvettes to be used) of DNA at 50 μg ml^{-1}, using 0.5× SSC to dilute the stock DNA preparations. Prepare a 10- to 20-ml volume of the standard DNA (e.g., the DNA of *E. coli* b, which has a G + C content of 51 mol% and a T_m of 90.5°C in SSC), because it will be used in each instrument run.

3. Dialyse all of the preparations overnight in the 0.5× SSC at 4°C. Change the buffer once after the first few hours of dialysis. After dialysis, return the DNA preparations to screw-cap tubes.

4. Determine the melting profiles with an automatic recording spectrophotometer having a sample chamber that is heated by a circulating bath containing ethylene glycol, or preferably one that has an electronically heated cuvette holder. A thermistor should be located in the cuvette chamber, the cuvette holder, or in the cuvette. Having the thermistor within the cuvette is theoretically the best, but it is very convenient when it is located in the cuvette holder and there appears to be very little difference in accuracy. The temperature variations within an electronically heated cuvette holder can be determined by placing samples of the same DNA preparation into each cuvette position and comparing the resultant T_m measurements. Even when the thermistor is located in the chamber, and the chamber is heated by passing ethylene glycol through the chamber walls, cuvettes holding about 1.0 ml of sample may be heated linearly at rates greater than 1.0°C min^{-1}, although the actual temperatures do lag behind those of the circulating fluid. It is important to make two or three measurements on each DNA preparation for improved accuracy (De Ley, 1970). Neutralise any temperature variations that there may be from one cuvette position to the next by running the samples at different cuvette positions.

5. Start at 60°C and determine the T_m values as shown in Fig. 3.

6. Calculate the mol% G + C of each sample DNA by using the following Marmur and Doty equation (1962), which has been modified for normalising the T_m values in 0.5× SSC to T_m values in SSC. This general procedure can also be used for other ionic strength buffers:

$$\text{Mol\% G + C} = \frac{[(A - B) + C] - 69.3}{0.41}$$

where $A = T_m$ of the DNA standard in SSC (for *E. coli*, 90.5); $B = T_m$ of the DNA standard in 0.5× SSC; and $C = T_m$ of the test DNA in 0.5× SSC.

The $A - B$ value is the difference between the T_m of the DNA standard in SSC and 0.5× SSC. Adding this value to the T_m of the test DNA normalises it to a T_m in SSC. The rest of the values are from the Marmur equation. Extrapolating to 100 mol%, A + T intercepts the temperature axis at 69.3°C and the slope of the line is 0.41.

(c) *Comments.* These analyses are usually determined using cuvettes that are from 0.3 to 3.0 ml in capacity. These cuvettes are satisfactory when plenty of DNA is available; however, for DNAs that are difficult to isolate in large quantities, such as plasmid, phage, or a recombinant DNA fragment, there would be an advantage in using smaller cuvettes. To this end, microcuvettes have been constructed so that less than 1 μg of DNA is needed for an analysis (Ansevin and Vizard, 1979; Karsten *et al.*, 1980).

This procedure is not particularly sensitive to RNA contamination. If there is still secondary structure in the RNA fragments, the base line will gradually go up before the DNA hyperchromic shift begins. The DNA fragment size does affect the T_m. The T_m of high-molecular-weight DNA, e.g., isolated by the Marmur procedure (1961), is 1–2°C higher than the same DNA preparation after it has been passed through a French pressure cell. Therefore, all of the DNA preparations being analysed, including the standard, should be prepared in the same manner. Also it should be noted, that the fragment sizes of several DNA preparations may not be of similar lengths, even if they are prepared in the same way. Nuclease activity during DNA isolation can cause this type of problem. We have found this to be most pronounced among low mol% G + C organisms, for example, clostridia and fusobacteria (Selin *et al.*, 1983).

3. High-performance liquid chromatography (HPLC)

As previously noted, all of our present indirect methods of estimating G + C content are based on the acid hydrolysis of DNA and quantitative paper chromatography of the bases. However, because of the extensive work involved, few of these chromatographic analyses have been done since the early 1960s. Recent advances in HPLC instrumentation and columns now enable investigators to make direct determinations of the nucleic acid bases, deoxyribonucleosides, and deoxyribonucleotides rapidly and easily. A schematic representation of HPLC instrumentation is shown in Fig. 4. The components include (1) a solvent reservoir for the mobile phase; (2) a pump to move the solvent at high pressure through the column; (3) a pressure gauge to measure pump pressure; (4) an injection device, usually a sample loop; (5) a column; (6) a detector (for nucleic acid components, a UV photometer); and (7) a potentiometric recorder for producing a written record of the analysis, the chromatogram. The instrument can be microprocessor controlled to generate solvent gradients automatically, to

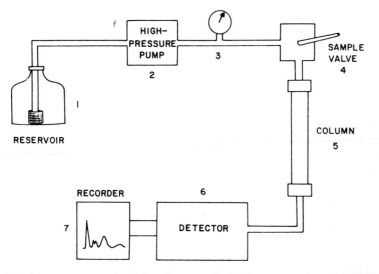

Fig. 4. Major components of a high-performance liquid chromatography system (HPLC). Reprinted from McNair (1980). © 1980 International Scientific Communications, Inc.

inject samples, and to quantitate the components by integration of the chromatograph peaks (McNair, 1980).

The UV detectors or monitors may be of a fixed wavelength (usually 254 nm), dual wavelength (254 and 280 nm), or variable wavelength (from about 190 to 300 nm). A photodiode array detector has recently been described which consists of 211 diodes covering a spectral range from 190 to 600 nm. This detector is able to take a spectrum in 10 ms (Elgass *et al.*, 1983). In addition to providing accurate analysis, it will allow for determining peak purity and peak identification. Other types of detectors that have been employed include fluorescence, conductivity, mass spectrometer, IR absorbance, refractive index (Wheals, 1982), voltametric, and polarographic (Štulík and Pacáková, 1983).

The three major types of HPLC columns used for the analysis of bases, nucleosides, and nucleotides are reversed phase, ion pairing, and ion exchange, either cation or anion (Zakaria and Brown, 1981). In reversed-phase chromatography (Caronia *et al.*, 1983; Ehrlich *et al.*, 1982; Gehrke *et al.*, 1978; Gehrke *et al.*, 1980; Hartwick *et al.*, 1979; Kraak *et al.*, 1981; Kuo *et al.*, 1980; Whitehouse and Greenstock, 1982; and Zumwalt *et al.*, 1982), which is used to separate both polar and non-polar compounds, hydrophobic interactions determine the extent of retention. A hydrophobic portion of the sample partitions into the hydrophobic surface of a chemically bonded material. The more polar or ionic solutes favour the

aqueous eluent and elute faster. In ion-pairing chromatography, compounds are added to the mobile phase which contain both a lipophilic
moiety that can interact with the non-polar reversed-phase stationary
phase and an ionic moiety that can pair with ionic compounds of an
opposite charge. This allows for greater retention of charged solutes on a
reversed-phase column. In ion-exchange chromatography, the stationary
phase contains fixed ions, anions for cation exchange, and cations for
anion exchange (Cashion *et al.*, 1977; Eksteen *et al.*, 1978; Floridi *et al.*,
1977; Ko *et al.*, 1977; and Van Boom and De Rooy, 1977). These interact
electrostatically with the ionic solutes which are thus retained. Some
examples of separations by different types of columns are shown in the
chromatograms in Figs. 5–10.

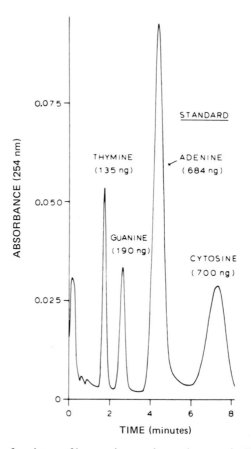

Fig. 5. Separation of a mixture of bases using a cation-exchange resin. Zipax-SCX stationary phase and 0.01 M NH$_4$H$_2$PO$_4$ (pH 5.56) as the mobile phase. From Ko *et al.* (1977).

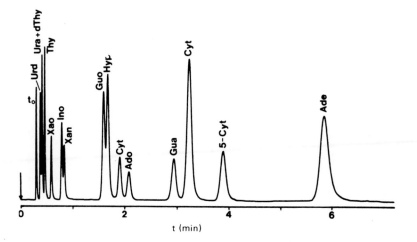

Fig. 6. Separation of a mixture of bases and nucleosides using a dynamic cation-exchange system with sodium dodecylsulphate (SDS) as an ion-pairing agent. Hypersil ODS stationary phase and 0.1 M HClO$_4$–ethanol (9:1, v/v) + 0.1% SDS mobile phase. Uracil (Ura), cytosine (Cyt), thymine (Thy), adenine (Ade), guanine (Gua), hypoxanthine (Hyp), xanthine (Xan), uridine (Urd), cytidine (Cyd), adenosine (Ado), guanosine (Guo), inosine (Ino), xanthosine (Xao), thymidine (dThy), and 5-methylcytosine (5-Cyt). From Kraak *et al.* (1981), *Journal of Chromatography* **209**, 369–376. © 1981 Elsevier Science Publishers, Amsterdam.

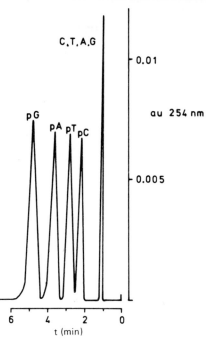

Fig. 7. Separation of four deoxyribonucleoside 5'-phosphates using a strong anion-exchange resin. Permaphase AAX stationary phase and 0.005 M KH$_2$PO$_4$ (pH 4.5) mobile phase. Standards: *d*-adenosine (A), *d*-cytosine (C), *d*-guanosine (G), *d*-adenosine 5'-phosphate (pA), *d*-cytosine 5'-phosphate (pC), *d*-guanosine 5'-phosphate (pG), and thymidine (pT). From Van Boom and De Rooy (1977).

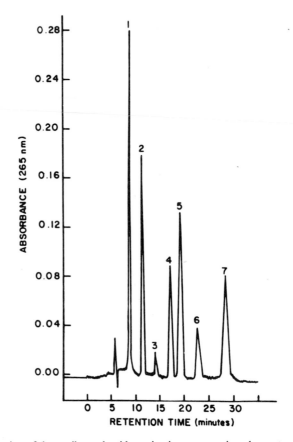

Fig. 8. Separation of deoxyribonucleotides using ion-suppression chromatography on a reversed-phase column. LiChrosorb RP-8 stationary phase and 0.6 M NH$_4$H$_2$PO$_4$ (pH 4.25) mobile phase. Standards: *d*-cytosine 5′-phosphate (1), *d*-uracil 5′-phosphate (2), impurity (3), *d*-guanosine 5′-phosphate (4), thymidine (5), *d*-bromouracil 5′-phosphate (6), and *d*-adenosine 5′-phosphate (7). Reprinted from Whitehouse and Greenstock (1982), p. 2090, by courtesy of Marcel Dekker, Inc.

Listed in Table I are the molecular weights, absorbance maxima, and molar extinction coefficients at the absorbance maxima (ε_{max}) for the major bases, deoxyribonucleosides, and deoxyribonucleotides that have been extracted from an extensive table compiled by Dunn and Hall (1975). Also included in Table I are 260 nm extinction coefficients (ε_{260}) from the Schwarz/Mann Biochemical Catalogue 1970/71. The relationship between the absorbance at x nm and the extinction coefficient at x nm (ε_x), concen-

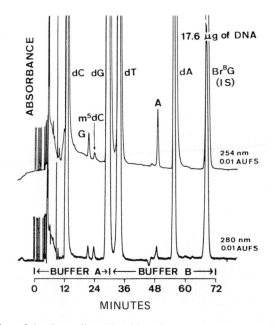

Fig. 9. Separation of the deoxyribonucleosides of *Xanthomonas oryzae* DNA using a reversed-phase column. Stationary phase, bondapak C_{18}, and mobile phase consisting of a two-step gradient. Buffer A: 0.01 M $NH_4H_2PO_4$, pH 5.3, containing 2.5% (v/v) methanol; buffer B: 0.01 M $NH_4H_2PO_4$, pH 5.1, containing 8% (v/v) methanol. Components: *d*-adenosine (dA), *d*-cytosine (dC), 5-methyl-*d*-cytosine (m⁵dC), *d*-guanosine (dG), thymidine (dT), guanine (G), adenine (A), and internal standard 8-bromoguanosine (Br⁸G). From Kuo *et al.* (1980).

tration in moles per litre (c) and distance of light path through the sample in centimetres (b) is shown below:

$$\text{absorbance}_x = \varepsilon_x c b$$

Although one should be able to measure concentrations directly from the absorbancies, instrument calibration differences, variation in light path distances, and variations in the liquid chromatography, buffer pH, and ionic strength make this difficult. The coefficients are best used for checking standard solutions and if one is going to use them in the calculations based on peak areas, one must determine them in the buffer that is being used for the chromatographic separation.

It is beyond the scope of this chapter to discuss the actual operation of the HPLC instruments and columns, since that information is best supplied by the various suppliers. What will be discussed is the preparation of

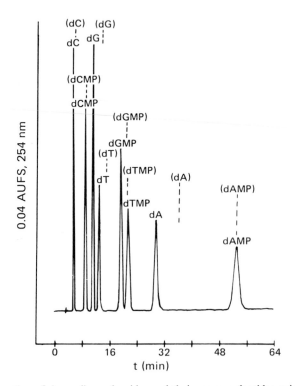

Fig. 10. Separation of deoxyribonucleosides and their mononucleotides using a reversed-phase column and tetrabutylammonium (TBA) hetaerons. Whatman Partisil 5 C8 stationary phase and 0.05 M phosphate buffer (pH 4.8), 2.0 mM TBA, and 7% methanol (v/v) mobile phase. Standards: d-adenosine (dA), thymidine (dT), d-cytosine (dC), d-guanosine (dG), d-adenosine 5'-phosphate (dAMP), thymidine 5'-phosphate (dTMP), d-cytosine 5'-phosphate (dCMP), and d-guanosine 5'-phosphate (dGMP). The dashed lines represent predicted retention times. Reprinted from Caronia *et al.* (1983) by courtesy of Marcel Dekker, Inc.

samples for HPLC analysis. The free bases are prepared by acid hydrolysis and the deoxyribonucleosides and deoxyribonucleotides are prepared by enzymatic hydrolysis.

(*a*) *Free bases.* Perchloric, hydrochloric, or formic acids have usually been used to hydrolyse DNA (Wyatt, 1955). Since formic acid causes the least destruction of the purine bases, it is employed in the following protocol.

1. Prepare reaction tubes out of 8- or 10-mm-diameter hard glass tubing by cutting it into lengths about 16 cm long and then cutting them in half with an oxygen flame.

TABLE I
Ultraviolet absorbancies of DNA components[a]

Component	MW	pH	Absorbance maximum	ε_{max} ($\times 10^{-3}$)	ε_{260}[b] ($\times 10^{-3}$)
Adenine	135.13	1	262.5	13.2	13.0
		7	260.5	13.4	13.3
		12	269	12.3	10.45
Deoxyadenosine	251.24	2	258	14.5	(14.6)[c]
		7	260	15.2	
		13	261	14.9	
Deoxyadenosine 5'- phosphate	331.22	2	258	14.3	14.7
		7			15.3
		12			15.4
Cytosine	111.1	1	276	10.0	6.0 (6.88)[c]
		7	267	6.1	5.55
		14	282	7.9	5.55
Deoxycytidine	227.22	1	280	13.2	6.15
		7	271	9.0	7.35
		13	271.5	9.1	7.05
Deoxycytidine 5'- phosphate	307.20	2	280	13.5	6.2
		7	271	9.3	7.4
		12			7.5
Guanine	151.13	1	248	11.4	8.0
			276	7.35	
		7	246	10.7	7.2
			276	8.15	
		11	274	8.0	
		14	274	9.9	6.4[d]
Deoxyguanosine	267.24	1	255	12.1	11.3 (11.8)[c]
		H_2O	254	13.0	11.75 (7)
		12	260	9.2	
Deoxyguanosine 5'- phosphate	347.23	1	255	11.8	11.5
		7			11.8
		12			11.5
Thymine	126.11	1	264.5	7.9	7.4
		7	264.5	7.9	7.4
		12	291	5.4	3.7
Thymidine	242.23	1	267	9.65	8.75 (9.43)[c]
		7	267	9.65	8.75
		13	267	7.4	6.65[c]
Thymidine 5'- phosphate	322.21	1	267	10.2	9.3
		7	267	10.2	9.3
		12			7.5

[a] From Dunn and Hall (1975).
[b] From the Schwarz/Mann 1970/71 Biochemical Catalogue.
[c] Anderson *et al.* (1963).
[d] Determined at pH 12.

2. Using highly purified DNA, for example, as prepared above (Section A,1), place 25 μg into a reaction tube. Freeze and then lyophilise the sample. Add 50 μl of 88% formic acid and then seal the top of the tube with a flame. Hydrolyse by autoclaving for 2 h (standard 15 psi, 123°C), or by heating in a 175°C oven for 1 h. Because of formic acid decomposition, there is considerable pressure in the reaction tubes which can be released by first heating the tip in a flame. Score and break the top off the tube, and remove the formic acid by lyophilisation.

3. Dissolve the bases in 50 μl of the HPLC buffer.

This procedure, of course, can be readily varied to fit the specific needs of an investigator. We have used screw-cap vials with Teflon-lined caps, and with the special precaution of retightening the caps, they have worked well (Ko *et al.*, 1977).

(*b*) *Deoxyribonucleosides and deoxyribonucleotides.* DNA can be readily converted to the 5'-nucleotides by several nucleases. The nucleotides are then converted to the nucleosides by digestion with phosphatase. Following is a variation of a protocol described by Kuo *et al.* (1980).

1. Place 25–50 μg of DNA into a 1.5-ml microcentrifuge tube. Add 0.1 volume of 3 M sodium acetate (pH 6.0), mix, and then add 2 volumes of −20°C 95% ethanol and mix again. Place the tube into a −20°C freezer for 1 h and then centrifuge for 10 min in a microcentrifuge. Carefully remove the buffer–ethanol with a Pasteur pipette and add 1.0 ml of −20°C 80% ethanol so as not to disturb the DNA pellet. Again place the tube in a freezer for 1 h, centrifuge, and remove the supernatant. Dry the pellet (0.5–1 h; a 37°C incubator works well).

2. Dissolve the DNA in 50 μl of 10 mM Tris–HCl, 4 mM MgCl$_2$ (pH 7.2) buffer. Add 2 μl of DNase I (1 mg ml^{-1} Tris–HCl, MgCl$_2$ buffer–50% glycerol; DP 200 Kunitz units mg^{-1}, Worthington). Incubate at 37°C for 18 h.

3. Add 100 μl of 30 mM sodium acetate, 0.5 mM ZnSO$_4$ (pH 5.2). Add 4 μl of P$_1$ nuclease (2 mg ml^{-1} sodium acetate, ZnSO$_4$ buffer–50% glycerol). Incubate at 37°C for 7 h. At this point, the DNA has been degraded to the 5'-deoxyribonucleotides.

4. Add 15 μl of *Escherichia coli* alkaline phosphatase (2.2 mg ml^{-1} 50 mM Tris–HCl, pH 8.0; 40 U mg^{-1} protein, type III, Sigma). Pre-incubate the phosphatase at 95°C for 10 min to inactivate contaminating deoxyadenosine deaminase. Incubate at 37°C for 16 h. If the samples are not to be analysed immediately, they can be frozen.

5. The digests can be injected directly into the chromatograph. Placing a 1- or 2-cm-long guard column between the injection port and the analyti-

cal column will prolong the life of the analytical column. The guard column will have to be replaced after 100–200 samples.

Kuo *et al.* (1980) have found this procedure to degrade 99.8% of the DNA to deoxyribonucleosides. Although the use of pure DNA preparations is preferred, Garrett and Santi (1979) have measured deoxyribonucleotides in cell extracts after destroying the ribonucleotides by treating the extract with periodate and methylamine.

4. DNA bromination

Bromine reacts with all of the DNA bases except adenine, leading to a disappearance of the absorbance of those bases in the 260- to 280-nm region (Wang and Hashagen, 1964). Measuring the absorbance change at 270 nm, resulting from the bromination reaction, enables one to measure the amount of adenine in the DNA preparation and thus calculate the mol% G + C. Although the method appears simple and straightforward (Wang and Hashagen, 1964; Wang, 1968), the procedure has not been widely used. This has probably been due to the requirement for pure DNA, since any RNA contamination could cause major inaccuracies. Also it would tend to be less sensitive for DNA preparations with high G + C content. Presented here is a variation of the Wang procedure that incorporates the use of NACS-52 purified DNA.

(*a*) *Reagents*
  Sulphuric acid (1 N): Dilute conc. H_2SO_4 1 : 36 with distilled water.
  N-Bromoacetamide solution (8 mM): Dissolve 110 mg in 100 ml of distilled water. The solution is unstable to both heat and light so the container should be wrapped with aluminium foil and stored in a refrigerator.

(*b*) *Protocol.* 1. Precipitate about 60 μg of DNA in a 1.5-ml microcentrifuge tube. Make the DNA solution 0.3 *M* with sodium acetate, pH 6.0, add 2 volumes of ethanol, mix, place at −20°C for 1 h, and centrifuge for 10 min. Wash the pellet once with 80% ethanol (see DNA isolation procedure). Allow the DNA to air dry in a 37°C incubator.

2. Dissolve the DNA by adding 1.25 ml of 1 *N* H_2SO_4 to the tube and mixing. Allow the tube to stand overnight with occasional mixing. Centrifuge the sample for 10 min to remove any particulate material.

3. Place 1.0 ml of the DNA solution in a semi-micro quartz cuvette fitted with a leakproof Teflon stopper. Determine the absorbancies at 270 and at 360 nm. The 360-nm reading will detect absorbance due to sample

turbidity. Subtract the 360-nm absorbance values from the 270-nm absorbance values. These represent the absorbance before bromination (ODB)
readings.

4. Add 25 μl of 8 mM N-bromoacetamide to each sample cuvette and
20 μl to the blank (containing 1 N H₂SO₄). Thoroughly mix contents of
each cuvette and place in the dark for 2 h.

5. Determine the absorbancies at 270 and at 360 nm. Subtract the 360-
nm absorbance values from the 270-nm values. These represent the absorbance after bromination (ODA) readings.

6. The percentage of absorbance remaining (P) is calculated as follows:

$$P = [ODA/(ODB \times 0.975)] \times 100$$

The addition of 25 μl of N-bromoacetamide to the cuvette dilutes the
sample by 2.5%, which is compensated for by the 0.975 value in the
equation.

7. Standard mol% G + C mixtures can be generated by mixing equal
molar mixtures of thymidylic acid (T) and deoxyadenylic acid (A) with

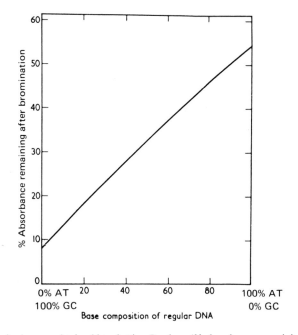

Fig. 11. Standard curve obtained by plotting P values (% absorbance remaining after bromination) against varying base compositions. From Wang and Hashagen (1964).

equal molar mixtures of deoxyguanylic acid (G) and deoxycytidylic acid (C). The P values for these mixtures can be calculated as follows:

$$P = \frac{100[x(ODA_A + ODA_T) + (100 - x)(ODA_G + ODA_C)]}{x(ODB_A + ODB_T) + (100 - x)(ODB_G + ODB_C)}$$

Again, the dilution factor should be taken into account when measuring the absorbance. The standard curve obtained by Wang and Hashagen (1964) is shown in Fig. 11.

(*c*) *Comments.* 1. The acid denatures and depurinates the DNA so that it is in a state similar to the four deoxyribonucleotides. The concentration of the N-bromoacetamide must be such that the bromium ions are in excess of the bases. Therefore, the bromination conditions must be re-checked if the DNA concentrations are altered. N-Bromoacetamide has a slight absorbancy in the 260- to 280-nm range and, since some of it is used up in the sample cuvette, adding the lesser amount to the blank cuvette compensates for this.
2. The method does require RNA-free DNA preparations.
3. The method is most accurate for low mol% G + C DNA preparations, since adenine is the only base being measured.

5. Depurination

The N-glycosidic linkages of purine nucleosides are less stable than those of pyrimidine nucleosides, thus mild acid hydrolysis (pH 1.6) results only in the removal of the purine bases, adenine and guanine (Tamm *et al.*, 1952). Huang and Rosenberg (1966) have used this property to devise a simple dialysis method for determining mol% G + C values.

(*a*) *Buffers and materials.* 1. Prepare SSC buffer adjusted to a pH of 1.58 with HCl.
2. Wash pieces of dialysis tubing ($\frac{1}{4}$ × 5 in.) as described in the thermal melting point section above.

(*b*) *Protocol.* 1. Prepare 0.5-ml DNA samples (100–500 μg ml^{-1}) in SSC.
2. Tie one end of a piece of dialysis tubing and place a DNA sample into it. After tying the other end, trim and rinse the tubing in SSC to remove any DNA that might be on the outside. Place the tubing into a screw-cap test-tube containing 5 ml of the pH 1.58 SSC. Tightly cap the tube and place it in a 37°C incubator with a reciprocal shaker and gently shake the tube for 24 h.

3. Measure the absorbance of the dialysate at 265 and at 280 nm. The mole fraction of G + C can be expressed as a function of the absorbance ratio at the two wavelengths:

$$A_{265}/A_{280} = R = \frac{(1 - X_G)a + X_G c}{(1 - X_G)b + X_G d}$$

where X_G = mole fraction of guanine; ε_X = molar extinction coefficient at X nm; $a = \varepsilon_{265}$ of adenine = 13.1 (12.98); $b = \varepsilon_{280}$ of adenine = 5.0 (4.76); $c = \varepsilon_{265}$ of guanine = 7.3 (7.19); $d = \varepsilon_{280}$ of guanine = 6.9 (6.95). The values for a, b, c, and d were experimentally determined at pH 1.58 by Huang and Rosenberg (1966). The values in parentheses were estimated from the 0.025 N HCl (adenine) and 0.1 N HCl (guanine) spectrum graphs of Beaven *et al.* (1955).

Solving for X_G in the above equation, one obtains

$$X_G = (a - bR)/[a - c + (d - b)R]$$

Inserting the values for a, b, c, and d,

$$X_G = (13.1 - 5.0R)/(5.8 + 1.9R)$$

The non-linear relationship between the absorbance ratio and the mol% G + C values is shown in Fig. 12.

(*c*) *Comments.* 1. Although some of the molar extinction coefficients obtained by Huang and Rosenberg (1966) differ somewhat from those obtained by Beaven *et al.* (1955) almost identical results are obtained using either of the two sets of values.

2. There is a discrepancy between the A_{265}/A_{280} ratios obtained using standard solutions of adenine and guanine and of those obtained using adenine and guanine depurinated from DNA preparations of known mol% G + C. It is not known whether guanine is slightly more resistant to depurination from intact DNA, if it is slightly more retained inside the dialysis tubing, or if there is some destruction.

3. Contamination of the DNA preparations with RNA would also affect the absorbance ratios. Assuming that ribosomal RNA (rRNA) would be the major contaminant, the 52–54 mol% G + C content of rRNA would cause an apparent increase in G + C content in low mol% G + C DNA and an apparent decrease in the G + C content for high mol% G + C DNA preparations. If there was rRNA contamination in the two high G + C content DNA preparations, that could account for the greater discrepancy at the high end of the mol% G + C scale.

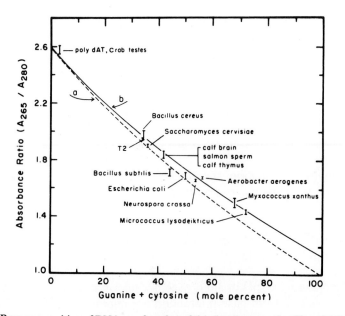

Fig. 12. Base composition of DNA as a function of the absorbance ratio of its acid dialysate. The broken line (a) represents the ratios of pure base-pair mixtures, and the solid line (b) represents the best line through 13 DNA samples whose mol% G+ C values had been obtained from the literature or unpublished (these determinations were by chemical analysis, T_m, or buoyant density). From Huang and Rosenberg (1966).

6. Absorbance ratios ($\varepsilon_{260}/\varepsilon_{280}$) at pH 3.0

While investigating the ultraviolet spectra of DNA and deoxyribonucleotides, Fredericq *et al.* (1961) observed that at pH 3.0, the absorption spectrum of DNA was very similar to that of the component nucleotides mixed in the same proportions. An $\varepsilon_{260}/\varepsilon_{280}$ ratio was characteristic of each mixture of deoxyribonucleotides and for each DNA preparation. The molar extinction coefficients of the deoxyribonucleotides at pH 3.0, which they determined, are listed below:

	Wavelength	
	260 nm	280 nm
Deoxyadenylic acid	13,800	2,950
Deoxyguanylic acid	11,500	8,000
Deoxycytidylic acid	5,800	12,500
Thymidylic acid	8,600	6,350

The absorbance ratio of any DNA preparation or mixture of deoxyribonucleotides, of equal molar ratios of A + T and G + C, or of known mol% G + C content, can be calculated from the following equation:

$$\varepsilon_{260}/\varepsilon_{280} = \frac{(X_{GC} \times 11{,}500) + (X_{GC} \times 5800) + [(1 - X_{GC}) \times 13{,}800] + [(1 - X_{GC}) \times 8600)]}{(X_{GC} \times 8000) + (X_{GC} \times 12{,}500) + [(1 - X_{GC}) \times 2950] + [(1 - X_{GC}) \times 6350]}$$

$$= [22400 - (5100 \times X_{GC})]/[9300 + (11200 \times X_{GC})]$$

Rearranging the equation, the mol% G + C of DNA preparations can be calculated from the $\varepsilon_{260}/\varepsilon_{280}$ ratios R:

$$\text{mol\% G + C} = [(22{,}400 - 9300R)/(5100 + 11{,}200R)]100$$

The non-linear relationship between the mol% G + C values and the $\varepsilon_{260}/\varepsilon_{280}$ is shown in Fig. 13.

(*a*) *Procedure.* A pH of 3.0 is used because at this pH the hypochromic effects resulting from base interaction are minimal and the molar absorbance at 260 nm approaches a maximum value. Also, at a pH of about 2.5, DNA becomes insoluble and the rate of depurination is greater. The pH 3.0 conditions are achieved with 0.1 N acetic acid.

1. Precipitate and wash 25–50 μg of DNA in a microcentrifuge as described above for the nuclease digestion of DNA in the HPLC section.
2. Dissolve the DNA in 0.6 ml of distilled water and then add 0.6 ml of 0.2 N acetic acid. Mix well and determine the absorbance at 260 and 280 nm using 0.1 N acetic acid as a blank. Make the measurements reasonably soon after preparing the samples to avoid depurination.

(*b*) *Comments.* 1. Accuracy using the method is dependent upon highly purified DNA; contamination with RNA will tend to skew the apparent mol% G + C values towards the 50% range.
2. Although the procedure has not been widely used, Bohácek et al. (1967, 1973) obtained results very similar to those determined by thermal melting point.

7. DNA base composition and concentration by UV absorbance

Skidmore and Duggan (1966, 1971) have described a method for estimating mol% G + C that is similar to that of Fredericq et al. (1961) except that the DNA samples and standard deoxyribonucleotide mixtures of

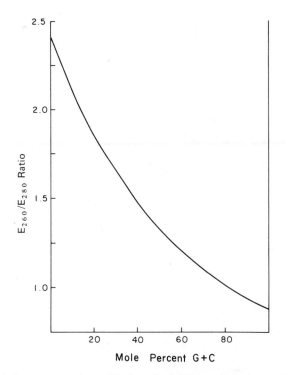

Fig. 13. Correlation between the mol% G + C of DNA and A_{260}/A_{280} ratios at pH 3.0. Calculated from Fredericq *et al.* (1961).

A + T and G + C are hydrolysed to the free bases with perchloric acid before being compared by UV absorbance.

The absorbance spectra of acid hydrolysed mixtures of deoxyadenosine 5'-phosphate plus thymidine 5'-phosphate (25 μM each) and deoxyguanosine 5'-phosphate plus deoxycytidine 5'-phosphate (25 μM each) and of a DNA preparation are shown in Fig. 14. The absorbance of each sample is recorded at the following wavelengths: 264 nm, the absorbance maximum of the A + T mixture; 273 nm, the isosbestic point for the A + T and G + C mixtures; and 286 nm, which is the wavelength at which the absorbance of the G + C mixture is the same as it is at 264 nm. The isosbestic point is important since this is the wavelength at which the molar extinction coefficient of any double-stranded DNA preparation is independent of the G + C content. The molar nucleotide concentration of a DNA sample can be determined from the absorbance at 273 nm (points *h* and *g* in Fig. 14) using the following equation:

$$h/g \times 50 \ \mu M = x \ \mu M \text{ DNA nucleotides}$$

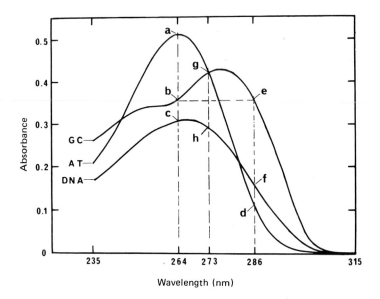

Fig. 14. Illustration of the use of multiple wavelengths for determining the mol% G + C and concentration of DNA, using the equation $[g/(a - d)] \times [(c - f)/h] \times 100 = $ mol% A + T. From Skidmore and Duggan (1966).

The equation for base composition is also based on proportionality. The absorbance difference for the DNA $(c-f)$ is normalised to that of the standards by multiplying it by g/h. The resulting number is arithmetically compared with the absorbance difference for the standards $(a-b) + (e-d)$. Since b and e are equal, this value reduces to $(a-d)$. The mol% A + T is as follows:

$$\text{mol\% A + T} = \left(\frac{g}{h}\right)(c - f) \times \frac{1}{(a - d)} \times 100$$

$$\text{mol\% G + C} = 100 - \text{mol\% A + T}$$

Procedure.   Prepare 50 μM solutions of each of the four deoxyribonucleotides in distilled water. Check their concentrations using the extinction coefficients. Since some of the absorbance maximum extinction coefficients used by Skidmore and Duggan (1971) are slightly different from those in Table I, they are listed below:

Deoxyadenosine 5'-phosphate	259 nm	15,200	pH 7
Thymidine 5'-phosphate	267 nm	9,600	pH 7
Deoxyguanine 5'-phosphate	253 nm	13,800	pH 7
Deoxycytidine 5'-phosphate	280 nm	13,200	pH 2

If any of the concentrations are off, adjust them. Working standards of the nucleotide pairs are prepared by mixing equal volumes of the two nucleotides.

1. Add 2.0-ml volumes of each of the deoxyribonucleotide mixtures and of DNA preparations (50 μM, about 16 μg ml^{-1}) to 16 × 125-mm screw-cap test-tubes. Add 0.2 ml of reagent-grade perchloric acid (70% w/v), mix, and put Teflon-lined caps loosely on the tubes. Place the tubes into a preheated 110–112°C oven or block heater for 15 min. Cool the tubes to room temperature in cold tap water.

2. Determine the spectra between 315 and 240 nm with a recording spectrophotometer, and record the absorbance values at 264, 273, and 286 nm. Calculate the concentrations and the mol% G + C values from the above equations.

8. Absorbance ratios of native DNA at pH 7.0

The spectral properties of both native and heat-denatured DNA at neutral pH have been investigated by Mahler et al. (1964) and Felsenfeld and Hirschman (1965). Felsenfeld and Hirschman (1965) also measured absorption spectra at various times during the denaturation profile. Methods for determining base compositions, based on the spectra of either native or denatured DNA, have been described by Hirschman and Felsenfeld (1966).

Ulitzur (1972) has described a method for determining mol% G + C values that is based on the correlation between the G + C content and absorbance ratios. The ratio between the absorbance at any wavelength in the region from 240 to 255 nm and any wavelength in the range from 265 to 280 nm, increases linearly with increasing mol% G + C. The most useful ratios were 240/275, 240/280, and 245/270. The method is unaffected by DNA concentration from 20 to 45 μg ml^{-1}, salt concentrations of SSC or 10× SSC, or temperatures from 5 to 35°C. High levels of protein contamination will affect the assay, but this effect can be minimised by using only the 245/270 ratio. The presence of RNA in the DNA preparations will cause inaccuracies. Accurate wavelength calibration of the spectrophotometer is also important. Meyer and Schleifer (1975) have used this procedure in conjunction with a rapid method for the isolation of DNA for the analysis of micrococci and staphylococci.

Procedure. 1. Using a protein- and RNA-free reference DNA sample of known mol% G + C content (two to four of varying G + C content would be even better), calculate the absorbance ratios at the following wavelengths (nm): 240/275, 240/280, and 245/270.

TABLE II

Numerical values and slope for absorbance ratios of 0 and 50 mol%
G + C DNA[a]

Absorbance ratio	Reference points for		Slope ($\times 10^{-3}$)
	0 mol% G + C	50 mol% G+ C	
240/280	0.720	1.100	7.60
240/275	0.580	0.868	5.76
240/270	0.460	0.725	5.30
240/265	0.410	0.640	4.60
245/280	1.016	1.395	7.58
245/275	0.810	1.095	5.70
245/270	0.680	0.915	4.70
245/265	0.555	0.825	5.40
250/280	1.425	1.697	5.44
250/275	1.140	1.330	3.80
250/270	0.917	1.100	3.66
250/265	0.818	0.970	3.04
255/280	1.640	1.862	4.44
255/275	1.315	1.460	2.90
255/270	1.090	1.215	2.50
255/265	0.960	1.072	2.24

[a] From Table II Ulitzur (1972).

2. Plot the values obtained in step 1 on a graph (ratios vs mol% G + C)
and draw lines through each of them using the pertinent slopes from Table
II. This step will normalise different spectrophotometers, which may differ in absolute values but the slopes will be the same.

3. Determine the absorbance ratios of the DNA samples that are being
analysed. Determine the mol% G + C values from each of the three
standard curves, and average the three values.

9. Other methods

A number of other methods have been described for the separation of
DNA bases, nucleosides, and nucleotides such as thin-layer chromatography and electrophoresis.

1. If a ^{32}P-labelled DNA preparation is hydrolysed to nucleotides and
separated by one of the above methods, the individual nucleotides can be
quantitated by measuring the radioactivity in each of the chromatographic
spots.

2. The electrophoretic mobility of DNA in composite polyacrylamide/agarose gels is influenced by the base composition (Zeiger *et al.*, 1972). Although the migration rates of higher mol% G + C DNA preparations are only slightly faster, Gilpin and Dale (1979) obtained a reasonable correlation between the migration distance and the mol% G + C values of several organisms.

3. The most recent approach to estimating the G + C content of DNA has been to double-stain bacterial cells with the fluorescent dyes chromomycin A3 and Hoechst 33258, which bind preferentially to DNA that is rich in G + C and A + T, respectively. The cells are then analysed in a dual-beam flow cytometer (Van Dilla *et al.*, 1983).

III. Biological and taxonomic significance

The DNA represents the genetic information contained within the cell or cells of an organism. The information is coded through the linear arrangement of the deoxyribonucleotides in the DNA strands. The DNA functions as a template to transcribe the information to RNA molecules from which the proteins of the cell are translated. Through replication of the DNA, this information is passed to the next generation of the organism. There is little change in the mol% G + C content in the DNA from one generation to the next, and as a result, it is characteristic for a species.

The mol% G + C content of DNA from the prokaryotes ranges from about 25 to 75% while for most eukaryotes, it is in the mid-forties. Since there are many species of prokaryotes, it is obvious that a given G + C content is not the preserve of a single species. The importance of the G + C content in bacterial taxonomy is that it can be an excluding characteristic. If two organisms have DNAs with different G +C content, they are different organisms. On the other hand, if two organisms have DNA with the same G + C content, all that can be said, using only this information, is that the organisms may be the same or similar. The other taxonomically important property of DNA is the order of the deoxyribonucleotides. In addition to having DNA with the same G + C content, the deoxyribonucleotides in the DNA of two organisms that are closely related are also in the same linear order. The linear order will be discussed in more detail in Chapter 2.

References

Anderson, N. G., Green, J. G., Barber, M. L., and Ladd, Sr. F. C. (1963). *Anal. Biochem.* **6**, 153–169.

Ansevin, A. T., and Vizard, D. L. (1979). *Anal. Biochem.* **97**, 136–144.
Ansevin, A. T., Vizard, D. L., Brown, B. W., and McConathy, J. (1976). *Biopolymers* **15**, 153–174.
Beaven, G. H., Holiday, E. R., and Johnson, E. A. (1955). *In* "The Nucleic Acids" (E. Chargaff and J. N. Davidson, Eds.), Vol. 1, pp. 493–553. Academic Press, New York.
Beidler, J. L., Hilliard, P. R., and Hill, R. L. (1982). *Anal. Biochem.* **126**, 374–380.
Boháček, J., Kocur, M., and Martinec, T. (1967). *J. Gen. Microbiol.* **46**, 369–376.
Boháček, J., Kocur, M., and Martinec, T. (1973). *Zentralbl. Bakteriol., Parasitenkd., Infektionler, Hyg., Abt. 1 Orig. Reihe A* **223**, 488–495.
Britten, R. J., Pavich, M., and Smith, J. (1970). *Carnegie Inst. Washington, Year Book* **68**, 400.
Caronia, J. P., Crowther, J. B., and Hartwick, R. A. (1983). *J. Liq. Chromatogr.* **6**, 1673–1691.
Cashion, P., Holder-Franklin, M. A., McCully, J., and Franklin, M. (1977). *Anal. Biochem.* **81**, 461–466.
Chargaff, E. (1955). *In* "The Nucleic Acids" (E. Chargaff and J. N. Davidson, Eds.), Vol. 1, pp. 307–371. Academic Press, New York.
De Ley, J. (1970). *J. Bacteriol.* **101**, 738–754.
Dunn, D. B., and Hall, R. H. (1975). *In* "Handbook of Biochemistry and Molecular Biology" (G. D. Fasman, Ed.), Vol. 1, pp. 65–215. CRC Press, Boca Raton, Florida.
Ehrlich, M., Gama-Sosa, M., Huang, L.-H., Midgett, R. M., Kuo, K. C., McCune, R. A., and Gehrke, C. W. (1982). *Nucleic Acid Res.* **10**, 2709–2721.
Eksteen, R., Kraak, J. C., and Linssen, P. (1978). *J. Chromatogr.* **148**, 413–427.
Elgass, H., Maute, A., Martin, R., and George, S. (1983). *Am. Lab.* **15**, 71–81.
Enquist, L. W., and Bradley, S. G. (1971). *Dev. Ind. Microbiol.* **12**, 225–236.
Felsenfeld, G., and Hirschman, S. Z. (1965). *J. Mol. Biol.* **13**, 407–427.
Ferragut, C., and Leclerc, H. (1976). *Ann. Microbiol.* (Paris) **127A**, 223–235.
Floridi, A., Palmerini, C. A., and Fini, C. (1977). *J. Chromatogr.* **138**, 203–212.
Fredericq, E., Oth, A., and Fontaine, F. (1961). *J. Mol. Biol.* **3**, 11–17.
Garrett, C., and Santi, D. V. (1979). *Anal. Biochem.* **99**, 268–273.
Gasser, F., and Mandel, M. (1968). *J. Bacteriol.* **96**, 580–588.
Gehrke, C. W., Kuo, K. C., Davis, G. E., Suits, R. D., Waalkes, T. P., and Borek, E. (1978). *J. Chromatogr.* **150**, 455–476.
Gehrke, C. W., Kuo, K. C., and Zumwalt, R. W. (1980). *J. Chromatogr.* **188**, 129–147.
Gibson, D. M., and Ogden, I. D. (1979). *J. Appl. Bacteriol.* **46**, 421–423.
Gilpin, M. L., and Dale, J. W. (1979). *Microbios Lett.* **12**, 31–35.
Igloi, G. L. (1983). *Anal. Biochem.* **134**, 184–188.
Hartwick, R. A., Assenza, S. P., and Brown, P. R. (1979). *J. Chromatogr.* **186**, 647–658.
Hirschman, S. Z., and Felsenfeld, G. (1966). *J. Mol. Biol.* **16**, 347–358.
Huang, P. C., and Rosenberg, E. (1966). *Anal. Biochem.* **16**, 107–113.
Johnson, J. L. (1981). *In* "Manual of Methods for General Bacteriology" (P. Gerhardt, Ed.-in-chief), pp. 450–472. American Society for Microbiology, Washington, D. C.
Karsten, H., Steger, G., and Riesner, D. (1980). *Anal. Biochem.* **101**, 225–229.
Knittel, M. D., Black, C. H., Sandine, W. E., and Fraser, D. K. (1968). *Can. J. Microbiol.* **14**, 239–245.
Ko, C. Y., Johnson, J. L., Barnett, L. B., McNair, H. M., and Vercellotti, J. R. (1977). *Anal. Biochem.* **80**, 183–192.
Kraak, J. C., Ahn, C. X., and Fraanje, J. (1981). *J. Chromatogr.* **209**, 369–376.
Kuo, K. C., McCune, R. A., and Gehrke, C. W. (1980). *Nucleic Acids Res.* **8**, 4763–4776.
Mahler, H. R., Kline, B., and Mehrotra, B. D. (1964). *J. Mol. Biol.* **9**, 801–811.

Mandel, M., and Marmur, J. (1968). *In* "Methods in Enzymology," Vol. 12, Part B, pp. 195–206. Academic Press, New York.

Mandel, M., Schildkraut, C. L., and Marmur, J. (1968). *In* "Methods in Enzymology," Vol 12, Part B, pp. 184–195. Academic Press, New York.

Mandel, M., Igambi, L., Bergendahl, J., Dodson, Jr. M. L., and Scheltgen, E. (1970). *J. Bacteriol.* **101**, 333–338.

Marmur, J. (1961). *J. Mol. Biol.* **3**, 208–218.

Marmur, J., and Doty, P. (1962). *J. Mol. Biol.* **5**, 109–118.

McNair, H. M. (1980). *Am. Lab.* **12**, 33–44.

Meyer, S. A. and Schleifer, K. H. (1975). *Int. J. Syst. Bacteriol.* **25**, 383–385.

Owen, R. J., Hill, L. R., and Lapage. (1969). *Biopolymers* **7**, 503–516.

Schildkraut, C., and Lifson, S. (1965). *Biopolymers* **3**, 195–208.

Schildkraut, C. L., Marmur, J., and Doty, P. (1962). *J. Mol. Biol.* **4**, 430–443.

Selin, Y. M., Harich, B., and Johnson, J. L. (1983). *Curr. Microbiol.* **8**, 127–132.

Skidmore, W. D., and Duggan, E. L. (1966). *Anal. Biochem.* **14**, 223–236.

Skidmore, W. D., and Duggan, E. L. (1971). *In* "Methods in Microbiology" (J. R. Norris and D. W. Ribbons, Eds.), Vol. 5B, pp. 631–639. Academic Press, London.

Štulík, K., and Pacáková, V. (1983). *J. Chromatogr.* **273**, 77–86.

Tamm, C., Hodes, M. E., and Chargaff, E. (1952). *J. Biol. Chem.* **195**, 49–63.

Ulitzur, S. (1972). *Biochim. Biophys. Acta* **272**, 1–11.

Van Boom, J. H., and De Rooy, J. F. M. (1977). *J. Chromatogr.* **131**, 169–177.

Van Dilla, M. A., Langlois, R. G., Pinkel, D., Yajko, D., and Hadley, W. K. (1983). *Science* **220**, 620–623.

Vizard, D. L., and Ansevin, A. T. (1976). *Biochemistry* **15**, 741–750.

Wang, S. Y. (1968). *In* "Methods in Enzymology," Vol. 12, Part B, pp. 178–184. Academic Press, New York.

Wang, S. Y., and Hashagen, J. M. (1964). *J. Mol. Biol.* **8**, 333–340.

Werner, H., Gasser, F., and Sebald, M. (1966). *Zentralbl. Bakteriol. Parasitenkd., Infektionskr. Hyg., Abt. 1: Orig.* **198**, 504–516.

Wheals, B. B. (1982). *In* "Techniques in Liquid Chromatography" (C. F. Simpson, Ed.), pp. 121–140. Wiley, New York.

Whitehouse, R. P., and Greenstock, C. L. (1982). *J. Liq. Chromatogr.* **5**, 2085–2093.

Wyatt, G. R. (1955). *In* "The Nucleic Acids" (E. Chargaff and J. N. Davidson, Eds.), Vol. 1, pp. 243–265. Academic Press, New York.

Zadrazil, S., Sastava, J., and Sormova, Z. (1973). *J. Chromatogr.* **91**, 451–458.

Zakaria, M., and Brown, P. R. (1981). *J. Chromatogr.* **226**, 267–290.

Zeiger, R. S., Salomon, R., Dingman, C. W., and Peacock, A. C. (1972). *Nature (London) New Biol.* **238**, 65–69.

Zumwalt, R. W., Kuo, K. C. T., Agris, P. F., Ehrlich, M., and Gehrke, C. W. (1982). *J. Liq. Chromatogr.* **5**, 2041–2060.

2

DNA Reassociation and RNA Hybridisation of Bacterial Nucleic Acids

JOHN L. JOHNSON

Department of Anaerobic Microbiology, Virginia Polytechnic Institute and State University, Blacksburg, Virginia, USA

I. Introduction

The elucidation of the structure and the physical properties of deoxyribonucleic acid (DNA) and ribonucleic acids (RNAs) has enabled investigators to compare directly the sequence similarities in these molecules from one organism with those from another. These comparisons, DNA reassociation or RNA hybridisation experiments, are measurements of the pairing of two DNA fragments or the pairing of an RNA molecule with a fragment of DNA. The specific pairings are between the base pairs adenine (A) and thymine (T) [or uracil (U) in RNA] and between guanine (G) and cytosine (C). As a result, the reassociation or hybridisation reaction is dependent upon the linear arrangement of the purine and pyrimidine bases along the nucleic acid strands and thus provides a comparative measurement of nucleotide sequence similarity.

33

METHODS IN MICROBIOLOGY
VOLUME 18

It has been more than 20 years since the introduction of nucleic acid homology experiments. Older methods are continually being improved and new methods introduced as a result of advances in recombinant DNA technology. These procedures have been described and updated in many excellent books over the years (Bendich and Bolton, 1968; Britten *et al.*, 1974; Church, 1974; De Ley, 1971; Gillespie, 1968; Johnson, 1981; and Midgley, 1971). In addition, the biochemistry and both general and specific information for a wide range of nucleic acid experiments can be found in Parish (1972) and Davidson (1975). Nucleic acid reassociation and hybridisation procedures have also been reviewed periodically (Kennell, 1971; McCarthy and Church, 1970; Moore, 1974; Wetmur, 1976). The purpose of this chapter is to update again the methods available for DNA and RNA homology experiments.

II. Methodology

The final success of the methods and procedures described in this chapter depends upon the initial growth of the bacterial cells and the isolation of DNA or RNA from them. Although techniques for following the growth of bacterial cultures are beyond the scope of this chapter, their importance cannot be overemphasised. Many organisms are fastidious and grow very slowly or to a low turbidity, others tend to die very rapidly after reaching stationary phase, while others may enter a sporulation phase.

A. Cell disruption

A major technical problem involved in the isolation of nucleic acids from bacterial cells is the disruption of the cell walls. Methods that have been used to disrupt cell walls include the addition of detergents, digestion with cell wall-specific enzymes, and physical. The method to be used in a specific case will depend upon the organism and the type or physical form of the nucleic acid being isolated.

Harvest cultures in the late logarithmic or early stationary phase of growth.

1. Gram-negative bacteria

Suspend the cells in a volume of saline–EDTA buffer (0.15 M NaCl, 0.01 M sodium ethylenediaminetetraacetate, pH 8.0) equal to about one-tenth to one-fortieth of the volume of the original culture, depending upon the method of isolation. Make sure that there are no cell clumps since in a clump only the outer cells will be lysed and the others will be lost in the isolation procedure. Add sodium dodecylsulphate (SDS) from a 20% (w/v) stock solution to give a final concentration of 1%. Swirl the flask in a 50–60°C water bath to increase the rate of lysis. Lysis will be indicated by a rapid increase in viscosity and a change in the suspension from turbid to opalescent.

For those bacteria that do not lyse sufficiently under these conditions, several alternatives can be tried. (1) Use a more dilute cell suspension; for some organisms a heavy suspension does not lyse as efficiently as a more dilute suspension. (2) Check the growth curve: make sure that the cells are still in the log phase. (3) Incubate the cells with enzymes that may weaken the cell walls such as lysozyme, pronase, or proteinase K. Check the effectiveness and determine time requirements by removing samples periodically during the incubation and testing them for lysis by SDS. (5) A detergent wash (Sarkosyl) coupled with a mild osmotic shock and lysozyme digestion has also been used successfully (Schwinghamer, 1980).

For Gram-negative bacteria that will not lyse by the above detergent procedures, one must resort to various physical procedures. These include sonication, passage through a French pressure cell, or shaking with glass beads. Cells of reasonable size can usually be disrupted by passage through a French pressure cell at 12,000–16,000 lb in.$^{-2}$. Ultrasonic oscillators are usually somewhat less effective. For organisms that cannot be disrupted by these methods, shaking with glass beads in a Bronwill cell homogeniser (Bronwill Scientific Co., Rochester, New York, distributor of B. Braun, Germany) will usually be effective. Operate the homogeniser at 4000 cycles min^{-1}, using equal volumes of cell suspension and glass beads (0.1 mm diameter). Remove the glass beads by suction filtration through a coarse-porosity sintered-glass filter. Wash the lysate free of the beads by mixing the beads with small amounts of the lysing buffer and then applying suction. Repeat until you have used a volume of buffer equal to the original cell suspension. Include SDS with the cells for the inactivation of nucleases.

2. Gram-positive bacteria

Several enzymes are available for rendering Gram-positive bacteria susceptible to lysis with SDS. The one most widely used is lysozyme which is an endoacetyl muramidase isolated from egg whites. Others include another *N*-acetylmuramidase, isolated from *Streptomyces globisporus*

(Miles Scientific); lysostaphin (Sigma), which is an endopeptidase isolated from *Staphylococcus* sp. K-6-WI and is specific for staphylococci; and achromopeptidase, TBL-1 (Wako Pure Chemical Industries, Tokyo, Japan), which is also active on the muramic acid cross-linking amino acids.

(*a*) *Lysozyme.* The following is a variation of a method described by Chassy and Giuffrida (1980). Suspend early stationary growth phase cells from 1 litre of culture in 50 ml of 0.01 M Tris–hydrochloride buffer, pH 8.2. Dilute and mix the suspension with 100 ml of 24% (w/v) 20 M PEG (Carbowax 20,000, Fisher) in water. Add 100 mg of lysozyme (estimating 100 μg mg dry wt.$^{-1}$ of cells) and incubate at 37°C until the cells are sensitive to SDS. Test by taking out aliquots at various times, add 0.1 volume of 10× NaCl–EDTA (NaCl–EDTA = 0.1 M NaCl, 0.01 M EDTA, pH 8.0), and add SDS to a final concentration of 1%. Mix and warm in a 50–60°C water bath. Lysis will be apparent by a clearing of the suspension and an increase in the viscosity. When the cells are SDS sensitive, treat all of the cells as described for the aliquots. Cultures that are resistant to lysis under these conditions can sometimes be made less intransigent by growing the cells in the presence of glycine (Yamada and Komagata, 1970), threonine, or lysine (Chassy and Giuffrida, 1980) or by adding penicillin at late log phase (about 0.5 h before harvesting).

(*b*) *N-acetylmuramidase.* This enzyme was isolated by Yokogawa *et al.* (1972) and although it has the same substrate specificity as lysozyme, the range of organisms that can be lysed by it is somewhat different (Yokogawa *et al.*, (1975). Suspend the cells in 25 ml of 50 mM Tris–malate–NaOH buffer, pH 6.5. Add enzyme to a level of about 5 μg ml^{-1} and incubate at 50°C. Again, remove aliquots and test for SDS sensitivity.

(*c*) *Lysostaphin.* This enzyme appears to be specific for the rather unique cross-linking peptides in the cell walls of *Staphylococcus* (Schindler and Schuhardt, 1964). Suspend the cells from 1 litre of culture in 25–50 ml of 50 mM Tris–0.145 M NaCl buffer, pH 7.5. Add 25 μg ml^{-1} of lysostaphin and incubate at 37°C. Remove aliquots and test for SDS sensitivity.

(*d*) *Achromopeptidase.* The *Achromobacter lyticus* strain that produces this enzyme was isolated at Takeda Chemical Industries, Ltd., Japan. The enzyme has been useful for plasmid isolation (Horinouchi *et al.*, 1977; Kahn and Pierson, 1983) and for routine DNA isolation in my laboratory. Suspend the cells from 1 litre of culture in 50–100 ml of 10 mM Tris–hydrochloride (pH 8.2 buffer). Add 50 μg ml^{-1} of achromopeptidase and incubate at 50°C until the cells become sensitive to lysis with

SDS. The pH and ionic strength requirements for this enzyme are very similar to those of lysozyme; perhaps they will complement one another.

B. DNA isolation

The most time-consuming part of a DNA homology study is the isolation of DNA, and including the cell disruption step, also the part that is most subject to technical problems. Two methods will be given for isolating DNA from 0.5 to 1.0 litres of culture.

1. Marmur method

The original procedure (Marmur, 1961) employs chloroform to remove protein from the lysate and is probably the most widely used DNA isolation procedure. The following is a variation of this method.

(a) *Reagents*
   Saline–EDTA buffer: 0.15 M NaCl, 0.01 M sodium EDTA, pH 8.0.
   SDS: 20% (w/v) solution.
   Sodium perchlorate: 5 M solution.
   Chloroform–isopentanol: 24 : 1 (v/v) mixture.
   Phenol–chloroform: (v/v) mixture of phenol (chromatography-grade liquid phenol equilibrated with saline–EDTA buffer, pH 8.0) and chloroform–isopentanol, containing 0.1% 8-hydroxyquinoline.
   SSC: 0.15 M NaCl, 0.015 M trisodium citrate, pH 7.0. Other concentrations of SSC are indicated in the text by a number such as 20× (20-fold concentration) or 0.1× (one-tenth the concentration).
   Ribonuclease (RNase): Bovine pancreatic RNase, 1.0 mg ml^{-1} in 0.15 M NaCl, pH 5.0. Heat the solution at 80°C for 10 min to inactivate any traces of deoxyribonuclease. RNase T_1, 1000 Sankyo units ml^{-1} of bovine pancreatic RNase. This is usually obtained as a solution and can be added just prior to use.

(b) *Procedure.*   1. Suspend the centrifuged cells in 50–100 ml of saline–EDTA buffer and disrupt them by one of the methods described above.
   2. Add sodium perchlorate to a final concentration of 1 M.
   3. Add one-third volume of phenol–chloroform, swirl the flask to mix the two phases together, and shake on a wrist-action shaker for 20 min. A ground glass-stoppered flask works best for this. Operate the shaker at an amplitude sufficient to maintain an emulsion, settings in the range of 2–5, depending on the size of the flask.
   4. Centrifuge the emulsion at 17,000 g for 10 min in a refrigerated centrifuge at 0–4°C.

5. Carefully decant and/or pipette the upper aqueous layer from each tube, being careful not to collect any of the white precipitate (protein) at the interface between the two phases. For pipetting, use an inverted 5- or 10-ml serological pipette, with the tip inserted into a rubber bulb or other mechanical pipetting device. The aqueous layer contains the DNA and is therefore very viscous. It is helpful to move the pipette continually back and forth in the tube to avoid collecting any of the protein at the interface.

6. Extract the lysate again by repeating steps 3 through 5. Repeat again, if there is still a substantial protein layer at the interface.

7. Place the aqueous phase in a beaker and slowly overlay with cold 95% ethanol (an amount equal to about 2 volumes of the aqueous phase). Collect the precipitated DNA with a glass stirring rod by gently stirring the two phases while spinning the rod. The DNA will adhere or "spool" onto the rod. Remove the excess ethanol by pressing the rod against the side of the beaker, and then place the rod in a test-tube containing cold ($-20°C$) 80% ethanol in water. Let it stand for 5–10 min to allow excess salt and other small molecules to diffuse away from the DNA. Stand the rod vertically (with the DNA end up) to allow it to drain and partially air dry.

8. Dissolve the spooled DNA in 10–20 ml of $0.1\times$ SSC. After the DNA is completely dissolved, adjust the SSC concentration to $1\times$ by adding a suitable volume of $20\times$ SSC.

9. Add RNase mixture, 1.0 ml per 20 ml of DNA solution, and incubate at 37°C for 1 h.

10. Add chloroform–isopentanol (5–10 ml) to the DNA solution and then shake the mixture on the wrist-action shaker. Centrifuge the emulsion and draw off the aqueous phase. Repeat the chloroform–isopentanol extractions until very little protein is observed at the interface between the two phases.

11. Precipitate the DNA with ethanol by "spooling," wash the spooled DNA in 80% ethanol, and dissolve in $0.1\times$ SSC. Repeat this step one or two more times to remove ribonucleotides. Adjust the SSC concentration to $1\times$ before each ethanol precipitation.

12. Finally, dissolve the DNA in $0.1\times$ SSC and store in a freezer at $-20°C$, or place two or three drops of chloroform in the tube and store in a refrigerator.

2. Hydroxylapatite method

Britten *et al.* (1970) were the first to isolate DNA routinely by adsorbing it to hydroxylapatite. Since then several modifications of the method have been described (Markov and Ivanov, 1974; Meinke *et al.*, 1974; Johnson, 1981). Following is an adaption of this method.

(*a*) *Reagents.*   See Marmur procedure above.
   Hydroxylapatite: Dry DNA grade.
   Phosphate buffer (PB): 1.0 *M*, pH 6.8 (prepare by mixing equal volumes of 1 *M* Na_2HPO_4 and 1 *M* NaH_2PO_4). This is used for preparing the lower concentrations as indicated in the text.

(*b*) *Procedure.*   1. Suspend the centrifuged cells in 25 ml of saline–EDTA buffer, and add 1.0 ml of RNase mixture.

2. Add 1.25 ml of SDS. Swirl the flask, and warm in a 50–60°C water bath until lysis is complete. For organisms that are not disrupted by detergent alone, see Section A.

3. Reduce the viscosity of the lysate by briefly subjecting it to sonic oscillation. Pronase b or proteinase K may be added at this time (50 μg ml^{-1}). For some organisms, proteinase digestion is not needed, while for others it is essential, for example, where there is a high level of nuclease activity or, as in some cases, where the DNA appears to be closely associated with cell protein and will be removed with the denatured protein if not first released with proteinase. If a proteinase is added, incubate it with the lysate at 50°C for 1 h.

4. Add 7 ml of phenol–chloroform, mix by hand to get the two phases well mixed, and then shake the flask on a wrist-action shaker for 20 min.

5. Centrifuge at 17,000 *g* in a polypropylene centrifuge tube, using a refrigerated centrifuge at 0–4°C. Carefully draw off the upper (aqueous) layer (see step 5 of the Marmur procedure), and return it to the flask.

6. Repeat steps 4 and 5.

7. After again returning the aqueous phase to the flask, add 1.0 *M* PB (0.1 ml ml^{-1} lysate). Add 2 g of hydroxylapatite (1 level measuring teaspoon full is convenient and sufficiently accurate), suspend well by hand, and then gently shake on a rotary or reciprocal shaker for 1 h at a speed sufficient to keep the hydroxylapatite from settling out.

8. Transfer the suspension to a 50-ml polypropylene centrifuge tube, and centrifuge for 2–3 min at 5000 *g* at room temperature. Return the supernatant layer (lysate) to the flask (this can be used for a second DNA adsorption cycle if it is needed).

9. To the sedimented hydroxylapatite add 8 ml of 0.10 *M* PB. Suspend the hydroxylapatite with the aid of a Vortex-type mixer. Immediately add an additional 24 ml of the PB (an automatic pipetter works well for these additions). The PB should be added with some force, so that the hydroxylapatite will become evenly distributed for maximum dilution of the

unadsorbed components such as nucleotides and phenol. Allow the hydroxylapatite to settle for 1–2 min, and then centrifuge for 2–3 min at 5000 g. Discard the supernatant. The low-speed centrifugation is for ease of resuspending the hydroxylapatite; if the tubes are not removed and decanted immediately after the centrifuge stops, the hydroxylapatite pellets will become loose.

10. Repeat step 9 six or seven times, or until the absorbance of the supernatant is less than 0.05 at 270 nm (the absorption maximum of phenol).

11. Suspend the hydroxylapatite in 5.0 ml of 0.5 M PB to desorb the DNA. Centrifuge as before, but for 5–10 min. Save the supernatant, which is the DNA preparation.

12. Filter the DNA preparation through a glass fibre filter to remove particles of hydroxylapatite.

13. Dialyse the DNA preparation against 0.1× SSC to remove the PB. Cut 13- to 15-cm lengths of dialysis tubing, and wash by heating it in a boiling solution of 2–5% sodium carbonate for a few minutes, thoroughly rinsing the lengths of tubing under running tap water, and then placing them in distilled water. The tubing can be stored in a refrigerator for 2–3 days. Dialyse in a 400 to 500 volume excess of the buffer for about 3 h, change the buffer, and continue the dialysis overnight.

14. Store the DNA preparations in a freezer, or add two or three drops of chloroform and store in a refrigerator.

(*c*) *Comments.*   The important feature of the hydroxylapatite procedure is that the RNA species are degraded to such an extent that they will not compete with DNA for adsorption sites on the hydroxylapatite. The lysate from some groups of organisms may inhibit the RNases; in these cases, dialyse the lysate before adding the RNase mixture.

Heavy suspensions of some groups of organism may not lyse; more dilute suspensions may help. The volume of the lysates will also increase if the organisms must be disrupted by glass beads. The increased volume will decrease the efficiency of DNA adsorption to the hydroxylapatite. For larger volumes of lysate, add the hydroxylapatite to part of the lysate and pour it into a 60-ml coarse-porosity sintered glass filter. After the hydroxylapatite has partially settled, let all of the lysate pass through it. Then transfer the hydroxylapatite to the centrifuge tube and complete the procedure as described above.

The major contaminant of DNA preparations, other than RNA, is polysaccharide. Many polysaccharides do not adsorb to hydroxylapatite, so this may be a useful alternative to the Marmur procedure when polysaccharide contamination is a problem. Some contaminants may preferentially adsorb to the hydroxylapatite. In these cases one may do a preliminary adsorption to remove the contaminant and then isolate the DNA with a second adsorption cycle.

If the lysate is digested with pronase b or proteinase K, a single phenol–chloroform extraction may be sufficient.

3. Other methods

Variations of the above methods have been described for the isolation of DNA from specific groups of organisms, such as mycobacteria, where the major changes are at the cell disruption step (Wayne and Gross, 1968). Some recent procedures for the isolation of DNA from fungi and higher plants (Specht et al., 1982; Bendich et al., 1979; Bowman and Dyer, 1982; Rawson et al., 1982) can also be adapted to the isolation of DNA from bacteria.

C. RNA isolation

The bulk of the nucleic acids within a bacterial cell is RNA [mostly ribosomal RNA (rRNA) and lesser amounts of transfer RNA (tRNA) and messenger RNA (mRNA)]. As a result it is easy to isolate large quantities of rRNA. From a taxonomic point of view there has not been as much interest in the tRNAs and the isolation of mRNA is rather difficult and beyond the scope of this chapter.

A major problem in the isolation of RNA is the ubiquitous presence of RNase within the cells, on the glassware, and in the buffers. Glassware can best be rendered free of RNase by baking in an ashing oven or a dry-heat oven. For plastic ware, autoclave it in a 1–2% solution of decontaminating detergent such as Isoclean (Isolabs Inc.). Rinse the plastic ware under hot tap water, and autoclave it again. Diethyl pyrocarbonate (0.2%) is added to the aqueous buffers and aqueous solutions, the flasks, or bottles and dispersed by rapid swirling (for additional information regarding diethyl pyrocarbonate, see Ehrenberg and Fedorcsak, 1976). Allow about 1 h for the diethyl pyrocarbonate to inactivate nucleases and decompose, and then autoclave the buffers and aqueous solutions.

1. Kirby method

Kirby (1968) and Kirby et al. (1967) have described in detail procedures for isolating RNA. The following is a variation of his procedure that works well for the isolation of bacterial rRNA that is essentially free of DNA and mRNA and contains very little tRNA.

(a) Reagents

Sodium naphthalene 1,5-disulphonate: 10% (w/v).

Phenol–cresol mixture: 550 ml of water-saturated chromatography grade phenol, 70 ml of *m*-cresol, and 0.5 g of 8-hydroxyquinoline.

Diethyl pyrocarbonate: This reagent is very unstable in water; add it directly to solutions or to cell suspensions just prior to lysing.

Sodium acetate: 3 *M* sodium acetate, pH 6.0.

SSC: (See DNA isolation section).

SDS: (See DNA isolation section). Use electrophoretic grade SDS for adding to RNA preparations.

(b) Procedure.

1. Harvest the cells by centrifugation, and wash them once with distilled water. Suspend the cells in approximately 20 ml of cold distilled water. Measure the volume of the suspension with a graduated cylinder.

2. Add 1.0 ml of 10% naphthalene disulphonate and 75 μl of diethyl pyrocarbonate. Swirl the cell suspension and disrupt the cells immediately by passage through a French pressure cell at 16,000 lb in.$^{-2}$, into a mixture containing 15 ml of phenol–cresol and 10 ml of 0.5% naphthalene disulphonate.

3. Shake the flask for 20 min on a wrist-action shaker. Place the mixture in a polypropylene centrifuge tube, and centrifuge at 17,000 g for 10 min in a refrigerated centrifuge at 0–4°C. Carefully draw off and save the upper (aqueous) layer.

4. Add 1.0 ml of 20× SSC to each 20 ml of aqueous layer, and SDS to a final concentration of 1%.

5. Add 15 ml of phenol–cresol, shake, and centrifuge as before. If there is a heavy protein layer at the interface, do another phenol–cresol extraction. Save the aqueous layer.

6. Add 2 volumes of −20°C 95% ethanol to the aqueous layer, mix, and allow to stand in a freezer for 30–60 min. Centrifuge at 4000 g in a refrigerated centrifuge for 10 min. Decant the supernatant, invert the bottle on a paper towel, and let it drain well.

7. Add 25 ml of cold (4°C) 3 *M* sodium acetate to the RNA pellet and homogenise it. A probe-type electric homogeniser works well, although a Dounce type will also work. Although not as efficient, one can simply stir the pellet around with a sterile glass stirring rod. Centrifuge the suspension at 4–5 g for 10 min, and save the pellet. This step will extract residual DNA, the mRNA, and most of the tRNA.

8. Dissolve the pelleted rRNA in 30 ml of SSC. Repeat the ethanol precipitation once or twice. Store the RNA in SSC containing 1% SDS at −20°C or colder.

2. Other methods

In addition to other RNA isolations described by Kirby (1968), Chirgwin *et al.* (1979) have described a procedure for the isolation of RNA from tissue containing a high level of RNase. A method for the partial purification of mRNA has been described by Chang *et al.* (1981), and a method for the isolation of polyadenylate-containing RNA from bacteria has been described by Gopalakrishna *et al.* (1981).

D. Nucleic acid quantitation and purity

1. Ultraviolet spectrophotometry

The most common method for determining the concentrations of nucleic acid solutions is to measure the absorbance at 260 nm. The molar extinction coefficient of native DNA is in the range of 6650–6700 (Chargaff, 1955; Le Pecq and Paoletti, 1966). Converting the DNA concentration to mg ml^{-1} and using a 1-cm light path cuvette, the following formula can be used: mg ml^{-1} = absorbance at 260 nm/20. Denatured DNA (single-stranded) or RNA has a higher extinction coefficient, for these use the following formula: mg ml^{-1} = absorbance at 260 nm/23. In practice, most stock DNA preparations will range from 0.2 to 2 mg ml^{-1}, and it is convenient to make a 1 : 20 dilution (0.1 ml of the stock DNA solution added to 1.9 ml of buffer). This is within the absorbance range of most UV spectrophotometers, and the absorbance then reads directly as milligrams of DNA per millilitre of the stock preparation. The same type of dilution can be done with preparations of denatured DNA or RNA.

Direct spectrophotometric assays were first used to detect nucleic acid contamination in protein preparations. Warburg and Christian (1942) used the 280 nm/260 nm absorbance ratio as a measurement of nucleic acid contamination. This ratio has also been used as an indicator of protein contamination in nucleic acid preparations; however, since the extinction coefficients of proteins are much lower than the coefficient for nucleic acids, it is very insensitive. More recently, Kalb and Bernlohr (1977) have recommended the use of the 230 nm/260 nm ratio as a more sensitive measurement of protein concentration in cell free lysates. The 230- to

234-nm absorbance is the minimal absorbance between the 208- to 210-nm and the 260-nm absorbance peaks of nucleic acids and is on the trailing end of the 208- to 210-nm absorbance peak of protein (most compounds absorb at 208 nm, so make sure that the buffer and salt concentrations are the same in both the blank and sample cuvette). When one is concerned about protein contamination in nucleic acid preparations, the 234/260 ratio is reasonably sensitive to protein contamination, as illustrated in Fig. 1. The following formulas of Kalb and Bernlohr (1977) provide reasonable estimates of nucleic acid and protein concentrations:

$$\mu g \ ml^{-1} \ \text{nucleic acid} = 49.1A_{260} - 3.48A_{230}$$

$$\mu g \ ml^{-1} \ \text{protein} = 183A_{230} - 75.8A_{260}$$

The major contaminant of DNA preparations is usually RNA, and the absorbance spectra of both are very similar. Both nucleic acids increase in absorbance (hyperchromic shift) when they are heated or degraded either chemically or enzymatically. The hyperchromic shift of DNA is sharp, occurring over a range of 10–15°C, and amounts to about 40% of the initial absorbance. [This is somewhat dependent on the mole percent guanine plus cytosine (mol% G + C), the hyperchromicity being a little higher for low mol% G + C DNAs.] The hyperchromic shift of RNA is broad, and amounts to about 28–30% of the initial absorbance. If the secondary structuring of RNA is destroyed by RNase, the hyperchromic shift can be used as a measurement of DNA concentration. Estimate the percent DNA in a nucleic acid preparation from the hyperchromic shift (H)

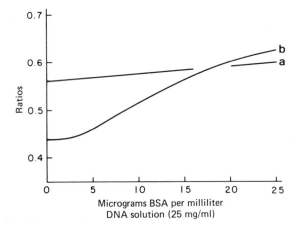

Fig. 1. Effect of a protein, bovine serum albumin (BSA), on the A_{260}/A_{280} (a) and A_{234}/A_{260} (b) ratios of DNA.

and the initial absorbance by the following formula: percentage DNA = $(2.5H/A_{260}) \times 100$.

2. Colorimetric and fluorimetric assays

Many colorimetric and fluorimetric assays have been developed for the measurement of nucleic acids.

(a) *Diphenylamine assay.* The diphenylamine reaction is specific for the deoxyribose in DNA. The Giles and Meyers (1965) variation of the Burton (1956) assay is given below:

1. Place 1.0 ml of the DNA sample in a 16 × 125-mm screw-capped tube and add 1.0 ml of 20% (w/v) perchloric acid.
2. Add 2.0 ml of glacial acetic acid containing 4% (w/v) diphenylamine.
3. Add 0.2 ml of 0.16% (w/v) solution of acetaldehyde.
4. Mix and incubate overnight at 30°C.
5. Read the absorbance at 595 and 700 nm and calculate the difference (the 700-nm reading will detect absorbance due to non-specific particulate material in the sample). Construct a standard curve using known amounts of a pure DNA preparation (salmon sperm DNA is very good and the initial concentrations can be determined by UV absorbance as described above.) Using the standard curve, estimate the DNA concentrations in the samples. The procedure works well for DNA solutions in the range of 5–50 μg ml^{-1}.

The diphenylamine procedure can also be used to quantitate the DNA bound to nitrocellulose membrane filters. If the membranes have been used in DNA homology or RNA hybridisation experiments, remove the membranes from the counting fluid (use a non-solubilising "cocktail" so that the membranes will not be dissolved or the DNA eluted), and dry them in air. Place each membrane (or if there are duplicates, both) in 2.0 ml of 10% perchloric acid, and heat in a boiling-water bath for 5 min. After cooling to room temperature, proceed with step 2 above.

(b) *Thiobarbituric acid assay.* Oxidation of 2-deoxyribose by periodate yields malonaldehyde, which when condensed with thiobarbituric acid forms a pink chromogen. Gold and Shochat (1980) have used this reaction for a sensitive assay for the determination of DNA concentrations, as follows:

1. Place 0.1-ml samples of DNA (0.5–500 μg) into 16 × 125-mm screw-capped (Teflon-lined caps) tubes.

2. Add 0.1 ml of 0.1 N H_2SO_4 to each tube. Heat at 100°C for 1 h. After cooling for 5 min, centrifuge the tubes at low speed to bring down any condensate on the tube walls.

3. Add 0.1 ml of periodate reagent (0.1 M sodium metaperiodate in 9.0 M phosphoric acid, prepared fresh every 2 weeks and stored at 4°C), mix well, and incubate at room temperature for 20 min.

4. Add 1.0 ml of arsenite solution [10% sodium arsenite (w/v) dissolved in 0.5 M sodium sulphate containing 0.1 N sulphuric acid] to destroy excess periodate. A yellow-brown discharge occurs, which upon vigorous mixing will disappear. Incubate for 10 min at room temperature.

5. Add 3.0 ml of thiobarbituric acid solution, mix well, put on the screw caps, and heat in a boiling-water bath for 15 min. Cool the tubes to room temperature.

6. Read the absorbance of the samples at 532 nm. If the samples contain less than 20 μg of DNA, the chromogen can be extracted into cyclohexanone. Add 1.0 ml of cyclohexanone, mix vigorously, and separate the phases by low-speed centrifugation. Read the 532-nm absorbance of the upper cyclohexanone phase.

(c) *Diaminobenzoic acid assay.* This assay, when read colorimetrically, has a range of 25–1000 μg of DNA, and when read fluorimetrically, has a range of 1–8 μg (Setaro and Morley, 1976, 1977).

1. Prepare DNA standards and samples in 0.1-ml volumes of 1 N perchloric acid.

2. Add 0.1 ml of 1.32 M diaminobenzoic acid. Make sure that all of the reaction components are at the bottom of the tube.

3. Cap and place the tubes in a 60°C water bath for 30 min.

4. Add 2.0 ml of 0.6 M perchloric acid, mix, and read at 420 nm. The reaction mixture, containing no DNA, is used as the blank.

For fluorimetric analysis, prepare the samples exactly as above, but using less DNA, and read, with an excitation of 420 nm at a fluorescence of 520 nm.

(d) *Mithramycin assay.* This assay is based on the fact that the antibiotic mithramycin binds to double-stranded DNA and fluoresces in direct proportion to the amount of DNA (Hill and Whatley, 1975). The antibiotic binds to a lesser extent to denatured DNA, and does not bind to DNase-treated DNA or to RNA. Prepare a mithramycin stock solution containing 200 μg ml^{-1} mithramycin in 300 mM $MgCl_2$. The buffer in which the DNA is dissolved is not critical, it may be a low ionic strength buffer, or phosphate-buffered saline.

1. To 1.9 ml of a DNA preparation (0.2–16 μ ml^{-1}) add 0.1 ml of mithramycin stock solution, mix, and determine fluorescence at 540 nm with an excitation wavelength of 440 nm.

E. Labelling of nucleic acids

Nucleic acids can be labelled *in vivo* by growing the organism in the presence of ^{3}H-, ^{14}C-, ^{32}F-, or ^{33}P-labelled precursors, or they can be labelled *in vitro* with ^{3}H-, ^{14}C-, ^{32}P-, ^{33}P-, or ^{35}S-labelled nucleotides or with ^{125}I by direct iodination. A recent substitution for radioactive labelling has been the incorporation of biotin-containing nucleotides, which can be detected by the binding of radioactive avidin or more often by the binding of a biotinated enzyme through an avidin bridge (Langer et al., 1981; Leary et al., 1983). The amount of enzyme bound can then be measured by a colorimetric assay.

1. In vivo labelling

The *in vivo* labelling of nucleic acids is dependent upon the medium required for culturing the organisms and also upon the nucleic acid precursors which the organism will take up. As a result, label incorporation is often inefficient and it is difficult to predict the specific activities of the nucleic acids. For these reasons, the use of *in vivo* labelled nucleic acids in DNA homology and RNA hybridisation experiments has decreased in recent years. The reader who is interested in the *in vivo* labelling of nucleic acids is referred to the article by Johnson (1981) for practical considerations.

2. In vitro labelling

Preparations of DNA can be labelled by "nick translation" or by iodination. Preparations of RNA can be labelled by iodination, end-labelled, or labelled RNA can be synthesised by *in vitro* transcription. Advantages of these procedures include less manipulation of radioactive components, less radioactive waste material, efficient incorporation of the labelled isotope, and predictable levels of specific activity.

(*a*) *"Nick translation."*   This procedure involves the generation of single-strand breaks (nicks) followed by nucleotide removal and replacement by *Escherichia coli* polymerase I (Chelm and Hallick, 1976; Kelly et al., 1970; Nonoyama and Pangano, 1973; Rigby et al., 1977). Kits patterned after the procedure of Rigby et al. (1977) are available from several companies. These come with complete instructions and are easy to use.

(*b*) *End labelling.* Nucleic acids can be labelled either at the 3' end by polynucleotide transferase, or at the 5' end by polynucleotide kinase. Again, kits are available for carrying out these reactions.

(*c*) *Iodination.* Iodine can be bound to cytosine in the presence of thallium chloride ($TlCl_3$) at pH 4.8 (Chan *et al.*, 1976; Commerford, 1971; Orosz and Wetmur, 1974; Tereba and McCarthy, 1973). Technical problems associated with sample preparation have been investigated by Selin *et al.* (1983). The following procedure incorporates aspects from each of the references above.

(*d*) *Reagents*
   Buffer: 0.2 *M* sodium acetate buffer, pH 4.8. Autoclave.
   Buffer–salt solution: 5.7 *M* sodium perchlorate, 7×10^{-5} *M* potassium iodide prepared in 0.05 *M* sodium acetate buffer, pH 4.8. Autoclave and filter through a 0.45-μm membrane filter. Thallium chloride solution: 1.0 mg ml^{-1} prepared in 0.05 *M* sodium acetate, pH 4.8. This is very unstable at room temperature, never let it become warm. Store in a refrigerator and keep on ice when it is out of the refrigerator. Under these conditions it is stable for a year or more.
   ^{125}I: sodium [^{125}I]iodide, pH 10. Usually supplied at 100 μCi μl^{-1}, which can be diluted to 40–50 μCi μl^{-1} with sterile distilled water. DNA preparations with $1–2 \times 10^6$ cpm μg^{-1} are obtained using about 40 μCi, and rRNA preparations having about 5×10^5 cpm μg^{-1} are obtained using about 100 μCi of ^{125}I.
   PB–mercaptoethanol: 0.5 *M* Na_2HPO_4–NaH_2PO_4 buffer, pH 6.8, 0.1 *M* mercaptoethanol.
   SSC: Standard saline citrate, 0.15 *M* NaCl, 0.015 *M* sodium citrate, pH 7.0.

(*e*) *Procedure.* 1. Place 10 μg of DNA (or for rRNA labelling, 100 μg of rRNA) into a 1.5-ml microcentrifuge tube. Make it 0.3 *M* with 3 *M* sodium acetate, mix, add 2 volumes of 95% ethanol, mix, and place in a -20°C freezer for 1 h.

2. Centrifuge for 10 min and carefully draw off the supernatant with a Pasteur pipette. Keep the tip of the pipette near the surface, so that liquid is not moving at high velocity near the nucleic acid pellet. Carefully add 1.0 ml of cold (-20°C) 0.2 *M* sodium acetate buffer–95% ethanol (1 : 4 v/v) so as not to dislodge the nucleic acid pellet. Place the tube in the freezer for 10 min.

3. Centrifuge for 5 min, carefully remove the supernatant, and let the tube dry in an incubator for 30 min.

4. Dissolve the DNA in 35 μl of buffer–salt solution (70 μl of 0.05 *M*

sodium acetate buffer, pH 4.8, for an rRNA sample). Transfer the sample to a 1.0-ml serum vial. In a fume hood, add ^{125}I (up to 2.5 μl for DNA or up to 5 μl for an rRNA sample). Immediately cap the vial and evacuate with a syringe (about 1 ml syringe volume). Mix the sample, add 12.5 μl of TlCl$_3$ solution (25 μl for rRNA), mix again, and incubate in a 70°C water bath for 20 min.

5. Remove from the water bath, cool in an ice-water bath, and add 0.05 ml PB–mercaptoethanol (0.1 ml for rRNA).

6. Return the vial to the 70°C water bath for an additional 20 min. Remove and cool.

7. Using a 1-ml disposable syringe, inject about 0.15 ml of SSC buffer into the vial, mix, and then using the syringe, remove all of the material from the vial and place on a PD10 column (this is a disposable Sephadex G-25 column, Pharmacia). Add SSC to 2.5 ml, place a collection tube under the column, and add an additional 2.0 ml of SSC. Add 30 μg of sheared salmon sperm DNA to each DNA preparation. Cap the column and place in radioactive waste.

8. Place the tube containing the labelled nucleic acid in the 70°C water bath for another 20 min and then cool.

9. Place the nucleic acid preparation on another PD10 column equilibrated against 0.1× SSC. After adding sample and buffer to 2.5 ml, collect 3.5 ml into a plastic 50-ml centrifuge tube. Add 0.35 ml of 3 M sodium acetate, mix, and add 8 ml of 95% ethanol. Mix and place in freezer for at least 1 h. Centrifuge at 12,000 g for 10 min; carefully decant the tubes and allow them to air dry. Dissolve in 3.3 ml of 0.1× SSC (3.0 μg ml^{-1} for DNA and 30 μg ml^{-1} for rRNA).

F. Immobilisation of nucleic acids

The ability to immobilise a nucleic acid (usually DNA) on a solid support has been one of the more important techniques for the comparison of nucleic acids. DNA was first successfully immobilised in an agar matrix (McCarthy and Bolton, 1963; Hoyer et al., 1964) which was soon replaced by nitrocellulose membranes (Gillespie and Spiegelman, 1965). More recently diazotised paper has been introduced for the covalent attachment of nucleic acids (Stark and Williams, 1979). Methods described here will deal primarily with nitrocellulose membranes.

1. Nitrocellulose fixation with NaCl

Denatured DNA will interact with nitrocellulose in the presence of about 1 M NaCl, and after baking becomes nearly irreversibly immobilised (Nygaard and Hall, 1963; Gillespie and Spiegelman, 1965). Under these conditions there is very little binding of RNA; however, RNA can be effectively bound if 3 M NaCl is used for immobilisation (Thomas, 1980).

The following is an adaptation of the Gillespie and Spiegelman (1965) procedure for the immobilisation of DNA. The procedure is for a 15-cm-diameter membrane with a surface filtering area of about 173 cm^2. The immobilisation of 4.35 mg of DNA will result in 25 μg cm^{-2}.

1. Dilute 4.35 mg of DNA to 50 μg ml^{-1} with 0.1× SSC (approximately 90 ml).

2. Denature by heating the solution (in a 125-ml flask) in a boiling-water bath for 10 min, and then cool it quickly by pouring it into 800 ml of ice-cold 6× SSC (to give about 5 μg of DNA ml^{-1}).

3. Float the 15-cm nitrocellulose membrane filter on distilled water so that the pores will be filled with water (if the membrane is submersed immediately, air pockets will form and cause uneven filtration).

4. Place the wet filter on the filtration device (Johnson, 1981), and wash the membrane with 500 ml of cold 6× SSC, using a flow rate of approximately 30 ml min^{-1}. Then pass the denatured DNA through the filter and wash again with 500 ml of 6× SSC.

5. Let the membrane dry at room temperature and then bake either in a 80°C vacuum oven for 2 h or in a 60°C plain oven overnight.

6. Label the membrane on the edge with a pencil, cut the filter in half, separate the halves with half of a sheet of the paper that is used to separate the membranes in the shipping box, and place them into an envelope. Store at room temperature over CaSO$_4$ in a desiccator.

Handle the membranes by the outside edge, and avoid touching the surface on which the DNA is bound. When cutting small membranes out of the large one, avoid getting too close to the edge.

2. Nucleic acid fixation on nitrocellulose membranes with NaI

In a series of recent papers, Gillespie and his collaborators have described the use of NaI for the immobilisation of nucleic acids on nitrocellulose membranes (Bresser et al., 1983a,b; Bresser and Gillespie, 1983; Gillespie and Bresser, 1983). The major advantages of using NaI include: (1) the chaotropic properties of NaI which allows DNA and RNA to become freed of proteins so that they can be immobilised directly on nitrocellulose membranes. (2) NaI promotes the helix-to-coil transition of nucleic acids (NaCl promotes the reverse), so that high concentrations of NaI keep the nucleic acids in the denatured state and promote their immobilisation on nitrocellulose. (3) The nucleic acid binding is immediate which allows for rapid filtration and the elimination of the baking step. (4) One can selectively bind either DNA or mRNA (at least for eukaryotic mRNA) to the membranes.

The following protocol for the nitrocellulose immobilisation of DNA from small cultures of bacteria without prior isolation of the DNA is a variation of the Bresser and Gillespie (1983) and Bresser *et al.* (1983b) procedure.

(a) Reagents

Supersaturated NaI: 250 g NaI in 100 ml of water. Store in the dark at 25°C as saturated solution over NaI crystals. Heat to 75°C to make it supersaturated. To prepare a NaI saturated lysate (12.2 M), add 0.813 volume of the supersaturated solution to each volume of cell lysate.

Tris–EDTA (TE) buffer: 10 mM Tris, 1 mM EDTA, pH 8.0.

Lysozyme: 10 mg ml^{-1} in TE buffer.

Proteinase K: 1.4 mg ml^{-1} in TE buffer.

RNase mixture: Use the same mixture as in the section on DNA isolation.

Acetic anhydride solution: 0.1 M triethanolamine to which acetic anhydride (25 μl in 10 ml) is added just before use.

70% ethanol: 70% ethanol in water.

Nitrocellulose membrane: Schleicher and Schuell (Keene, New Hampshire) 0.45-μm porosity BA85 nitrocellulose membranes.

(b) Protocol.

The procedure is designed for a filter with approximately a 2-cm-diameter surface. Assuming 1 × 10^9 cells ml^{-1} and genome size of 2.5 × 10^9 Da, a culture will yield about 4 μg ml^{-1}. Therefore, 5 ml of culture will result in a reasonable amount of DNA per square centimeter of membrane. The volume can be adjusted for greater or less growth and differences in genome size and copies per cell.

1. Harvest a 5-ml culture at late log phase by centrifugation. Resuspend the cells in 2.5 ml of TE buffer, add 0.5 ml of lysozyme, and incubate at 37°C for 10 min.

2. Add 0.35 ml of RNase mixture. Freeze–thaw three times using an ethanol–dry ice bath. Add 0.25 ml of proteinase K and incubate for 30 min in a 50°C water bath. (For organisms that are not lysed by the above treatment, add an additional 0.5 ml of TE in place of the lysozyme, and after adding the RNase mixture, disrupt the cells by passage through a French pressure cell at 15,000 psi. Then add the proteinase K and pro-

ceed. If the cell disruption is not complete, remove the intact cells by centrifugation.)

3. Add 3.0 ml of supersaturated NaI, mix, and heat the lysate at 90°C for 5–10 min.

4. Cool the lysate to 50°C and pass it through a nitrocellulose filter at that temperature.

5. Wash the membrane free of NaI by soaking it with three changes of 70% ethanol at room temperature. Soak it for 5 min in each change.

6. Place the membrane in fresh acetic anhydride solution for 10 min at room temperature.

7. The membranes may be used immediately or they may be dried and stored in a desiccator.

3. Other supports

In addition to nitrocellulose, several other support membranes have been used for immobilising nucleic acids. These include nylon membranes such as Genescreen (New England Nuclear, Boston, Massachusetts), Biodyne (Pall Corporation, Glen Cove, New York), and Zetapor (AMF Cuno, Meriden, Connecticut) and diazotised paper such as Transabind (Schleicher and Schuell, Keene, New Hampshire). Although these membranes and paper can be used for the direct immobilisation of nucleic acids they have been used mostly in conjunction with the transferring of nucleic acids from agarose or polyacrylamide electrophoresis gels. Protocols for the use of these materials are available from the manufacturers.

G. Reassociation and hybridisation kinetics

Similarities in the linear arrangement of nucleotides along DNA (or RNA) strands from different organisms are determined by measuring the extent to which DNA (or RNA) from one organism forms DNA duplexes (reassociation) or RNA hybrids (hybridisation) with DNA from another, or by comparing the thermal stability of the heterologous DNA duplexes or RNA hybrids with homologous duplexes or hybrids. These reactions may be carried out either in free solution or with one of the reactants immobilised on a filter. The major parameters that affect the rates of these reactions include (1) the concentrations of the reactants, (2) the ionic strengths (salt concentrations) of the buffers, (3) the temperature of reassociation or hybridisation, (4) the concentration of organic solvents such as formamide, dimethyl sulphoxide, or dextran sulphate, (5) the extent of nucleotide sequence mismatch in heterologous reassociation, and (6) the genetic complexity of the nucleic acid. The free solution reactions can be described by either second-order kinetics (DNA reassociation or the hybridisation of RNA with an excess of DNA) or pseudo-first-order kinetics (DNA with an excess of RNA). Although immobilising one of the nucleic acids on a membrane makes the kinetic description of the reaction more difficult, it is affected by the above parameters in much the same manner as the free solution reactions.

The kinetic properties of DNA reassociation (Britten and Davidson, 1976; Galau et al., 1977a; Smith et al., 1975) and RNA hybridisation (Bishop, 1969; Smith et al., 1975; Galau et al., 1977a,b; Melli et al., 1971) have been investigated extensively.

The second-order reassociation of DNA (Britten and Kohne, 1968; Wetmur and Davidson, 1968) is illustrated by the so-called C_0t curve. In this example, the usual form of the equation [Eq. (1)] is rearranged [Eq. (2)], where C is the concentration of single-stranded DNA at time t in moles of nucleotides per litre, C_0 is the concentration of single-stranded DNA at time zero in moles of nucleotides per litre, t is the time in seconds; and k is the reassociation rate constant:

$$1/C - 1/C_0 = kt \tag{1}$$

$$C/C_0 = 1/(1 + kC_0t) \tag{2}$$

Milligram per millilitre DNA concentration can be expressed as moles per litre by dividing it by 331 (the average molecular weight of the deoxyribonucleotide sodium salts). Equation (2) represents a hyperbolic curve for which the general formula is

$$y = 1/(1 + ax) \tag{3}$$

When $y = 1/2$ in Eq. (3), $x = 1/a = x_{1/2}$ and $a = 1/x_{1/2}$. If the logs of the x values are plotted, when $y = 1/2$, $\log a = 1/x_{1/2} = -\log x_{1/2}$. A general log C_0t plot is illustrated in Fig. 2. There is an almost linear region on the curve that extends for about 2 logs. This is because any significant change in C/C_0 occurs between $1/(1 + 0.1)$ and $1/(1 + 10)$, i.e., when kC_0t is between 0.1 and 10. The rate constant $k = 1/C_0t_{1/2}$. The units for the rate constant are litres $mol^{-1} t^{-1}$.

One can now consider the effects of some of the parameters listed above on the reassociation kinetics. The y values represent the fraction of DNA fragments that are still single-stranded. The x values (either logs of the values or the values plotted on semi-log paper) are dependent upon the interrelationship of the DNA concentration and the reassociation time.

The x values are dependent upon the genome size. If the genome of an

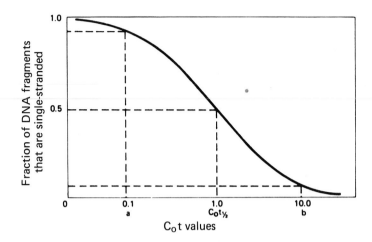

Fig. 2. Generalised C_0t plot. At values less than or equal to a, $1/(1 + C_0t)$ values are 0.9 or greater. At C_0t values equal to or greater than b, $1/(1 + C_0t)$ values are less than 0.1. From Johnson (1981).

organism is 10 times larger than the one illustrated in Fig. 2, the curve would move up the scale 1 log, i.e., the $C_0t_{1/2}$ would $= 10$. If the genome of the organisms was 10 times smaller, the curve would move down the scale 1 log. Since most bacteria do not differ greatly in genome size, there will be little if any detectable differences with DNA from one organism to the next.

The effect of salt concentration and reassociation temperature is shown for *Bacteroides vulgatus* DNA in Fig. 3. The reassociation rate in 6× SSC is three times faster than in 2× SSC and about 1.5 times faster than in 4× SSC. Salt concentrations less than 2× SSC would result in much slower rates, although concentrations greater than 6× SSC do not significantly increase the rate (Marmur and Lane, 1960). Temperatures 25°C below the midpoints of the melting temperatures (T_m) are noted by the arrows and the optimal temperatures are between 20 and 25°C below the T_m, as originally observed by Marmur and Doty (1961). Britten *et al.* (1974) have presented in table form the influence of phosphate buffer concentration on the reassociation rate of DNA.

Several organic compounds depress the thermal stability of DNA duplexes and RNA hybrids. These include formamide (Bonner *et al.*, 1967; McConaughy *et al.*, 1969; Schmeckpeper and Smith, 1972; Hutton, 1977), urea (Kourilsky *et al.*, 1971; Hutton, 1977), and dimethyl sulphoxide (Legault-Demare *et al.*, 1967). These allow one to do reassociation and hybrid-

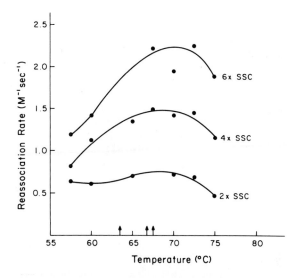

Fig. 3. Effect of salt concentration on the reassociation rate of *Bacteroides vulgatus* DNA (41% G + C content). The arrows indicate temperatures that are 25°C below the melting temperatures of the DNA at these salt concentrations.

isation experiments at lower temperatures so that there is less thermal scission of the nucleic acid strands, particularly RNA strands (Gillespie and Gillespie, 1971). Formamide has been more extensively used than the others and will be the only one discussed in this chapter.

The effects of formamide and SSC concentration on the T_m's of *E. coli* and *Clostridium perfringens* DNA are shown in Fig. 4. Within the range of SSC concentrations usually used for reassociation experiments, the effect of salt concentration is rather minimal. Formamide, however, lowers the T_m 0.61°C per each percent for *E. coli* DNA and 0.50°C per each percent for *C. perfringens* DNA. The effect of formamide on the thermal stability of RNA hybrids is not as pronounced as on DNA duplexes; thus, at high concentrations of formamide, 50–80%, conditions can be obtained in which the hybridisation of rRNA proceeds in the absence of DNA reassociation (Vogelstein and Gillespie, 1977; Casey and Davidson, 1977). In both instances, *E. coli* nucleic acids were employed in the experiments, where the G + C content of both the DNA and rRNA are similar. Therefore, conditions for these experiments can be met for organisms with DNA having a G + C content of 50 mol% or less, but may not be for organisms having DNA with a G + C content significantly greater than 50 mol%.

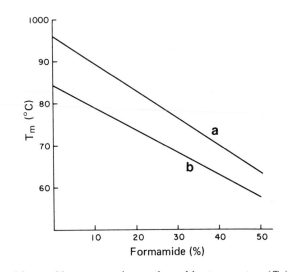

Fig. 4. Effect of formamide concentration on the melting temperature (T_m) of (a) *Escherichia coli* and (b) *Clostridium perfringens* DNA.

The addition of 10% dextran sulphate 500 (Pharmacia) to reassociation solutions can increase the rates 10-fold (Wetmur, 1975). Similar or even greater rate increases have been observed for hybridisation or reassociation with immobilised DNA (Wahl *et al.*, 1979) and with cytological preparations (Lederman *et al.*, 1981). The effect of dextran sulphate appears to be the tying up of water, such that the effective concentrations of the reactants become greater and thus increase the rates.

The reassociation rates that have been considered above are for the homologous reactions; however, in homology experiments most of the reactions are heterologous. Under optimal reassociation conditions, for heterologous duplexes that have diverged to the extent of lowering the thermal stability 10–15°C below that for the homologous duplexes, the association rate is about two times slower than for the homologous duplexes (Bonner *et al.*, 1973; Marsh and McCarthy, 1974). If the reassociation C_0t value is 40–50 times greater than the homologous $C_0t_{1/2}$ value, heterologous duplexes capable of forming will have been formed.

Although an organism's genome size has a profound effect on the reassociation rate, the genome size variation among bacteria is not great enough to require special consideration of this for most homology experiments. There appear to be 8–10 rRNA genes in most bacteria, so reassociation involving these genes would be faster than for the bulk of the genes. Reassociation experiments involving bacterial phage or plasmids will, of course, have much lower C_0t values.

H. DNA reassociation

The reassociation of DNA in solution can be measured optically by use of an ultraviolet recording spectrophotometer, or if a probe (either radioactive or biotin labelled) is used, the reassociation of the probe with excess unlabelled DNA can be followed by resistance to S_1 nuclease or by adsorption to hydroxylapatite.

The procedures work best with highly purified DNA and the concentrations should be accurately determined. It is important that the DNA in all of the preparations be uniformly fragmented. This is best accomplished by passing each preparation ($0.2–0.5$ mg ml^{-1}) two or three times through a French pressure cell at 15,000 psi. This will result in an average fragment size of $4–5 \times 10^5$ Da. If a French pressure cell is unavailable, a sonicator will also work; however, because of variations in energy and tip diameters, sonicators should be calibrated in terms of energy setting and the length of sonication required under a standard set of conditions.

1. Optical

The reassociation of DNA may be measured spectrophotometrically by the initial rate of reassociation (Wetmur and Davidson, 1968; De Ley *et al.*, 1970) or by determining the time and DNA concentrations at which one-half ($C_0 t_{1/2}$) of the DNA has reassociated (Britten and Kohne, 1968; Seidler and Mandel, 1971; Kurtzman *et al.*, 1980).

The most popular method for measuring homology between DNAs from two organisms has been the initial reassociation rate method described by De Ley *et al.* (1970). With this procedure one compares the rates of reassociation of preparations of DNA from organism A (V_A) and from organism B (V_B) with that of an equal mixture of the two DNA preparations (V_M). The percentage homology can be calculated from the following formula (De Ley *et al.*, 1970):

$$\text{percentage homology} = \frac{4\,V_M - (V_A + V_B)}{(V_A\,V_B)^{1/2}}\,(100)$$

With this formula, one makes the assumption that the reassociation rate constants for the two organisms are the same (i.e., they have the same genome size), which will be true for bacterial and most double-stranded DNA viruses, although one may need to be cautious when comparing virus and plasmid DNA preparations. Also the formula compensates for differences in DNA concentrations in the preparations from organisms A and B by averaging the two rates. As a result, only arbitrary rates (i.e., change in absorbance) per unit time are needed for DNA homology calculation.

An ultraviolet spectrophotometer with an automatic sample positioner and cuvette temperature control is required for these experiments. The best type of temperature control is an electronically heated cuvette holder which holds small (0.2- to 0.3-ml) Teflon-stoppered quartz cuvettes. The samples can be readily heated to denaturing temperature and then rapidly cooled to a reassociation temperature.

The DNA preparations should be of similar concentration, fragmented to similar fragment lengths (see above), and dissolved in $0.1 \times$ SSC buffer. A reassociation mixture should contain about 75 μg ml^{-1} DNA, a 4 or 6$\times$ SSC concentration, and formamide concentration in the range of 25–50%. The major parameter affecting the conditions selected will be the mol% G + C content of the organisms being compared. The following protocol is for DNA reassociation mixtures containing 4$\times$ SSC and 25% formamide.

1. Dilute fragmented DNA preparations to 150 μg ml^{-1} with $0.1 \times$ SSC. Prepare 1.0 ml and use an aliquot to obtain an accurate concentration measurement; if needed, adjust the concentration.

2. Prepare a solution containing 8$\times$ SSC and 50% deionised formamide [deionise formamide by mixing 5–10 g of mixed bed ion-exchange resin such as AG 501-$\times$8 (Bio-Rad Laboratories) per 100 ml and after thorough mixing, allow the resin particles to settle to the bottom of the bottle or filter them out].

3. Mix 0.75 ml of each DNA preparation with 0.75 ml of the 8$\times$ SSC–50% formamide solution. This will result in preparations containing 75 μg ml^{-1} DNA, 4$\times$ SSC buffer, and 25% formamide. Prepare DNA mixtures (i.e., DNA from organism A with DNA from organism B) by combining 0.50-ml volumes of the DNA–4$\times$ SSC–25% formamide mixtures. As a blank, prepare a mixture containing no DNA.

4. Fill the cuvettes, making sure to leave an air space for thermal expansion, with the buffer blank mixture, with DNA from organism A, with DNA from organism B, and with the DNA mixture from organisms A and B.

5. Set the spectrophotometer at 270 nm (formamide does not interfere with absorbance at this wavelength) and place the cuvettes in the cuvette holder set at the reassociation temperature. Allow the temperature of the samples to equilibrate for about 5 min and then increase the temperature at 1°C min^{-1} to 5°C above the end of the melting profile. After 5 min at this temperature, rapidly lower the temperature to the reassociation temperature, which should be 22°C below the T_M in the same buffer system (this takes about 1.5–2.5 min). Follow the reassociation for about 20 min.

6. From the initial linear portion of each reassociation curve, measure the absorbance change per unit time. Calculate the DNA homology value by inserting these values in the equation above.

2. S_1 nuclease and hydroxylapatite

The S_1 nuclease and hydroxylapatite procedures represent two methods for assaying the extent of reassociation of a small quantity of labelled (usually radioactively labelled) probe DNA with an excess of unlabelled DNA. S_1 nuclease is specific for single-stranded DNA, and therefore any unreassociated probe DNA or segments of probe DNA fragment that is not in a duplex structure (either a loop or tail) will be degraded by the nuclease. By using the proper ionic strength, 0.14 M sodium phosphate buffer, pH 6.8 (0.14 M PB), double-stranded DNA will adsorb to hydroxylapatite and single-stranded DNA will not. As a result, the reassociated fragments of probe DNA can be separated from those unreassociated, but fragments that are only partially duplexed will be selected as a fully duplexed fragment. Thus, the two methods do not result in identical results, but results can be readily correlated (Grimont *et al.,* 1980).

The procedures described are for the use of ^{125}I-labelled probe DNA, with a specific activity of at least 1×10^6 cpm μg^{-1}. Some of the special equipment that will be referred to in the procedures has been described and illustrated in detail (Johnson, 1981).

(a) Reagents and components. 1. *Concentrated SSC or NaCl equivalent:* Prepare and dilute these with 10^{-3} M (pH 7.0) HEPES buffer. Reassociation experiments have traditionally been carried out in a SSC salt solution, but to reduce the chelating effect of citrate during S_1 nuclease digestion, a NaCl solution that is ionically equivalent to SSC has also been used. The concentrates are prepared such that the addition of 25 μl to a total reassociation volume of 110 μl will result in the desired salt concentration.

Reaction concentration	Concentrate
6× SSC	26.4× SSC (3.94 M NaCl, 0.394 M sodium citrate, 10^{-3} M HEPES, pH 7.0)
6× SSC equivalent	5.15 M NaCl, 10^{-3} M HEPES, pH 7.0
4× SSC	17.6× SSC (2.64 M NaCl, 0.264 M sodium citrate, 10^{-3} M HEPES, pH 7.0)
4× SSC equivalent	3.43 M NaCl, 10^{-3} M HEPES, pH 7.0

2. *Unlabelled DNA preparations.* Fragment DNA preparations (dissolved in 0.1× SSC) by passing each two or three times through a French

pressure cell at 15,000 psi. Denature the preparations by heating in a boiling-water bath for 5–10 min, then quickly cool by placing the tubes in ice water. Centrifuge the preparations for 10 min at 17,000 g to remove any particulate debris. Adjust each DNA preparation to 0.4 mg ml^{-1} (other concentrations can also be used as long as one compensates for the reassociation time and labelled DNA concentration). Also use these preparations for the *in vitro* iodination. This assures that the labelled fragments are of the same size range as the unlabelled DNA.

3. *Formamide:* Prepare deionised formamide as described above.

(*b*) *Procedure.* 1. The reassociation vials that we use are constructed out of 6 mm O.D. × 24 mm lengths of hard glass tubing. One end is sealed with a flame and the other is capped with a serum vial stopper (Johnson, 1981). Other small vials, such as microcentrifuge tubes, should also work well, as long as they can be incubated submerged without leaking.

2. Place in each vial 10 μl of labelled DNA (30,000 cpm, 0.03 μg or less of DNA), 50 μl of unlabelled DNA (0.4 mg ml^{-1} = 20 μg per vial), 25 μl of salt concentrate (26.4 or 17.6× SSC or NaCl equivalent), and 25 μl of deionised formamide. Mix on a vortex mixer, stopper, and incubate submerged in a water bath at a temperature 25°C below the T_M of the reference DNA (i.e., labelled DNA) measured in the salt-formamide mixture used in the experiment.

(*c*) *Comments.* 1. Each experiment requires two controls. The first is for measuring the self-reassociation of the labelled DNA. For this, 50 μl of salmon sperm DNA (0.4 mg ml^{-1}, fragmented but not denatured) is substituted for the denatured bacterial DNA. The second control is for measuring the homologous reassociation. Have four replicates of each control and duplicates of each DNA preparation that is being compared.

2. The length of time required for reassociation will depend upon the salt and the unlabelled DNA concentrations. For example, 50 μl of 0.4 mg ml^{-1} unlabelled DNA results in 20 μg 110 μl^{-1} or 182 μg ml^{-1} or 5.5 × 10^{-4} M nucleotides. A reassociation time of 24 h results in a C_0t value of (5.5 × 10^{-4} M nucleotides) (8.64 × 10^4 s) = 47.5 M s. The $C_0t_{1/2}$ of *B. vulgatus* DNA in 6× SSC (Fig. 3) is about 0.5; therefore, the reassociation is about 10 times longer than the time required for the total reassociation of the homologous DNA. Using 4× SSC will result in a final C_0t value about two-thirds that of 6× SSC. Using one-half the amount of unlabelled DNA will cut the final C_0t value by one-half. Lower DNA concentrations can be compensated for by longer reassociation times. If the reassociation times are extremely long, one may have to consider the self-reassociation of the labelled DNA. Using smaller amounts of higher specific activity DNA will

eliminate this as a possible problem. The major consideration is that heterologous reassociation may be two to four times slower than homologous reassociation, so one does not want to terminate the reassociation reaction prematurely.

(*d*) *Duplex analysis.* The extent of duplex formation can be measured by either adsorption to hydroxylapatite or by S_1 nuclease digestion of the unreassociated DNA.

Hydroxylapatite procedure. The hydroxylapatite (HA) procedure separates double-stranded DNA from single-stranded DNA by selective adsorption. Double-stranded DNA will adsorb to the HA in 0.14 *M* phosphate buffer, whereas single-stranded DNA will not (Brenner *et al.,* 1969).

1. Place 0.5 g of dry HA into a 13 × 100-mm test-tube, add 3 ml of 0.14 *M* phosphate buffer (PB; equimolar NaH_2PO_4 and Na_2HPO_4) containing 0.4% SDS, suspend the HA with a Thermomix mixer, and warm the tube in a 60°C water bath.

2. Add the reassociation mixture from the vial (see above), and rinse the vial twice with 100 μl of 0.14 *M* PB–0.4% SDS. Mix each tube and return all of them to the water bath. Allow the temperature to equilibrate for 5 min, mix each tube, and immediately return it to the bath for an additional 5 min.

3. Centrifuge the tube at 5000 *g* for 3–4 min, decant the buffer into a 15.8 × 100-mm polypropylene test-tube (we use this size because it will fit directly into our gamma scintillation counter), and add another 3 ml of PB–0.4% SDS and return the tubes to the water bath. At 5 min mix each tube and at 10 min repeat the centrifugation. Repeat this washing procedure four times, pooling two washes per scintillation tube.

4. Add 3 ml of 0.28 *M* PB, mix, and centrifuge. Repeat the elution step twice and pool these into one scintillation tube.

5. The summation of counts in all three scintillation tubes represents the total radioactivity in the reassociation vial and the total count in the third scintillation tube represents the duplexed labelled DNA adsorbed to the HA.

Calculate homology values by dividing the amount of heterologous reassociation by the amount of homologous reassociation and multiplying by 100.

Comments. This is only one of several ways that HA can be used for measuring DNA reassociation. It was originally developed as a column procedure (Miyazawa and Thomas, 1965), and Brenner *et al.* (1969) then described an effective batch procedure in which larger amounts of DNA are used. Lachance (1980) has described the use of microcolumns, and Sibley and Ahlquist (1981) have described an apparatus for the automated thermal elution of DNA from multiple HA columns.

S_1 *nuclease procedure.* This procedure is based on the observation by Ando (1966) that S_1 nuclease from *Streptomyces oryzae* will, under carefully controlled conditions, hydrolyse single-stranded nucleic acid but not double stranded. Therefore, the extent of duplex formation between radioactively labelled DNA fragments and an excess of unlabelled DNA fragments can be determined by measuring the amount of S_1-resistant (i.e., acid-precipitable or binding to DEAE paper; Maxwell *et al.*, 1978; Popoff and Coynault, 1980) radioactivity. The procedure described here is a variation of the one described by Crosa *et al.* (1973).

Titration of S_1 nuclease. Determine the nuclease activity of the stock S_1 nuclease in the following manner.

1. Place 50-μl amounts of a 0.6 mg ml^{-1} sheared, denatured bacterial DNA (''spiked'' with ^{125}I DNA) preparation (dissolved in 0.02 M NaCl–1 mM HEPES buffer, pH 7.0) into a series of polypropylene tubes containing 1.0-ml assay buffer [0.05 M sodium acetate–0.3 M NaCl–0.5 mM ZnCl$_2$, pH 4.6; this is a variation of Vogt's buffer A (Vogt, 1973)]

2. Add 50-μl amounts of twofold dilutions of the S_1 nuclease using the assay buffer for making the dilutions. Mix the contents of each tube, and incubate at 50°C for 1 h.

3. Add 50 μl of sheared salmon sperm DNA (0.6 mg ml^{-1}) to each tube and acid precipitate by adding one-fifth volume of 1 N HCl–1% sodium pyrophosphate (Na$_4$P$_2$O$_7$ · 10H$_2$O)–1% monosodium phosphate (NaH$_2$PO$_4$) to each tube (Rigby *et al.*, 1977), mix, and place at 4°C for 1 h. Collect the precipitates on nitrocellulose or GF/F glass fibre filters, washing the filters with cold acid wash solution (the acid stock solution diluted one-fifth with water).

Perform the homology assays using a dilution of the enzyme that is twice as concentrated as that required to hydrolyse the 30 μg of titrating DNA effectively. Dilute only the amount of enzyme that is required for each experiment. The enzyme is very stable when stored at -20°C, so only if an enzyme preparation is kept for longer than 6–8 months is a second titration needed.

Assay the 110-μl volume reassociation mixtures (as described above) with S_1 nuclease; proceed as follows:

1. Quantitatively transfer the reaction mixture from each vial into a polypropylene test-tube containing 1.0 ml of assay buffer (0.05 M sodium

acetate–0.3 M NaCl–0.5 mM ZnCl$_2$, pH 4.6), and 50 μl of 0.5 mg ml^{-1} denatured salmon sperm DNA (do not add to control tubes that will not receive S$_1$ nuclease, i.e., tubes for estimating total radioactivity per reaction vial). The reaction vials can be rinsed with a couple of 100-μl volumes of assay buffer. Mix the tube contents on a Thermomix mixer, add 50 μl of diluted S$_1$ nuclease, and mix again. Incubate the tubes in a 50°C water bath for 1 h.

2. Remove the tubes from the water bath, add 50 μl of 0.6 mg ml^{-1} fragmented native salmon sperm DNA, and 0.4 ml of the acid solution (1 N HCl, 1% sodium pyrophosphate). Mix the contents of each tube and place at 4°C for 1 h.

3. Collect the precipitates on nitrocellulose or GF/F glass fibre filters, dry the filters, place them into scintillation vials, and measure the radioactivity.

Calculate the homology values by first subtracting the self-reassociation of the labelled DNA (i.e., labelled DNA reassociated in the presence of native salmon sperm DNA) from each reassociation mixture, and then dividing the counts per minute of the heterologous S$_1$-resistant DNA by the counts per minute of the homologous S$_1$-resistant DNA, and multiplying by 100.

Thermal stability procedures. The extent of base-pair mismatching can be estimated by comparing the thermal stability of heterologous duplexes with the stability of homologous duplexes. Thermal stability measurements can be determined using either the HA or S$_1$ nuclease assay procedures.

Thermal stability measurements using HA differ from the assay procedure above in that after the non-duplexed labelled DNA has been washed from the HA, the temperature of the water bath is increased in increments (usually 5.0°C), and the washing steps continued until all of the labelled DNA is denatured and eluted from the HA. This can readily be done using batch or column procedures (Brenner *et al.,* 1969; Lachance, 1980; and Sibley and Ahlquist, 1981).

When S$_1$ nuclease is used for determining thermal stability values, the denaturation step (i.e., temperature increases) is done directly on the reassociation mixture, which is then assayed for remaining S$_1$ nuclease-resistant duplexes.

1. Prepare a multiple volume (12–15 times larger) reaction mixture which is then reassociated in the usual manner, but in a screw-cap test-tube.

2. After reassociation one might wish to increase the volume of the

reaction mixture by adding additional reassociation buffer or additional formamide (to lower the duplex stabilities) so that larger and perhaps more accurate aliquots can be removed.

3. Place the tubes in a circulating water bath. Maintain the bath water level just below the cap to avoid concentration changes in the tubes due to evaporation and condensation on the tube walls.

4. Start the denaturation profile 30°C below the expected $T_{m(e)}$ of the homologous duplexes. Increase the water bath temperature by 5°C increments. Allow 5 min to attain each temperature (may need a little longer at the higher temperatures) and maintain each temperature for 5 min. At the end of each temperature cycle, remove an aliquot for S_1 nuclease digestion (see above), using a standard pipette or a positive displacement micropipette (other micropipettes will be inaccurate as a result of the temperature variations).

Plot the S_1 nuclease-resistant counts per minute versus the temperatures. The temperature at which 50% of the duplexes become dissociated and thus digested by S_1 nuclease is called the $T_{m(e)}$ (named for the dissociation and subsequent elution from an HA column). The difference between the homologous and heterologous $T_{m(e)}$ values is the $\Delta T_{m(e)}$ value.

3. Immobilised DNA procedures

An immobilised DNA procedure represents the first DNA or RNA homology method that was amenable to routine measurements of DNA similarity. Originally high-molecular-weight and denatured unlabelled DNA was immobilised in an agar gel matrix (McCarthy and Bolton, 1963), which was later replaced by the immobilising of DNA on nitrocellulose membrane filters (Gillespie and Spiegelman, 1965; Denhardt, 1966).

These experiments can be done by directly measuring the amount of duplex formation between labelled DNA fragments in solution and various unlabelled DNA preparations, or indirectly by the inhibition (competition) of the homologous labelled DNA and immobilised DNA duplex formation with excess amounts of heterologous DNA fragments in solution. Because of subtle differences from one immobilised DNA preparation to the next, the direct binding measurement is more qualitative, while the competition procedure, which usually involves the use of small membranes cut from a large membrane containing immobilised DNA, results in more quantitative measurements. In recent years, the membrane competition procedure has been largely replaced by either the HA or S_1 nuclease procedures because of its requirement for rather large amounts of DNA.

For additional information on membrane competition experiments, see Johnson (1981). Following are methods suitable for the rapid qualitative detection of specific DNA sequences.

(*a*) *Labelled probes.*   The radioactively labelled DNA or RNA (probe) will determine the uniqueness of a particular experiment. The probe may represent an RNA, total bacterial DNA, plasmid DNA, or specific sequences of bacterial or plasmid DNA (cloned DNA).

(*b*) *Preincubation of the DNA-containing membranes.*   The Denhardt (1966) preincubation mixture (0.02% bovine serum albumin, fraction V; 0.02% polyvinylpyrrolidone; and 0.02% Ficoll 400) prepared in 6× SSC is used to cover the remaining DNA binding sites on the membrane filters.

1. In most cases, the membranes used in an experiment will be small ones, cut from a larger one prepared as described above. We use a Keysort punch that makes 3 × 9-mm membranes; unfortunately these punches (McBee Systems) are no longer available. Other punch shapes will also work, although one may have to alter the reaction volumes (see below).

2. Preincubate the membranes at the incubation temperature of the experiment for 1–2 h. Agitate the membranes several times during the preincubation period.

3. Remove the membranes, place on a paper towel to blot off excess liquid, and then place into the reaction vials.

(*c*) *Reassociation.*   The reassociation volume should be the minimum that will completely cover the membrane. The following protocol is for 3 × 9-mm filters and 110-μl reassociation volumes.

1. Place into each reaction vial 10 μl of labelled DNA and 100 μl of 6.6× SSC prepared in 55% formamide. Mix each vial on a Thermomix.

2. Place the membranes in the vials, stopper, and incubate at a temperature 15°C below the T_m of the reference DNA (i.e., labelled DNA) in 6× SSC–50% formamide. For bacterial DNA, an overnight reassociation period is usually sufficient; however, a preliminary time-course experiment to determine the time required for a specific system may be well worthwhile.

3. Remove the membranes from the vials and place in a washing chamber (Johnson, 1981). Wash in 2× SSC at the incubation temperature. Use two 300-ml volumes of washing buffer, and wash for 5 min in each. Move the washing chamber back and forth in the beaker several times during the washing period to assure that the buffer passes through each compart-

ment. Shake the washing chamber free from buffer, and remove the membrane filters, and dry on a paper towel under a heat lamp.

4. Determine the radioactivity on the membranes using an appropriate scintillation counter.

5. Calculate the percentage homology by dividing the counts bound by the heterologous DNA membrane by the counts bound by the homologous DNA membrane and multiplying by 100.

(*d*) *Thermal stability of duplexes.*   Follow the general procedure above but reassociate at 25°C below the T_m of the reference DNA and do not dry the membranes after the washing step. This procedure is described for using ^{125}I-labelled DNA.

1. Remove the membrane filters from the washing chamber, skewer on a stainless-steel insect pin near the head end, and insert the pin into the end of a rubber stopper core plugging a short piece of glass tubing (Johnson, 1981), which is used as a handle to transfer the membranes from one tube to the next. Each membrane (or membranes) is then placed into the first elution tube.

2. Into each elution tube (we use 15 × 84-mm polypropylene tubes which fit directly into the gamma counter) put 1.0 ml of 4× SSC–1 mM HEPES buffer–50% formamide. Start the elution profile at about 35°C below the $T_{m(e)}$ of the homologous duplexes and increase the bath temperature at 5°C increments as described for the HA and S_1 nuclease thermal stability profiles above. At each temperature, move the membranes up and down a couple of times during the incubation period and touch against the side of the tube when transferring them from one tube to the next to drain off most of the buffer.

3. The elution profiles are calculated in the same manner as for the HA procedure described above.

I. RNA hybridisation

The bacterial cell (as do eukaryotic cells) contains several classes of RNA. These include messenger RNA (mRNA), ribosomal RNA (rRNA), and transfer RNA (tRNA). Since all of the RNA molecules are transcribed from one of the DNA strands, there are no associations between RNA molecules, although there is a lot of secondary structure within the tRNA and rRNA molecules.

The mRNA is transcribed from the largest fraction of the bacterial genome, and as a result total mRNA homology results are very similar to

DNA homology results (McCarthy and Bolton, 1963). Since there is only a very small amount of mRNA in the cell, mRNA is rapidly turning over, and is very difficult to separate from rRNA; mRNA has not been used in homology studies. Most investigations involving mRNA have dealt with the synthesis and stability of specific mRNA molecules. The tRNA molecules are very small (about 80 nucleotides) and represent about 10% of the RNA in a bacterial cell. Although many of these molecules from *E. coli* have been sequenced and Brenner *et al.* (1970) have shown that these molecules are conserved among the enteric bacteria, hybridisation studies have been limited.

Most comparative RNA hybridisation studies have been performed with the 16 S and/or 23 S rRNA molecules. These two types of molecules account for about 80% of the nucleic acid in a bacterial cell and can be readily isolated. In addition, the oligonucleotides resulting from the T_1 ribonuclease digestion of 16 S rRNA have been compared for a wide range of organisms by Woese and collaborators (Fox *et al.*, 1980).

The G + C content of rRNA is in the 52–54 mol% range, and as a result requires reasonably high hybridisation temperatures of 65–70°C in 2× SSC (Gillespie and Spiegelman, 1965; Johnson and Francis, 1975), which causes thermal degradation of the rRNA (Gillespie and Spiegelman, 1965; Gillespie and Gillespie, 1971). Including formamide in the buffer system allows one to use lower hybridisation temperatures which alleviates the rRNA degradation problem (Gillespie and Gillespie, 1971).

An additional interesting property of formamide is that it does not depress the thermal stability of RNA hybrids to the same extent as DNA duplexes. At higher formamide concentrations, RNA hybrids are progressively more stable than comparable DNA duplexes, and the maximum hybridisation rate occurs at a higher temperature relative to DNA duplex formation (Birnstiel *et al.*, 1972; Schmeckpeper and Smith, 1972). Using this property, Vogelstein and Gillespie (1977) and Casey and Davidson (1977) have optimised conditions for the hybridisation of *E. coli* rRNA, while at the same time totally repressing DNA duplex formation. These conditions may be difficult to attain when the average G + C content of the DNA is in the 60–70 mol% range.

1. RNA hybridisation with immobilised DNA

Following are procedures for direct binding, thermal stability, and competition experiments (Johnson and Harich, 1983). The procedures are described for the use of 3 × 9-mm membranes (Johnson, 1981); other membrane sizes may require volume adjustments.

(a) *Reagents*

^{125}I-*labelled rRNA:* Use 16 S plus 23 S rRNA labelled as described above. The specific activity should be in the range of $2-6 \times 10^5$ cpm μg^{-1} and at a concentration of 30 μg ml^{-1} in 1 × SSC–1 mM HEPES (pH 7)–0.5% SDS.

Concentrated SSC: Prepare 17.6× SSC–1 mM HEPES (pH 7) buffer. When 25 μl of this is used per reaction vial, a 4× SSC concentration will result.

Competitor rRNA: 2.0 mg total RNA ml^{-1} as isolated above in 1× SSC–1 mM HEPES (pH 7)–0.5% SDS. Heat each preparation for 5 min in a boiling-water bath at the time of preparation to disrupt secondary structure. Store at $-20°$C.

RNase mixture: Prepare concentrated mixture as described above. The final concentrations should be 10 μg RNase A ml^{-1} and 0.25 units of RNase T$_1$ in 2× SSC.

Immobilised DNA: Prepared as described above with about 25 μg DNA mm^{-2} of membrane.

(b) *Direct binding procedure.* 1. Preincubate membranes for 1–2 h at 50°C in the Denhardt solution prepared in 2× SSC. Remove from the preincubation mixture and blot on a paper towel just prior to placing them into the reaction vials.

2. Reaction mixtures consist of 10 μl of labelled rRNA (0.3 μg), 25 μl 1× SSC–1 mM HEPES, 25 μl 17.6 × SSC, and 25 μl of deionised formamide. Mix the contents of each vial on a Thermomix mixer.

3. Place the DNA-containing membranes into the vials; incubate the vials at 50°C for 16–20 h.

4. After completion of the hybridisation reaction, place the membranes into a washing chamber (Johnson, 1981) and wash them twice in 300 ml of 2× SSC warmed to 50°C. The washing chamber is placed in the RNase mixture and incubated for 1 h at 37°C. The membranes are then washed a final time in 300 ml 2× SSC at room temperature.

5. Dry the membranes and measure their radioactivity in a gamma scintillation counter.

Although this procedure has been used for the measurement of rRNA homology, the problems of the extent of hybridisation to heterologous DNA as compared to homologous DNA, the variability of the amount of DNA from one membrane to the next, and the variable number of rRNA gene copies per genome result in only qualitative measurements.

(c) *Thermal stability profiles.* 1. Follow the direct binding procedure above, but do not dry the membranes after the final wash.

2. Follow the procedure as outlined for DNA duplexes above. Start the profile at 35°C and increase to 85°C at 5°C increments.

(*d*) *Competition experiments.*   This procedure is also very similar to the direct binding experiment, except that only membrane-immobilised DNA homologous to the labelled rRNA is used and an excess of unlabelled rRNA is included in most of the reaction vials. The experiment measures the ability of a heterologous unlabelled rRNA to compete with the hybridisation of the labelled rRNA relative to that of unlabelled homologous rRNA. Therefore, the experiment must include a measurement of the direct binding of the labelled rRNA in the absence of unlabelled rRNA, in the presence of an excess of homologous unlabelled rRNA, and in the presence of the various heterologous rRNA preparations. We routinely use eight replicates for direct binding and for homologous competition and four replicates for each heterologous competition.

1. Follow the procedure as outlined for the direct binding experiment, except add 25 μl of competitor rRNA in place of the 25 μl of 1× SSC in those vials requiring competitor rRNA.

(*e*) *Comments.*   All of the experiments described above are in the presence of added 4× SSC. When the 55 μl of 1× SSC, added with the rRNA preparations is also calculated in, the final SSC concentration in the reaction vials is 4.5× SSC. We have found that adding 6× SSC has very similar results (Johnson and Harich, 1983), so at this salt concentration range, small variations in salt concentration probably have little effect on the reaction. The major factors influencing the specificity of the reaction are the combined effects of the formamide concentration and the hybridisation temperature. The 45% formamide concentration is readily attained, and the hybridisation temperatures can be lowered to a range where thermal degradation of the rRNA is not a problem (Gillespie and Gillespie, 1971). The hybridisation temperature of 50°C, which is about 20°C below the homologous $T_{m(e)}$ in this buffer system, allows for the detection of a wide range of homology values. Increasing the hybridisation temperature to 60°C, for example, depresses the homology values in the 40–50% range down to background (Johnson and Harich, 1983). Also, under these experimental conditions, there is a high correlation between rRNA homology or thermal stability values and oligonucleotide similarity (S_{AB}) values.

If one is interested in using rRNA hybridisation for identifying specific organisms, then the stringency of the hybridisation reaction must be increased to the maximum. Although these classic membrane hybridisation procedures may not be the best approach, increasing the hybridisation temperature in the above systems to the 60–65°C range will provide the stringency required.

2. RNA hybridisation to DNA in solution

A major problem with experiments involving the hybridisation of RNA with DNA in solution is the simultaneous reassociation of the DNA. This results in low hybridisation values, or requires vast amounts of DNA (Melli *et al.*, 1971). The following protocol is a variation of the methods described by Vogelstein and Gillespie (1977) and Casey and Davidson (1977).

(*a*) *Reagents*

   SSC–formamide: Mix 2.0 ml of 25× SSC–1 mM HEPES (pH 7) with 8.0 ml of deionised formamide. This solution will be 5× SSC.

   Labelled rRNA: Prepare ^{125}I-labelled rRNA with a specific activity of about 2×10^6 cpm μg^{-1}. Adjust concentration to 0.5 μg ml^{-1} in 1× SSC–1 mM HEPES.

   DNA: The DNA can be isolated by either of the methods given above or by small-scale adsorption chromatography (see Chapter 1). The DNA must be rRNA free, so an alkaline denaturation step is needed (0.5 M NaOH or 15 min at 50°C; neutralise and ethanol precipitate).

   RNase mixture: 0.5 mg RNase A ml^{-1} and 2500 units RNase T$_1$ ml^{-1} prepared as described in Section B, diluted 1:5 with 1× SSC–1 mM HEPES.

(*b*) *Procedure.* 1. Place 10–20 μg of DNA into a 0.5-ml microcentrifuge tube. Add 3.0 M sodium acetate (pH 6) to a final concentration of 0.3 M and precipitate with 2 volumes of −20°C 95% ethanol. Wash the pellet with 250 μl of cold 80% ethanol and allow the pellet to dry for about 0.5 h in an incubator.

   2. Dissolve the DNA in 50 μl of the 5× SSC–80% formamide mixture. Mix to dissolve fully.

   3. Add 10 μl of labelled rRNA, place a piece of Teflon tape over the top, and close the cap. Mix the contents on a Thermomix mixer and centrifuge briefly to bring the contents to the bottom of the tube. Incubate at 50°C for 2–4 h.

   4. Add 50 μl of RNase mixture to each vial, mix, and incubate at 37°C for 1 h.

   5. Acid precipitate the rRNA hybrids and collect them on nitrocellulose or GF/F glass fiber filters.

III. Biological and taxonomic significance

The genetic information residing within an organism (bacterial cell) is contained in the linear arrangement of the deoxyribonucleotides of the DNA. The DNA functions as a template to transcribe the information to RNA molecules, which in turn are translated for the synthesis of proteins having specific amino acid sequences. The RNA molecules will be complementary to one of the DNA strands. Through the replication of the DNA, the genetic information is passed on to the next cell generation. Differences in the nucleotide sequences between organisms are the result of changes, such as inversions, transitions, deletions, and additions, that have occurred in the past and as a result reflect phylogenetic relationships.

The methods described in this chapter result in comparative measurements of nucleotide sequence similarity, ranging from a value which is the same as for the homologous reaction (100% homology) to no measurable similarity (0% homology). Complementary strands that can form DNA duplexes or RNA hybrids probably do not have base mismatches greater than 10–20% (Brenner *et al.*, 1969; Ullman and McCarthy, 1973). As a result, there is a range of similarity values that cannot be measured by these experiments. In the case of specific cistrons, these comparisons have been done by nucleotide sequencing, but these procedures are at present not applicable to long stretches of DNA.

The major contribution of DNA homology studies has been to provide a more unifying concept of a bacterial species. Although the specific level of homology at which organisms are considered to belong to the same species is arbitrary, similar homology clusters have been found for all groups of bacteria that have been investigated and homology levels for species inclusion have been suggested (Johnson, 1973, 1984).

Comparisons of rRNA cistrons by rRNA homology experiments and by 16 S rRNA oligonucleotide catalogue similarities are providing data from which a more unifying phylogenetic concept for higher bacterial taxa is possible.

References

Ando, T. (1966). *Biochim. Biophys. Acta* **114**, 158–168.

Bendich, A. J., and Bolton, E. T. (1968). *In* "Methods in Enzymology" (L. Grossman and K. Moldave, Eds.), Vol. 12B, pp. 635–640. Academic Press, New York.

Bendich, A. J., Anderson, R. S., and Ward, B. L. (1979). *In* "Genome Organization and Expression in Plants" (C. J. Leaver, Ed.), pp. 31–33. Plenum, New York.

Birnstiel, M. L., Sells, B. H., and Purdom, I. F. (1972). *J. Mol. Biol.* **63,** 21–39.

Bishop, J. O. (1969). *Nature (London)* **224,** 600–603.

Bonner, J., Kung, G., and Bekhor, I. (1967). *Biochemistry* **6,** 3650–3653.

Bonner, T. I., Brenner, D. J., Neufeld, B. R., and Britten, R. J. (1973). *J. Mol. Biol.* **81,** 123–135.

Bowman, C. M. and Dyer, T. A. (1982). *Anal. Biochem.* **122,** 108–118.

Brenner, D. J., Fanning, G. R., Rake, A. V., and Johnson, K. E. (1969). *Anal. Biochem.* **28,** 447–459.

Brenner, D. J., Fournier, M. J., and Doctor, B. P. (1970). *Nature (London)* **227,** 448–451.

Bresser, J., and Gillespie, D. (1983). *Anal. Biochem.* **129,** 357–364.

Bresser, J., Doering, J., and Gillespie, D. (1983a). *DNA* **3,** 243–254.

Bresser, J., Hubbel, H. R., and Gillespie, D. (1983b). *Proc. Natl. Sci. U.S.A.* **80,** 6523–6527.

Britten, R. J., and Davidson, E. H. (1976). *Proc. Natl. Sci. U.S.A.* **73,** 415–419.

Britten, R. J., and Kohne, D. E. (1968). *Science* **161,** 529–540.

Britten, R. J., Pavich, M., and Smith, J. (1970). *Carnegie Inst. Washington, Year Book* **1968,** 400–402.

Britten, R. J., Graham, D. E., and Neufeld, B. R. (1974). *In* "Methods in Enzymology" (L. Grossman and K. Moldave, Eds.), Vol. 27E, pp. 363–406. Academic Press, New York.

Burton, K. (1956). *Biochem. J.* **62,** 315.

Casey, J., and Davidson, N. (1977). *Nucleic Acids Res.* **4,** 1539–1552.

Cashion, P., Holder-Franklin, M. A., McCully, J., and Franklin, M. (1977). *Anal. Biochem.* **81,** 461–466.

Chan, H. C., Ruyechan, W. T., and Wetmur, J. G. (1976). *Biochemistry* **15,** 5487–5490.

Chang, S. H., Majumdar, A., Dunn, R., Makabe, S., Rajbhandary, U. L., Khorana, H. G., Ohtsuka, E., Tanaka, T., Taniyama, Y. O., and Ikehara, M. (1981). *Proc. Natl. Acad. Sci. U.S.A.* **78,** 3398–3402.

Chargaff, E. (1955). *In* "Nucleic Acids" (E. Chargaff and J. N. Davidson, Eds.), Vol. 1. Academic Press, New York.

Chassy, B. M., and Giuffrida, A. (1980). *Appl. Environ. Microbiol.* **39,** 153–158.

Chelm, B. K., and Hallick, R. B. (1976). *Biochemistry* **15,** 593–599.

Chirgwin, J. M., Przybyla, A. E., MacDonald, R. J., and Rutter, W. J. (1979). *Biochemistry* **18,** 5294–5299.

Church, R. B. (1974). *In* "Molecular Techniques and Approaches in Developmental Biology" (M. J. Chrispeels, Ed.), pp. 223–301. Wiley, New York.

Commerford, S. L. (1971). *Biochemistry* **10,** 1993–2000.

Crosa, J. H., Brenner, D. J., and Falkow, S. (1973). *J. Bacteriol.* **115,** 904–911.

Davidson, J. N. (1975). "The Biochemistry of Nucleic Acids." Academic Press, New York.

De Ley, J. (1971). *In* "Methods in Microbiology" (J. R. Norris and D. W. Ribbons, Eds.), Vol. 5A, pp. 301–329. Academic Press, New York.

De Ley, J., Cattoir, H., and Reynaerts, A. (1970). *Eur. J. Biochem.* **12,** 133–142.

Denhardt, D. T. (1966). *Biochem. Biophys. Res. Commun.* **23,** 641–646.

Ehrenberg, L., and Fedorcsak, I. (1976). *Prog. Nucleic Acid Res. Mol. Biol.* **16,** 189–262.

Fox, G. E., Stackebrandt, E., Hespell, R. B., Gibson, J., Manniloff, J., Dyer, T. A., Wolfe, R. S., Balch, W. E., Tanner, R. S., Magrum, L. J., Zablen, L. B., Balkemore, R., Gupta, R., Bonen, L., Lewis, B. J., Stahl, D. A., Leuhrsen, K. R., Chen, K. N., and Woese, C. R. (1980). *Science* **209,** 457–463.

Galau, G. A., Britten, R. J., and Davidson, E. H. (1977a). *Proc. Natl. Acad. Sci. U.S.A.* **74,** 1020–1023.

Galau, G. A., Smith, M. J., Britten, R. J., and Davidson, E. H. (1977b). *Proc. Natl. Acad. Sci. U.S.A.* **74,** 2306–2310.

Giles, K. W., and Meyers, A. (1965). *Nature (London)* **206**, 93.

Gillespie, D. (1968). *In* "Methods in Enzymology" (L. Grossman and K. Moldave, Eds.), Vol. 12B, pp. 641–668. Academic Press, New York.

Gillespie, D., and Bresser, J. (1983). *Biotechniques* **1**, 284–292.

Gillespie, D., and Spiegelman, S. (1965). *J. Mol. Biol.* **12**, 829–842.

Gillespie, S., and Gillespie, D. (1971). *Biochem. J.* **125**, 481–487.

Gold, D. V., and Shochat, D. (1980). *Anal. Biochem.* **105**, 121–125.

Gopalakrishna, Y., Langley, D., and Sarkar, N. (1981). *Nucleic Acids Res.* **9**, 3545–3554.

Grimont, P. A. D., Popoff, M. Y., Fromont, F., Coynault, C., and Lemelin, M. (1980). *Curr. Microbiol.* **4**, 325–330.

Hill, B. T., and Whatley, S. (1975). *FEBS Lett.* **56**, 20–23.

Horinouchi, S., Uozumi, T., Beppu, T., and Arima, K. (1977). *Agric. Biol. Chem.* **41**, 2487–2489.

Hoyer, B. H., McCarthy, B. J., and Bolton, E. T. (1964). *Science* **144**, 959–967.

Hutton, J. R. (1977). *Nucleic Acids Res.* **4**, 3537–3555.

Johnson, J. L. (1973). *Int. J. Syst. Bacteriol.* **23**, 308–315.

Johnson, J. L. (1981). *In* "Manual of Methods for General Microbiology" (P. Gerhardt, Ed.), pp. 450–472. American Society for Microbiology,

Johnson, J. L. (1984). *In* "Bergey's Manual of Systematic Bacteriology" (N. R. Krieg, Ed.), Vol. 1, pp. 8–11. Williams & Wilkins, Baltimore.

Johnson, J. L., and Franics, B. S. (1975). *J. Gen. Microbiol.* **88**, 229–244.

Johnson, J. L., and Harich, B. (1983). *Curr. Microbiol.* **9**, 111–120.

Kahn and Pierson (1983). *Am. Soc. Microbiol. Abstr. H92.*

Kalb, V. F. Jr., and Bernlohr, R. W. (1977). *Anal. Biochem.* **82**, 362–371.

Kelly, R. B., Cozzarelli, N. R., Murray, M. P., Deutscher, P., Lehman, I. R., and Kornberg, A. (1970). *J. Biol. Chem.* **245**, 39–45.

Kennell, D. E. (1971). *Prog. Nucleic Acid Res. Mol. Biol.* **11**, 259–301.

Kirby, K. S. (1968). *In* "Methods in Enzymology" (L. Grossman and K. Moldave, Eds.), Vol. 12B, pp. 87–99. Academic Press, New York.

Kirby, K. S., Fox-Carter, E., and Guest, M. (1967). *Biochem. J.* **104**, 258–262.

Kourilsky, Ph., Manteuil, S., Zamansky, M. H., and Gros, F. (1971). *Biochem. Biophys. Res. Commun.* **41**, 1080–1087.

Kurtzman, C. P., Smiley, M. J., Johnson, C. J., Wickerham, L. J., and Fuson, G. B. (1980). *Int. J. Syst. Bacteriol.* **30**, 208–216.

Lachance, M.-A. (1980). *Int. J. Syst. Bacteriol.* **30**, 433–436.

Langer, P. R., Waldrop, A. A., and Ward, D. C. (1981). *Proc. Natl. Acad. Sci. U.S.A.* **78**, 6633–6637.

Leary, J. J., Brigati, D. J., and Ward, D. C. (1983). *Proc. Natl. Acad. Sci. U.S.A.* **80**, 4045–4049.

Lederman, L., Kawasaki, E. S., and Szabo, P. (1981). *Anal. Biochem.* **117**, 158–163.

Legault-Démare, J., Desseaux, B., Heyman, T., Seror, S., and Ress, G. P. (1967). *Biochem. Biophys. Res. Commun.* **28**, 550–557.

Le Pecq, J.-B., and Paoletti, C. (1966). *Anal. Biochem.* **17**, 100–107.

McCarthy, B. J., and Bolton, E. T. (1963). *Proc. Natl. Acad. Sci. U.S.A.* **50**, 156–164.

McCarthy, B. J., and Church, R. B. (1970). *Annu. Rev. Biochem.* **39**, 131–150.

McConaughy, B. L., Laird, C. D., and McCarthy, B. J. (1969). *Biochemistry* **8**, 3289–3304.

Markov, G. M., and Ivanov, I. G. (1974). *Anal. Biochem.* **59**, 555–563.

Marmur, J. (1961). *J. Mol. Biol.* **3**, 208–218.

Marmur, J., and Doty, P. (1961). *J. Mol. Biol.* **3**, 585–594.

Marmur, J., and Lane, D. (1960). *Proc. Natl. Acad. Sci. U.S.A.* **46**, 453–461.

Marsh, J. L., and McCarthy, B. J. (1974). *Biochemistry* **13**, 3382–3388.
Maxwell, I. H., Van Ness, J., and Hahn, W. E. (1978). *Nucleic Acids Res.* **5**, 2033–2038.
Meijs, W. H., and Schilperoort, R. A. (1971). *FEBS Lett.* **12**, 166–168.
Meinke, W., Goldstein, D. A., and Hall, M. R. (1974). *Anal. Biochem.* **58**, 82–88.
Melli, M., Whitfield, C., Rao, K. V., Richardson, M., and Bishop, J. O. (1971). *Nature (London) New Biol.* **231**, 8–12.
Midgley, J. E. M. (1971). *In* "Methods in Microbiology" (J. R. Norris and D. W. Ribbons, Eds.), Vol. 5A, pp. 331–360. Academic Press, New York.
Miyazawa, Y., and Thomas, C. A. (1965). *J. Mol. Biol.* **11**, 223–237.
Moore, R. L. (1974). *Curr. Top. Microbiol. Immunol.* **64**, 105–128.
Nonoyama, M., and Pagano, J. S. (1973). *Nature (London)* **242**, 44–47.
Nygaard, A. P., and Hall, B. D. (1963). *Biochem. Biophys. Res. Commun.* **12**, 98–104.
Orosz, J. M., and Wetmur, J. G. (1974). *Biochemistry* **13**, 5467–5473.
Parish, J. H. (1972). "Principles and Practice of Experiments with Nucleic Acids." Wiley, New York.
Popoff, M., and Coynault, C. (1980). *Ann. Microbiol. (Paris)* **131 A**, 151–155.
Rawson, J. R. Y., Thomas, K., and Clegg, M. T. (1982). *Biochem. Genet.* **20**, 209–219.
Rigby, P. W., Dieckmann, J. M., Rhodes, C., and Berg, P. (1977) *J. Mol. Biol.* **113**, 237–251.
Schindler, C. A., and Schuhardt, V. T. (1964). *Proc. Natl. Acad. Sci. U.S.A.* **51**, 414–421.
Schmeckpeper, B. J., and Smith, K. D. (1972). *Biochemistry* **11**, 1319–1326.
Schwinghamer, E. A. (1980). *FEMS Microbiol. Lett.* **7**, 157–162.
Seidler, R. J., and Mandel, M. (1971). *J. Bacteriol.* **106**, 608–614.
Selin, Y. M., Harich, B., and Johnson, J. L. (1983). *Curr. Microbiol.* **8**, 127–132.
Setaro, F., and Morley, C. D. G. (1976). *Anal. Biochem.* **71**, 313–317.
Setaro, F., and Morley, C. D. G. (1977). *Anal. Biochem.* **81**, 467–471.
Sibley, C. G., and Ahlquist, J. E. (1981). *In* "Evolution Today" (G. G. E. Scudder and J. L. Reveal, Eds.), Proceedings of the Second International Congress of Systematic and Evolutionary Biology, Carnegie-Mellon University, Pittsburgh, Pennsylvania, pp. 301–335.
Smith, M. J., Britten, R. J., and Davidson, E. H. (1975). *Proc. Natl. Acad. Sci. U.S.A.* **72**, 4805–4809.
Specht, C. A., DiRusso, C. C., Novotny, C. P., and Ullrich, R. C. (1982). *Anal. Biochem.* **119**, 158–163.
Stark, G. R., and Williams, J. G. (1979). *Nucleic Acids Res.* **6**, 195–203.
Tereba, A., and McCarthy, B. J. (1973). *Biochemistry* **12**, 4675–4679.
Thomas, P. S. (1980). *Proc. Natl. Acad. Sci. U.S.A.* **77**, 5201–5205.
Ullman, J. S., and McCarthy, B. J. (1973). *Biochim. Biophys. Acta* **294**, 416–424.
Vogelstein, B., and Gillespie, D. (1977). *Biochem. Biophys. Res. Commun.* **75**, 1127–1132.
Vogt (1973). *Eur. J. Biochem.* **33**, 192–200.
Wahl, G., Stern, M., and Stark, G. R. (1979). *Proc. Natl. Acad. Sci. U.S.A.* **76**, 3683–3687.
Warburg, O., and Christian, W. (1942). *Biochem. Z.* **310**, 384–421.
Wayne, L. G., and Gross, W. M. (1968). *J. Bacteriol.* **95**, 1481–1482.
Wetmur, J. G. (1975). *Biopolymers* **14**, 1517–1524.
Wetmur, J. G. (1976). *Annu. Rev. Biophys. Bioeng.* **5**, 337–361.
Wetmur, J. G., and Davidson, N. (1968). *J. Mol. Biol.* **31**, 349–370.
Yamada, K., and Komagata, K. (1970). *J. Gen. Appl. Microbiol.* **16**, 215–224.
Yokogawa, K., Kawata, S., and Yoshimura, Y. (1972). *Agric. Biol. Chem.* **36**, 2055–2065.
Yokogawa, K., Kawata, S., and Yoshimura, Y. (1975). *Agric. Biol. Chem.* **39**, 1533–1543.

3

16 S Ribosomal RNA Oligonucleotide Cataloguing

E. STACKEBRANDT,*,1 W. LUDWIG,* and G.E. FOX†

Institut für Botanik und Mikrobiologie, Technische Universität München, Munich, Federal Republic of Germany

†*Department of Biochemical and Biophysical Sciences, University of Houston, University Park, Houston, Texas, USA*

I. Introduction

16 S ribosomal RNA is currently the subject of intense study. As a major component of the small ribosomal subunit, elucidation of 16 S rRNA secondary and tertiary structures is a key objective in the ongoing efforts to understand the structure of ribosomes (Woese *et al.*, 1983). As a

---

1 Present address: Institut für Allgemeine Mikrobiologie, Universität Kiel, Kiel, Federal Republic of Germany.

METHODS IN MICROBIOLOGY
VOLUME 18

readily isolatable and ubiquitous macromolecule, 16 S rRNA is ideal for studies in molecular systematics. Indeed, in this latter context, studies of 16 S rRNA have become a major factor in the revolution in microbial systematics which is now underway (Fox *et al.*, 1980; Stackebrandt and Woese, 1981). Progress in both areas is dependent on the availability of sequence data from 16 S rRNAs isolated from a large variety of microbial species. Unfortunately, even with modern RNA/DNA sequencing techniques this is a tedious and time-consuming task. To circumvent this difficulty, procedures for rapid characterisation of 16 S rRNA by complete analysis of the oligonucleotide fragments produced by digestion with ribonuclease T_1 have been developed. Over 400 16 S rRNAs have been characterised in this way with considerable success.

In this review we discuss the two techniques that have been employed and describe one of them, the thin-layer chromatography method, in full detail. This includes description of the experimental procedures as well as the deduction process used to establish the sequences of the individual oligonucleotides. Next we describe the data analysis procedures and show how the oligonucleotide catalogue data can be used to elucidate relationships between species and to evaluate structural proposals. Finally, in the last section we discuss practical considerations for workers who may wish to begin 16 S rRNA cataloguing.

II. Basis of 16 S rRNA cataloguing

In 16 S rRNA cataloguing the RNA is digested completely with a ribonuclease of known base specificity (usually T_1, which is specific for guanosine). The resulting oligonucleotide products are separated in two dimensions and visualised by autoradiography. The photographic image of the separation is referred to as the primary fingerprint. The oligonucleotides in each spot on the fingerprint are next sequenced in their entirety. The catalogue then is the tabulation of sequences of all the oligonucleotides that comprise the original RNA. The underlying idea is that a comparison of such catalogues can give quantitative information of the same type as that obtained by actual determination of the 16 S rRNA sequences. This is illustrated in Fig. 1 where oligonucleotide sequence data have been used to superimpose schematically the *Escherichia coli* 16 S rRNA fingerprint with fingerprints from *Aerobacter aerogenes* and *Yersinia pestis,* respectively (Woese *et al.*, 1974; Zablen *et al.*, 1975). In these comparisons white spots represent oligomers which are present in both RNAs while the dark or shaded spots are unique to one or the other RNA.

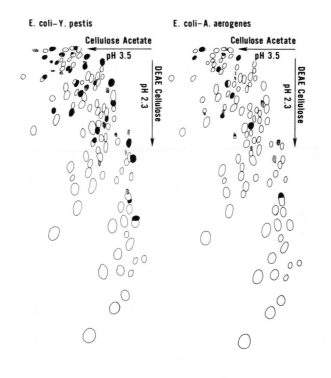

Fig. 1. Schematic comparison of the HVE version of the *E. coli* 16 S rRNA fingerprint with *Yersinia pestis* and *Aerobacter aerogenes*, respectively. Open spots are common to both fingerprints, dark spots are unique to *E. coli,* and cross-hatched spots are unique to either *Y. pestis* or *A. aerogenes*. The superposition was accomplished by utilising the oligonucleotide sequence data in conjunction with double-label experiments in which the *E. coli* RNA was labelled with tritium (Woese *et al.,* 1974; Zablen *et al.,* 1975). If the reader makes an examination of the number of dark and cross-hatched spots in the *E. coli/Y. pestis* pair and contrasts this with the number seen in the *E. coli/A. aerogenes* comparison it is easy to understand how the cataloguing method can give quantitative insight into systematic relationships. The diagrams also illustrate nicely the isopleth groupings which characterise the fingerprints when they are run in the HVE format.

By simply counting the number of unique or identical spots one quickly obtains quantitative insight into which two strains are most closely related. This counting procedure has been formalised by use of an association coefficient, S_{AB}, which allows for differing lengths among the oligomers. This association coefficient is defined as follows (Fox *et al.,* 1977):

$$S_{AB} = 2N_{AB}/(N_A + N_B)$$

where N_A is the total number of residues of at least length L in catalogue A; N_B, the total number of residues of at least length L in catalogue B, and N_{AB}, the total number of residues represented by all the coincident oligomers between the two catalogues, A and B, of at least length L. As will be discussed in a later section the usual value of L is six. The values of the association coefficient can be calculated for every pair of organisms for which data exist. The resulting matrices of S_{AB} values are readily utilised in any of a number of different clustering algorithms to produce dendrograms. As new catalogues are determined they can, like complete sequences, be compared to all previous catalogues regardless of laboratory origin. This continuously growing data base provides a significant advantage over hybridisation methods which rely on comparison with key organisms. Intercomparisons with hybridisation results are possible, and this has been recently discussed in some detail (Johnson and Harich, 1983).

III. Alternative methods for 16 S rRNA cataloguing

Primary ribonuclease T_1 fingerprints have typically been obtained in one of four ways. In the original methodology the separation was achieved by two-dimensional high-voltage electrophoresis (HVE) on cellulose acetate paper in the first dimension and DEAE paper in the second dimension (Sanger *et al.*, 1965). Difficulties in separating very large oligonucleotides led to modification of these procedures. Woese and colleagues introduced significant improvements that made the original Sanger method usable for 16 S rRNA cataloguing (Uchida *et al.*, 1974; Woese *et al.*, 1976). This modified Sanger method will herein be referred to as the HVE method. An alternative two-dimensional polyacrylamide gel separation procedure (gel method) was developed (De Wachter and Fiers, 1972) that is now widely used with viral genomic RNA and in which the pattern of spots alone is used to characterise the relationship between various viral strains. Brownlee and Sanger (1969) improved the separation of oligonucleotides in the second dimension by use of homochromatography (homochromatography method) on thin-layer DEAE plates. It was difficult to obtain a complete catalogue of all the digestion products with this methodology, so currently ammonium formate gradient thin-layer chromatography (TLC method) on DEAE plates is used in the second dimension (Domdey and Gross, 1979).

At present, only fingerprints produced by the HVE method and TLC method are used in the determination of complete 16 S rRNA catalogues.

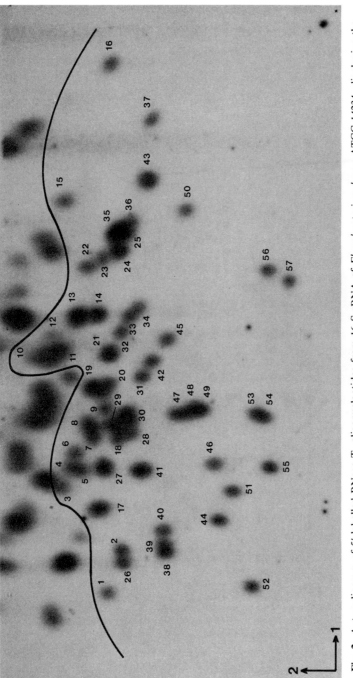

Fig. 2. Autoradiogram of 5'-labelled RNase T₁ oligonucleotides from 16 S rRNA of *Flavobacterium breve* ATCC 14234, displaying the position of oligonucleotides (1–57) of size hexanucleotide and larger. These oligonucleotides are used in the analysis of phylogenetic relationships. Sequences are listed in Table I. The solid line indicates the division between hexanucleotides and the smaller fragments.

The HVE method utilises RNA that is uniformly labelled *in vivo*. Individual spots are isolated from the primary fingerprint (Fig. 1) and fragmented a second time with a variety of enzymes. Still unidentified fragments are digested a third time. Once all the information is available the sequences are reconstructed by logical arguments. The procedures and ideas needed for interpretation of the results in the HVE method have been described in detail previously (Uchida *et al.*, 1974; Woese *et al.*, 1976; Fox *et al.*, 1977) and will be omitted here. In contrast the TLC method uses RNA in which the products of the original ribonuclease T_1 digestion have been 5'-end labelled with T_4 polynucleotide kinase and $[\gamma\text{-}^{32}P]ATP$ *in vitro*. Spots containing the oligonucleotide fragments of interest are removed from the fingerprint (Fig. 2) and subjected to mobility shift analysis. Since the procedures currently used in the TLC method are only partially documented in the literature this method will be described in detail.

IV. Advantages and disadvantages of the HVE and TLC methods

The most fundamental distinction between the methods is the mode of introducing the label. Because the HVE method utilises an *in vivo* label, its use is restricted to organisms that grow readily in the presence of high levels of ^{32}P. In contrast the TLC method can be applied to any organism for which a few grams of cells can be obtained. As a result it has been possible to study many important organisms with the TLC method that could not be approached with the HVE method. TLC on the other hand utilises a much smaller separation area than the traditional Sanger fingerprints. Thus the small oligomers are allowed to run into the buffer to optimise the separation of the oligomers of size six and larger. This results in the loss of some useful information because the small posttranscriptionally modified oligomers cannot be identified. Larger oligomers containing posttranscriptional modifications cause interpretation difficulties in both methods though these are most severe with TLC. Because of these problems neither method is readily used with highly modified, e.g., primarily eukaryotic, ribosomal RNAs. The efficiency of the end-labelling step in the TLC method is variable so it is not, in general, possible to establish the quantitation of individual oligomers. This is of minimal consequence because multiple occurrence oligomers of length six and larger are rare. One noteworthy exception is the oligomer A*A*CCUG (A* = 2,6-dimethyladenine). This sequence is quite common among Gram-negative bacteria but has never been seen by the TLC method, apparently because it is not labelled by polynucleotide kinase (Stackebrandt *et al.*,

1982). Both methods are reliable and each has been extensively used and these slight differences are taken into account in the data analysis procedures so that the catalogues are readily integrated into a single data base.

V. Detailed description of the TLC method

A. Isolation and purification of 16 S rRNA

Many techniques are available today for isolation and purification of ribosomal RNA. Typically, rRNA is isolated either from ribosomal subunits or from deproteinised cell lysates. The latter is usually faster but may cause problems when large amounts of high-molecular-weight DNA inhibit the subsequent purification of RNA by polyacrylamide gel electrophoresis. Whatever method is used, precautions against ribonucleases are a prerequisite for success (see D'Alessio, 1982, for recommendations). It is impossible to give a general recipe for the isolation of 16 S rRNA, since many factors, including cell shape, age of the cells, composition of the cell envelope, cell density, and endonuclease activity, influence the various steps in the isolation of a pure rRNA. Our experience with more than 400 bacterial species from almost all major groups of prokaryotes, however, has shown that the following steps will lead in most cases to a satisfactory amount and purity of 16 S rRNA.

Cell pellets (2–4 g wet wt.) are resuspended in buffer (0.04 M Tris, 0.02 M Na acetate, 0.001 M Na$_2$ EDTA, pH 7.2) at a ratio of 1–1.5 ml g^{-1} of cells and opened by passage through a precooled French pressure cell at 1.7 GPa in the presence of a few drops of sodium dodecylsulphate (10%). An equal volume of water-saturated, redistilled phenol is added to the lysate, vortexed for 10 s, and centrifuged at 20,000 g for 20 min. The phenol phase and the protein interface material are discarded and the extraction is repeated (two to four times) until the proteinaceous interface has disappeared. The nucleic acids are then precipitated by 3 volumes of cold ethanol for at least 2 h at −20°C. Separation of rRNA species is achieved by one-dimensional polyacrylamide slab gel electrophoresis. The electrophoresis apparatus, consisting of two glass plates (20 × 20 cm), separated from each other by greased plastic spacers (0.4 cm) is basically the same as described by De Wachter and Fiers (1982), who in addition describe the general experimental procedure in detail. A total of 0.8 μl ml^{-1} N,N,N',N'-tetramethylethylenediamine and 1.6 μl ml^{-1} ammonium peroxodisulphate (50%) is added to 150 ml of a degassed polyacrylamide solution (2.8% PA, 0.02% N,N'-methylene bisacrylamide, in 0.04 M Tris, 0.02 M Na acetate, 0.001 M Na$_2$EDTA, pH 7.2) and poured into the gel mold. Gels are allowed to stand for 1 h to complete polymerisation. Pre-electrophoresis is carried out for 1 h at 70 mA. The nucleic acid precipitate is resuspended in up to 1.5 ml electrophoresis buffer (0.04 M Tris, 0.02 M Na acetate, 0.001 M Na$_2$EDTA, 0.2% SDS, pH 7.2) and mixed with 10 μl bromophenol blue (saturated in distilled water); glycerol is added to a total of 10% and the mixture is loaded on top of the gel. Electrophoresis is at 30 mA for the first hour and at 70 mA for an additional 6–8 h. Following electrophoresis the plates are separated and the RNA is visualised by placing the gel on a thin-layer plate containing a fluorescent indicator (254 nm, Merck, Darmstadt) which has been covered with plastic wrap. Under UV light (254 nm) nucleic acids appear as dark bands on the fluorescent background (Hassur and Whitlock, 1974).

The gel region containing 16 S rRNA is removed with a razor blade and embedded in a cylindric (1-cm diameter) agarose gel (0.5% agarose in 0.1 M NH$_4$ formate). The 16 S rRNA is collected in 2–3 ml of elution buffer (0.1 M NH$_4$ formate), which is separated from the reservoir by a dialysis membrane (boiled in 5% Na$_2$CO$_3$ for 20 min prior to use). Electrophoresis is towards the anode at a current of 10 mA per tube for 5–7 h. To avoid the loss of rRNA which accumulates on the dialysis membrane the voltage is reversed for 3–5 min. The rRNA-containing solution is subjected to two phenol extractions to remove SDS and traces of contaminants. Traces of phenol are removed by three subseqent extractions with 4 volumes of ether and the elution is precipitated by the addition of 3 volumes of ethanol (96%). Alternatively, the solution is precipitated by 3 volumes of ethanol, stored at -20°C for 3 h, redissolved in 1 ml distilled water, made up to 0.4 M by the addition of 100 μl 4 M NaCl, precipitated again with 3 volumes of cold ethanol, and stored at -20°C. The precipitate is then dried in a desiccator over phosphorus pentoxide and redissolved in 100 μl sterile water checked for purity, and the yield is determined spectrophotometrically at 260 nm.

B. RNase T$_1$ digestion and *in vitro* labelling of 16 S rRNA

Forty micrograms 16 S rRNA are completely digested with RNase T$_1$ (2.4 U, Calbiochem–Behring, La Jolla, California) in 80 μl 12 mM Tris buffer (pH 8.0) at 37°C for 3 h, boiled for 2 min, and stored frozen. In contrast to methods published previously (Stackebrandt *et al.*, 1981, 1982) the removal of 3'-terminal phosphates is omitted, since in the following *in vitro* labelling of 5' termini, phosphatase-free polynucleotide kinase is used (Cameron *et al.*, 1979). Two micrograms of the digested RNA are used for enzymatic 5'-end-group labelling. The reaction mixture is as follows: 5 μl RNase T$_1$ digest, 1 μl mercaptoethanol (0.3 M), 0.5 μl spermidine

(0.34 M), 0.5 μl MgCl$_2$ (0.4 M),1 μl unlabelled ATP (1 mM), 1 μl Tris–HCl buffer (1 M, pH 8.0), 10 μl of [γ-^{32}P]ATP (NEN, 1 mCi, 1000–3000 Ci mmol^{-1}), and 1.2 μl of phosphatase-free T$_4$ polynucleotide kinase (NEN, 8000 U ml^{-1}, 6000 U mg^{-1} protein). The mixture is incubated at 37°C for 30 min and the reaction stopped by boiling for 2 min (Simsek *et al.*, 1973; Raba *et al.*, 1979; Silberklang *et al.*, 1979). The mixture is then frozen, dried over P$_2$O$_5$ in a vacuum desiccator, and redissolved in 2.4 μl of water.

C. Sequence determination

Separation of 5′-labelled RNase T$_1$-resistant oligonucleotides is in two dimensions using a combination of HVE in the first dimension (Sanger *et al.*, 1965) and TLC in the second (Silberklang *et al.*, 1977, 1979; Domdey and Gross, 1979; Diamond and Dudock, 1983). These methods have been improved to facilitate sequencing of the more than 60 oligonucleotides of lengths 6–20, which make up a 16 S rRNA catalogue. The changes include the replacement of the "Homomix" used for separation of fragments in the second dimension (Brownlee and Sanger, 1969) by NH$_4$ formate gradient chromatography, the utilisation of alkaline hydrolysis rather than enzymatic cleavage to produce fragments, the introduction of tertiary analysis for comigrating oligonucleotides, whose sequences cannot be unambiguously determined by secondary analysis alone, and the money- and time-saving separation of several oligonucleotides on a single cellulose acetate strip or DEAE cellulose thin-layer plate in the secondary and tertiary analyses.

1. Experimental preparation of the primary fingerprint

For separation in the first dimension, a cellulose acetate strip (3 × 100 cm, Schleicher and Schuell, Dassel, Federal Republic of Germany) is presoaked in electrophoresis buffer (5% acetic acid, 0.05% pyridine, 5 mM EDTA, pH 3.5), and an area 5–7 cm from one end of the strip is blotted on both sides with absorbent paper to remove surplus buffer. This region of the strip is supported on two (0.7-cm) glass rods that are 2 cm apart. One microlitre of the radioactive sample is then applied in a line 1.2 cm long across the width of the strip, and 6 cm from the end of the strip, using a fine glass capillary. On either side of the sample, a marker mix containing 1% xylene cyanol, 2% fuchsin, and 1% orange C, is applied. After the sample has been absorbed, excess buffer is removed from the rest of the strip, which is then placed in a high-voltage electrophoresis tank containing the same buffer. Varsol or a comparable solvent is used as hydrophobic phase and cooling fluid. Electrophoresis is at 5000 V and the xylene cyanol marker is allowed to migrate 28 cm. The fuchsin and orange G markers move approximately 2 and 2.5 times as quickly, respectively, and the oligonucleotides of interest lie between the blue and the orange markers.

In preparation for the second dimension, labelled fragments are transferred from the strip to the long side of a 20 × 40-cm DEAE cellulose thin-layer glass plate (Cel DEAE HR-Mix-20, 0.2-mm layer, Macherey-Nagel, Düren, Federal Republic of Germany). The strip is laid lengthwise on a glass rod (0.4 cm) and strips of water-soaked Whatman 3MM filter-paper (5 × 45 cm), are put on either side to overlap the strip by 3 mm in the region between the blue and the yellow markers. The thin-layer plate is placed onto the strip so that the line of contact between the strip and plate runs 2 cm from the long edge of the plate (Stackebrandt *et al.*, 1982). The glass plate is held in place by weights. Oligonucleotides are transferred by capillary action and bound to the DEAE cellulose (Southern, 1974, as modified by Silberklang *et al.*, 1977), within a period of about 20 min. After the transfer is complete, this part of the DEAE cellulose plate is rinsed with 150 ml of ethanol to remove the high-voltage buffer salts, and allowed to dry.

Oligonucleotides are separated in the second dimension by NH_4 formate TLC, carried out at 70°C. This technique replaces the earlier homochromatography procedure (Silberklang *et al.*, 1977) which caused severe problems in the two-dimensional separation of fragments during tertiary analysis (see below). The conditions used for the separation of oligonucleotides in the second dimension depend to some extent on the particular lot of DEAE TLC plates. While in some batches a 0.3–0.4 *M* NH_4 formate gradient will achieve good separation, others will require a higher (or lower) molarity gradient as indicated by a test run.

Prechromatography in water is carried out for approximately 30 min, i.e., until the water has migrated about 15 cm, followed by the transfer of the plate to a solution containing 0.3 *M* NH_4 formate, 7 *M* urea, and 1 m*M* Na_2EDTA. An equal volume of a solution of 0.5 *M* NH_4 formate in 7 *M* urea and 1 m*M* Na_2EDTA is added dropwise over a period of about 3 to 4 h, resulting in a concentration gradient of 0.3–0.4 *M* NH_4 formate. To ensure good mixing of the solvents the plate is supported on notched plastic rods and the solution in the chromatography tank is stirred. Because of the prechromatography, the front runs ahead of the smallest oligonucleotides, and is absorbed by a filter paper wick (three sheets of 20 × 40-cm, Whatman 3MM) fastened onto the top of the chromatography plate by two metal binder clips. Chromatography is terminated when the orange G marker has migrated 18 cm. Small oligonucleotides including some pentamers have by then migrated into the wick. Since hexamers are the smallest oligonucleotides used in the data analysis, this is not a problem. After removing the wick, plates are dried and marked with radioactive ink, and the positions of the labelled oligonucleotides are located by autoradiography for 10–30 min.

2. Characteristics of the primary fingerprint

In the first dimension, separation of RNase T_1-resistant oligonucleotides depends on three major factors. These are the average pK value of the bases comprising each oligonucleotide at the pH value of the high-voltage electrophoresis buffer used, the fragment size, and the fragment base composition. A detailed analysis is presented by Brownlee (1972). The most prominent factor affecting the separation of oligonucleotides of a given fragment size is the number of uridine residues. Using a pH 3.5 buffer, there is a direct correlation between uridine content and the migration distance. Thus uridine-rich fragments move faster toward the anode than do fragments with fewer uridine residues. Oligonucleotides of the same size, same uridine content, and a larger number of adenine residues will migrate slightly faster than those with lower adenine content.

In the second dimension, separation occurs according to size and base composition. In particular, for oligonucleotides of the same size class, pyrimidine-rich fragments migrate faster than purine-rich ones. The net effect of the two-dimensional separation then is that oligonucleotides of a distinct size but different composition are separated in the shape of a triangle in which the corners are occupied by fragments consisting of one kind of nucleotide only (with its 3-terminal Gp). C_nG and U_nG are found at the corners of the long side of the triangle at the left and right side, respectively, while A_nG runs faster than C_nG in the first but markedly slower in the second dimension. Isomeric oligonucleotides are located within this triangle according to their gross nucleotide composition (Fig. 3): Fragments of a given size with the same uridine content are arranged along diagonal lines, with cytidine-rich fragments moving slower in the first and faster in the second dimension than adenine-rich fragments. Oligonucleotides of the same size and gross composition that differ in sequence usually occupy essentially the same position in the fingerprint. There are instances, however, where the order of bases can cause substantial differences in fragment migration. An example is UUCCCG which migrates faster the first and slightly slower in the second dimension than CCUUCG, resulting in a comigration of UUCCCG with fragments of

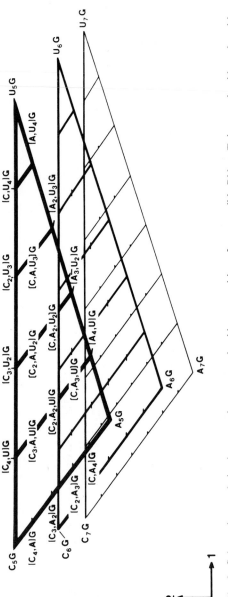

Fig. 3. Schematic correlation between the gross nucleotide composition of any possible RNase T_1 hexanucleotide and position of hexa-, hepta-, and octanucleotides after two-dimensional NH_4 formate gradient chromatography. The reader should note the high degree of overlap, resulting in the comigration of oligonucleotides of different sizes and uridine content.

the composition $(U_2C_2A)G$. Similarly, UACACACCG runs slightly faster in the first dimension than CUACACACG and CUCACCAAG.

In general, triangles of oligonucleotides of size X_nG and $(X_{n+1})G$ are not separated completely from each other but overlap (Fig. 3). Domdey *et al.* (1978) show a detailed schematic diagram, including triangles of dimers up to decamers. The degree to which the overlapping occurs depends to some extent upon the molarity of the NH_4 formate solution used in the second dimension. As can be seen in Fig. 6B, in one case a heptamer containing two uridine and three adenine residues comigrates with octamers, containing three uridine and one adenine residue. Figure 2 shows a fingerprint of *Flavobacterium breve* 16 S rRNA. Indicated are those spots (1–57) which were isolated for sequence determination as described below. The results of these determinations are listed in Table I. As seen in Fig. 3, the distinction between hexamers and smaller oligonucleotides is not immediately apparent from the fingerprint. Indeed, experience is necessary to recognise this boundary, and it is recommended that a neophyte do the extra work associated with the isolation of the neighbouring pentamers rather than risk the loss of fast-migrating hexamers.

This separation pattern is in considerable contrast to that seen in the primary fingerprint generated by two-dimensional high-voltage electrophoresis (Sanger *et al.*, 1965; Uchida *et al.*, 1974). There fragments with the same uridine content are arranged in distinct wedge-shaped isopleths (see Fig. 1). The limited space available on a 40×20-cm thin-layer plate (as compared to the DEAE paper of size 35×100 cm) causes the smaller oligonucleotides to migrate close to each other. On the other hand, large oligonucleotides are usually better separated than in the two-dimensional HVE method. Moreover, the difference in appearance between the TLC and HVE fingerprints is further aggravated by the fact that in the HVE method it is usual to retain the small oligomers on the fingerprint while in the TLC case these are allowed to run into the buffer. Whichever method is used, knowing the rules that govern the separation of the oligonucleotides is essential if one is to obtain reliable results. Thus the reasonableness of the gross composition of a sequence determined is routinely tested by comparison with the sequences of neighbouring spots and by looking at the position of the spots in the original fingerprint.

3. *Isolation and controlled hydrolysis of oligonucleotides*

Individual spots believed to contain labelled oligonucleotides of length six or larger are isolated from the DEAE cellulose plate (Domdey *et al.*, 1978; Stackebrandt *et al.*, 1982). Each spot is wet with 96% alcohol, scraped

TABLE I

Oligonucleotide catalogue of *Flavobacterium breve* ATCC 14234[a]

Sequence	Spot no.	Sequence	Spot no.	Sequence	Spot no.
CACAAG	1	AAUCUCG	20	AUUUAUUG	37
AAACAG	2	CAUAAUG	20	CUCACCAAG	38
UCCACG	3	UAACUAG	21	CUACACACG	39
CCAAUG	4	UAAUACG	21	UACACACCG	40
AUCCAG	5	CCUUAUG	13	CUAACUCCG	41
UACCAG	6	UUAUCCG	22	ACUCCUACG	41
AAUACG	7	AUACUUG	23	ACUCUAAUG	42
ACAAUG	8	AAAUUG	24	UCUAUAUUG	43
UAAACG	8	AAAUUUG	24	UUUAAUUCG	43
AAUAAG	9	UUAUAAG	25	AAACUCAAAG	44
UAAAAG	9	UUUAAAG	25	CAUCAUUUAG	45
UUCCCG	10	CCAUCCCG	26	AACACCAAUUG	46
CCAUUG	10	CCCUUACG	27	UAUCCCACCUG	47
CUACUG	10	CCCUACUG	27	AACAUCUCAUG	48
ACUUCG	11	CAUAUCAG	28	CUUAACACAUG	48
AUCCUG	11	UCAUCACG	29	AAAUCCUAUCG	49
UACCUG	12	AUACCCUG	29	AUAUUCUUCG	50
AUACUG	13	CUACAAUG	30	AACCUUACCAAG	51
UAAUCG	13	UCACUAAG	30	AACCCCACCACUG	52
AAUUAG	14	UAAAACUG	31	AAAACUUAUCUCAG	53
UUUAAG	15	ACUCUAUG	32	AUAAACCUACUUACG	54
UAUUUG	16	CUUAAAUG	33	CAACCCCUAUCAUUAG	55
CAACUCG	17	AUAUUACG	34	UAAACUACUUUUAUCUG	56
UAACAAG	18	AUCUUUAG	35	AUUAAUACUUUAUAAAUAG	57
CAACUUG	19	AUUUAUCG	35		
CUAAUUUG	19	CUAAUUUG	36		

[a] Numbers refer to those in Fig. 2, indicating the position of the respective oligonucleotide in the 16 S rRNA "fingerprint." The 3' end of the molecule was not found.

from the plate, and sucked by vacuum into the open end of a drawn-out cotton-plugged glass tube (0.4 × 3–4 cm). Oligonucleotides are then eluted from the DEAE powder by three washes with 80 μl 1 M NaCl. These washings are collected by centrifugation through the glass tubes into Eppendorf test tubes with a swinging-bucket rotor (Beckman TJ 6, 750 g for 2 min). RNA fragments are next precipitated by 3 volumes of ethanol at −20°C for at least 2 h. Following centrifugation at 1000 g for 20 min, the supernatant is discarded and the precipitate is dried in a vacuum desiccator over P_2O_5. Controlled alkaline hydrolysis of the oligonucleotides is conducted in 10 μl 50 mM NaHCO$_3$, adjusted to pH 9.2 with 50 mM NaOH (Wengler et al., 1979) at 100°C for 20 min, followed by chilling in ice. To remove any cyclic phosphate groups, 1.2 μl of 2.5 N HCl is added and the samples are dried in vacuo.

4. Secondary analysis

To determine the sequences of individual oligonucleotides the alkali-generated fragments are again separated in two dimensions. This technique, known as "mobility-shift" analysis (Jay et al., 1974; Silberklang et al., 1977; Gross et al., 1978), involves essentially the same procedures as described earlier for the generation of the primary fingerprint. Each alkaline hydrolysate is dissolved in 2.4 μl distilled water, and a 1-μl aliquot is applied to a cellulose acetate strip for high-voltage electrophoretic separation in the first dimension. The quality of the separation patterns of chemically generated fragments depends to a great extent on the careful application of the samples to the cellulose acetate strips. On the other hand, rapid work is necessary to prevent drying of the presoaked strips. Thus the number of samples that can be applied during one session is limited. The following arrangement allows two co-workers rapidly to apply 30 of the 60 samples to 10 strips which can then be run simultaneously in one electrophoresis tank. Three samples are applied to one 55-cm strip, 6, 18, and 30 cm from the end. The xylene cyanol marker is allowed to migrate 6 cm. An alternative application method prevents any drying of the strips by immersing them in a shallow tray containing the hydrophobic cooling fluid. Here the separation of six samples per 100-cm strip, applied 6, 18, 30, 50, 62, and 74 cm from the end, results in a satisfactory resolution of the fragments. For application, the respective region of the strip is brought out of the hydrophobic fluid, the fluid is removed by a stream of air, and the strip is submersed after the sample has been dried. This alternative procedure allows the separation of all samples isolated from a 16 S rRNA fingerprint in one operation by one person. Depending on which method is used, the three or six samples separated on a strip are transferred to one or two adjacent DEAE cellulose plates. For this purpose, thin-layer foils (20 × 40 cm, Polygram, Cel 300 DEAE/HR-2/15, 0.1 mm layer, Macherey-Nagel, Düren) on glass plates are used. Transfer of oligonucleotides is as described for the primary fingerprint.

The second dimension is developed in a NH_4 formate gradient. Up to 25 plates with a wick fastened to the top are placed in a suitable tank. The plates are supported on notched rods to allow circulation. Excess water not used during prechromatography is removed by vacuum when the water front reaches the middle of the plates. The water is replaced by 1 litre of buffer (0.28 M NH_4 formate in 7 M urea, 1 mM Na_2EDTA, 1 M boric acid, pH 4.3 adjusted with formic acid), and an more concentrated NH_4 formate solution (2 $M \rightarrow 0.5$ M NH_4 formate as indicated above, 85 ml each) is added dropwise over a period of 3–4 h at 70°C, up to a final concentration of 0.44 M NH_4 formate. This procedure establishes an approximately asymptotic increase in the NH_4 formate concentration in the tank, resulting in a better resolution of fragments than is obtained with a linear gradient. By the time the xylene cyanol dye has almost reached the upper edge of the plates and the water front has been absorbed by the wick, the wicks are removed, the foils are dried, marked with radioactive ink, and wrapped in cellophane, and the positions of 5'-labelled fragments are visualised by autoradiography (12–24 h).

5. Characteristics of the mobility-shift patterns

Detailed descriptions of the two-dimensional mobility-shift analysis of enzymatically cleaved DNA fragments have been presented previously (Jay *et al.,* 1974; Bambara *et al.,* 1974). Silberklang *et al.* (1979) obtained similar results for RNA fragments separated by homochromatography and provided R_f values for representative intermediary fragments. These earlier descriptions do not apply exactly to the patterns found for alkali-cleaved intermediates, separated by gradient chromatography. Thus, a detailed description of these patterns will be given here. In general, the separation follows the same rules as discussed previously for the primary fingerprint. The relative mobilities of the intermediates in the first dimension depend upon pK values, size, and base composition, while in the second dimension intermediates are separated by size and base composition. Using controlled conditions, alkali cleaves oligonucleotides randomly to give a family of all possible intermediate products. Since oligonucleotides are end labelled, the only intermediates seen after autoradiography of the two-dimensionally separated digest are those with a ^{32}P-labelled 5' terminus. Knowing the influence the addition of a mononucleotide to a mono- or oligonucleotide has on its migration pattern, the cleavage series of an oligonucleotide can be directly read as angular mobility shifts between successively longer fragments.

As depicted schematically in Figs. 4 and 5 on a series of five octamers with different uridine content, the following rules can be used to interpret these shift patterns. The addition of Cp to [32]pCp or [32]pAp, or to intermediates ending in Cp or Ap causes a distinct mobility shift to the left, the addition of Up causes a shift to the right (towards the anode), while the addition of Ap causes a shift that is either slightly to the left or almost vertical. The addition of Cp or Ap to either [32]pUp or to an intermediate ending in Up causes the longer intermediates to shift to the left, with [32]UpCp and [32]pXpUpCp running shorter in the first and faster in the second dimension than [32]pUpAp or [32]pXpUpAp. The addition of Up makes the dimer run faster in the first dimension but the shift is markedly reduced when more Up residues are added. Similarly, uridine-caused shifts to the right get shorter with increased oligonucleotide length.

In longer sequences containing Cp and Ap only, both nucleotides are easily distinguished by mobility differences in the second dimension because purines cause a larger shift than pyrimidines. Long Cp stretches (>5 bases) turn almost vertical, while long Ap stretches at the end of a long oligonucleotide (>10 bases) tend to give a shift to the right. The uncleaved, original material, ending in Gp, also shows a shift to the right (Fig. 5).

With the introduction of phosphatase-free T_4 polynucleotide kinase, the uncleaved oligonucleotides do not have to be dephosphorylated prior to

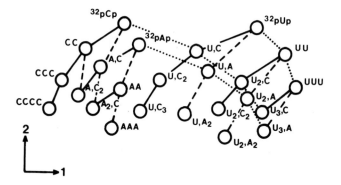

Fig. 4. Schematic correlation between gross nucleotide composition of oligonucleotides and their separation according to the mobility-shift method. Shifts are caused by the extention of a given fragment by a mononucleotide. The effects of the addition of Cp, Ap, and Up are indicated by solid, dashed, and dotted lines, respectively.

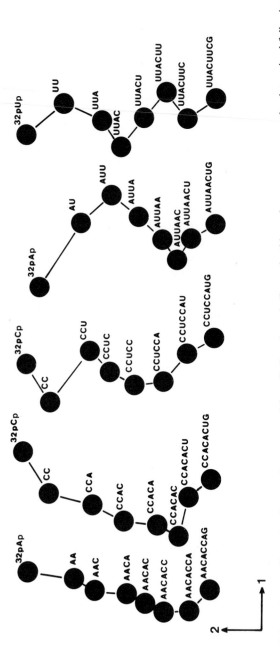

Fig. 5. Schematic, two-dimensional separation pattern of five alkaline-cleaved oligonucleotides, whose sequences can be determined following the characteristic mobility shifts depicted in Fig. 4.

the labelling step. As a result, the separation pattern seen (Fig. 5) is slightly different from those published previously (Stackebrandt et al., 1981, 1982). In that earlier work G_{OH}-ending oligomers ran slower in the first dimension and faster in the second dimension. This resulted in an overlap of intermediary spots by the non-cleaved material in certain cases, causing severe difficulties in the interpretation of sequences. The use of the new labelling procedure has eliminated these problems in all cases except the 3'-terminal oligonucleotide which lacks the 3'-phosphate group. In oligonucleotides with more than 10 bases, the larger intermediates (towards the 3' terminus) are usually insufficiently resolved. In such cases a sample of the oligonucleotide is separated in two dimensions as described above, but the NH_4 formate gradient is replaced by "Homomix" $(10:20:30:50:75$ mM KOH $= 0.5:1:1:1:1$, containing 1 M boric acid). "Homomix" is prepared by hydrolysis of a 3% solution of yeast RNA in 7 M urea by KOH (10, 20, 30, 50, and 75 mM) at 65°C for 20 h. The pH is adjusted to 4.7 with glacial acetic acid (Jay et al., 1974; Silberklang et al., 1977). Homochromatography produces a better separation of the long fragments than the gradient method. Since fragments up to heptanucleotides are allowed to migrate into the wick (the yellow marker dye, orange G, has just left the thin-layer plate), the longer fragments are well resolved enough to allow proper interpretation.

6. Tertiary analysis

There are cases in which the sequence cannot be determined unambiguously from secondary analysis alone. In these instances it is necessary to perform a tertiary analysis by one or a combination of several procedures (Fowler et al., 1984). The determination of the 5'-terminal nucleotide is necessary whenever its identity is not obvious from the secondary analysis. This is the case when a spot in the fingerprint contains only one oligonucleotide beginning with [32]pCp or with [32]pAp, or when two or more oligonucleotides with different 5' termini run as a single spot on the fingerprint. In this latter case, the determination of the 5' terminus of selected intermediates helps to follow the migration pattern of the individual oligonucleotides (Fig. 6). Experimentally the 5' terminus of complete oligonucleotides or fragments thereof can be determined by scraping the respective material from the thin-layer foil, and isolating the RNA as described earlier. The precipitated and dried material is completely digested by 4 μl of nuclease P_1 (25 ng ml^{-1} in 0.05 M NH_4 acetate, pH 4.5) at 37°C for several hours. The 5'-labelled nucleotides are identified by thin-layer chromatography in one dimension on cellulose plates (20 × 20 cm)

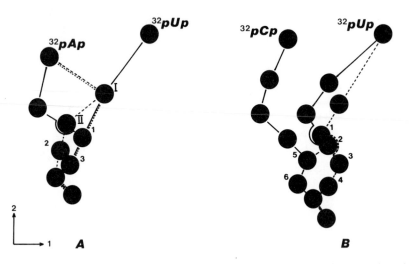

Fig. 6. Sequence determination by tertiary analysis. In those cases in which several comigrating sequences cannot be determined unambiguously, the relevant intermediate fragments are isolated and their sequences are determined as described for the secondary analysis, by the identification of their 5'-labelled termini. (A) Without tertiary analysis, the following sequences could be present: AAUACG, AAUCAG, AUAACG, AUCAAG, AUACAG, UAAACG, UAACAG, and UACAAG. The determination of 5'-labelled end groups of intermediate I reveal the presence of A and U; that for II, A only. The sequence determination of 1 gave UAA, AUA, and AAU, that of 2 gave AUCA, and that of 3 gave UAAA, AUAA, and AAUA. The presence of the following four oligonucleotides could therefore be proven: AAUACG, AUCAAG, AUAACG, and UAAAC. (B) The following sequences could be present: CCCUUAUG, CCUUCAUG, CCUUUACG, CCUCUAUG, UCCCUAUG, UCCUCAUG, UCCUUACG, and UAAUACG. The sequence determination of intermediates gave the following results: 1, UAA and traces of UCCU; 2, UCCU and traces of UAA; 3, UAAU; 4, UAAUA; 5, CCCUU and UCCUC; and 6, CCCUUA and UCCUCA. Intermediate 2 did not contain the sequence CCUU, 3 did not contain CCUUU or UCCUU, and 5 did not contain CCUCU, CCUUC, or UCCCU, excluding the presence of the versions CCUCUAUG, CCUUCAUG, CCUUUACG, UCCUUACG, and UCC-CUAUG. Note that the spot contains a mixture of octamers containing three uridine and one adenine residue, together with a heptamer with two uridine and three adenine residues (see Fig. 3).

in tertiary butanol : H_2O : conc. HCl (14 : 3 : 3) (Domdey *et al.*, 1978). In this solvent system the motility is as follows: uridine ≫ cytidine > adenosine. Autoradiography requires from one to several days.

Even knowing the 5' termini of two or more comigrating oligonucleotides it is sometimes impossible to determine their sequence. This is so when coinciding oligonucleotides or those from overlapping spots in the fingerprint share common intermediates of alkaline hydrolysis. In such cases fragments are isolated from the foil as described above and subjected to a second controlled alkaline hydrolysis. The resulting tertiary fragments are again separated in two dimensions as detailed for the secondary analysis. Because of the smaller size of the fragments, four samples can be applied to each cellulose acetate strip, 6, 15, 24, and 33 cm from one end and the xylene–cyanol marker allowed to migrate 4.5 cm. The low activity of these fragments requires autoradiography times of 2–3 weeks which can be reduced to 1 week by using preflashed film, intensifying screens, and storage at −80°C.

Difficulties are also sometimes encountered when discriminating between adenine and cytidine in the middle or at the end of uridine-rich stretches. Such sequences, as well as oligonucleotides up to 14 bases in length, can be elucidated and verified by one-dimensional thin-layer chromatography (Jay *et al.*, 1974) of the alkaline hydrolysate. As explained previously, the differences in the migration behaviour allow easy distinction between pyrimidines and purines in one dimension. Since guanosine occurs only at the 3′ terminus and the presence of uridine can be determined from the two-dimensional separation pattern, adenosine and cytidine can be easily distinguished in this way. The hydrolysate of an oligonucleotide is dissolved in 100 μl distilled water and the rRNA precipitated in 3 volumes of ethanol to remove salts. After 2 h at −20°C the precipitate is dried, redissolved in 2.4 μl of water, and the sample applied to a DEAE cellulose thin-layer plate (20 × 30 cm) and separated in one dimension by homochromatography (equal volumes of 20, 30, 50, and 75 mM KOH "Homomixes") at 70°C until the blue marker dye has almost reached the upper edge of the plate. Autoradiography of 1–3 days is required.

7. Posttranscriptional modifications

The methods described here are only partially effective in detecting posttranscriptional modifications. The presence of such modifications, if not unravelled, may cause errors in the interpretation of mobility-shift patterns or, in the case of hypermodified oligonucleotides, may prevent their detection altogether. On the other hand, less than 1% of the total number of nucleotides in eubacterial (Woese *et al.*, 1975) and most archaebacterial (Balch *et al.*, 1979) 16 S rRNAs are modified and only three to four of the RNase T_1-resistant oligonucleotides of hexamer size and larger carry modifications. The only exceptions are found in the 16 S rRNAs of thermoacidophilic archaebacteria (*Thermoplasma, Thermoproteus, Sulfolobus, Pyrodiction,* and others), which contain a much higher number of posttranscriptional modifications. Most of the modified oligonucleotides are highly conserved in sequence and occur in other organisms in their unmodified versions. This finding can be used to decide the correct nucleotide sequence of an otherwise incompletely modified oligonucleotide, although neither is the sequence proven nor the nature of the modification determined. Additional insight is obtained from the analysis of modified oligonucleotides, elucidated by DEAE paper chromatography. This method is able to detect modifications in both *in vivo* and *in vitro* labelled fragments (Silberklang *et al.*, 1979; Brownlee, 1972).

Of the two possible targets of modification, base and ribose, the base modifications are the most difficult to detect by the mobility-shift method because the modification does not markedly influence the migration pattern as compared to the pattern obtained with unmodified fragments. The only nucleotides of *in vitro* labelled rRNA oligonucleotides that can be easily tested for modification are the 5′ termini. Total digestion with nuclease P_1 followed by one-dimensional chromatography on cellulose thin-layer plates as described above immediately indicates the presence of such modifications. The situation is different with ribose-2-*O*-methyl

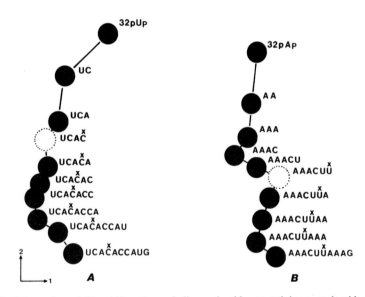

Fig. 7. Schematic mobility-shift pattern of oligonucleotides containing a nucleotide with a posttranscriptional 2-*O*-methyl group at the ribose moiety. Since alkali does not cleave the phosphodiester bond between this modified nucleotide and the 3′ neighbouring nucleotide, the pattern contains an unusually large gap in (A) and an unusual angle in (B). The position of the fragment with the modified nucleotide at its 3′ terminus is indicated by a white circle. Neither of the two sequences is strictly proven, but both are highly likely to be correct because close relatives of the respective organisms contain the equivalent oligonucleotides in an unmodified version.

modifications. Alkali is not able to cleave the phosphodiester bond between a ribose with a methyl group and its 3′ neighbouring nucleotide, resulting in a gap in the mobility-shift pattern (Fig. 7). Without support from the equivalent unmodified oligonucleotide the exact migration pattern cannot be interpreted (see legend to Fig. 7). Even more severe difficulties arise when an intermediate of a comigrating fragment occupies this gap and hence increases the probability that the presence of a modified oligonucleotide will be overlooked. These cases occur rather frequently in the highly modified rRNA of thermophilic archaebacteria, which makes the use of additional methods, e.g., the use of snake venom phosphodiesterase and nuclease P_1 (Silberklang *et al.*, 1977) or DEAE paper electrophoresis as indicated by Silberklang *et al.* (1979), for an unambiguous determination of the exact sequence absolutely necessary.

VI. Data Analysis

A. Data handling procedures

Like macromolecular sequence data, oligonucleotide catalogues obtained by different workers by different methods at different times can be usefully intercompared. Each catalogue is thus part of an ever-expanding data base. At present, this catalogue data is derived from the two methods described herein and by reduction of complete small subunit RNA sequences to catalogues. Currently the data base consists of approximately 27,000 individual oligonucleotides of length six and larger from over 350 strains of eubacteria and 35 strains of archaebacteria. It is clear that electronic data processing is needed to keep track of and to manipulate this data base. Programs for this purpose have been developed and these are available to interested researchers (Sobieski *et al.*, 1984).

Beyond the task of getting the raw data accurately entered into the data base it is necessary to ensure that data from the different methods can be intercompared. To this end certain conventions are followed and the potential user of the data base needs to be aware of these. Oligomers of less than five nucleotides are not recorded unless they contain posttranscriptional modifications because they typically occur numerous times in each sequence and their correspondence between two catalogues is thus meaningless unless the experiments produce reliable values for copy number, which they do not.

Oligomers carrying modifications are frequently detected. One usually knows with some confidence the identity of the original unmodified base but the nature of the modification is obscure except that one can distinguish base modifications from ribose methylations. In catalogue comparisons it is usual to consider two sequences as identical irrespective of the precise nature of the modification as long as the underlying sequence is the same. In the data base it is usual (for obscure historical reasons) to simply indicate a modified nucleotide by the letter *D*. More detailed information can be obtained from publications or manual records.

Pentamers are readily determined in the HVE method and are useful in obtaining perspective during the data interpretation phase. They have some value in evaluating very close phylogenetic relationships and are thus included in the data base. As evolutionary distances exceed the generic level the probability that a pentamer coincidence between two catalogues is due to chance rather than homology becomes unacceptably high. One can readily demonstrate this statistically and it is intuitively obvious from the large number of multiple-copy pentamers in a typical catalogue. In the TLC method, determination of the pentamer sequences would materially affect the time and cost required and they are thus ignored since they have only dubious value anyway.

By the time one reaches oligomers of length six the number of possibilities (243) is large and the number of occurences (approximately 30 in a typical catalogue) is small so that each oligomer is typically unique. At this stage identity argues strongly for homology since even random sequences of length 1600 are expected to share only two identical hexamers (Pechman and Woese, 1972). The data base convention of using the letter *D* to indicate modifications is dropped for these larger oligomers unless confusion with a nonhomologous oligomer is known to be possible.

As the oligomers get larger they in general become more difficult to sequence. In the HVE method this sometimes leads to situations in which the correct sequence must be one of a small number of alternative possibilities. When the number of such possibilities is three or less all of them are entered into the data base. The presence of these anomalies is controlled by keeping a separate exception list. Another source of exceptions is microheterogeneity. Since there are typically several ribosomal RNA genes it is best to regard purified 16 S rRNA as a population of nearly identical molecules. Occasionally sequence analysis will reveal that two large oligomers (10 or more bases) differ in sequence by only one base. In addition close examination of the fingerprint reveals that both are found in low yield. Taken together, this evidence strongly implies that the two oligomers are arising from a single location in the sequence. Although both versions are entered into the data base, they are treated as if they were alternative sequence possibilities and are again included on the exception list. The reader should realise, however, that most microheterogeneity is not readily documented by 16 S rRNA cataloguing.

B. Dendrogram construction

The goal in 16 S rRNA cataloguing is most often to obtain insight into the phylogenetic position of the strains being studied. The ideal way to do this would be to calculate the actual percentage of homology between the various 16 S rRNAs. It would then be possible to apply the methodologies usually used with macromolecular sequence data to produce evolutionary trees. Unfortunately, the number of oligonucleotide identities between two catalogues is not uniquely related to this true homology (Fox, 1984) and no reliable method for estimating the actual homology from catalogue data alone is currently available. Probabilistic attempts (Pechman and Woese, 1972; Bonen and Doolittle, 1975) to devise such a relation were based on the assumption that each position in the 16 S rRNA was equally mutable. Markov simulations of the cataloguing procedure (Young et al., 1981; Aaronson et al., 1982; Hori et al., 1982) made the same assumption. That this is not a correct assumption is revealed by comparisons of the actual number of oligomers of each length with the number expected (Pechman and Woese, 1972). A simulation in which known conserved positions and established secondary structural features were taken into account (Chen, 1980) demonstrated that when the assumption of equal mutability was dropped the predicted values of the actual homology were significantly changed. All things considered, the best hope for predicting actual sequence homologies from catalogue data lies in the development of correlations based on the growing number of 16 S rRNA sequences.

In the absence of reliable estimates of the underlying sequence homology it is somewhat risky to apply the usual cladistic methods because they assume that the similarity or difference measurement used has the same meaning (i.e., is linear) throughout the range of application. If the measure does not behave this way the branch lengths generated would quite possibly be meaningless. The usual approach, then, has been to apply average linkage cluster analysis using the association coefficient S_{AB}, defined in the introductory section as the measure of similarity. This similarity measure is one of several that might have been employed and should not be taken as an indicator of the actual homology, which it in fact underestimates by a wide margin. It does offer the advantage that it is readily calculated. The dendrograms generated represent the relationship between extant strains and should not strictly be interpreted as evolutionary trees. Since average linkage clustering is in fact used to obtain a first approximation in many cladistic methods it is not unreasonable to attempt to draw evolutionary inferences and this liberty has been freely taken.

A dendrogram is actually generated by applying the average linkage clustering algorithm to a matrix of association coefficients (S_{AB} values) which contains all possible pairwise comparisons for the group of strains under consideration. The algorithm and a very useful computer program that implements it are described in detail elsewhere (Anderberg, 1973). Briefly, the lower diagonal matrix is scanned for entries which are maxima in both their column and row. The organisms represented by such two-way maxima are linked together in the dendrogram and the two rows and columns representing these strains are merged into a single row and column. The merged combination is produced by taking the arithmetic average of the rows and columns of the two newly linked strains. The process is continued until all the strains have been linked into the dendrogram.

The actual assembly of the matrix must be done with some care, especially if catalogues obtained from both the TLC and HVE are included as is normally the case. It is essential that the same data be used in all the S_{AB} calculations. Thus it is usual to omit the pentamers and small post-transcriptionally modified oligomers which are not determined in the TLC method and which have little clustering value in any event. Likewise, the oligomer A*A*CCUG must be deleted from all catalogues that have it since it is not detectable by the TLC method. An especially difficult problem is presented by the 3'-terminal oligomer which is typically AUCACCUCCUUU or a related sequence. It is obvious that all 16 S rRNAs have a 3' terminus but it is not always detected in the fingerprint data. That it is easily missed (it lacks the terminal guanosine and terminal phosphate group and thus behaves in an aberrant fashion in both methods) has been demonstrated on numerous occasions where it has been found on a second attempt. How then does one handle these situations? If one chooses not to include this oligomer in the analysis then its considerable phylogenetic value is lost. On the other hand its absence from some organisms distorts needlessly their S_{AB} values. The usual practice then is to look at the close phylogenetic relatives of the organisms where it was missed and to include the most likely version in the data base. The inclusion of this speculative entry is then noted on the exception list. The 3'-end presents a second challenge in that length differences between terminal oligomers (e.g., AUCACCUCCUUU vs AUCACCUCCUUU) of two organisms are more likely to represent differences in the processing of the 16 S rRNA precursor rather than actual differences in the underlying genes. This problem is resolved by only considering two 3'-terminal sequences to be different when no combination of bases can be added to the shorter version to make it identical to the larger. A program which implements the calculation of S_{AB} values is available (Sobieski *et al.*, 1984) for use with the data base. The small number of oligomers (usually hexamers) that occur twice in more than one organism give four identities instead of the correct two in a straightforward searching comparison. The program automatically corrects for this. It is necessary, however, to correct manually a small number of S_{AB} values due to instances in which the same multiple versions have been included in two or more organisms due to sequencing problems. The exception list indicates which these are. The dendrograms themselves are, of course, sensitive to experimental errors and these can affect branching orders, especially among closely related organisms. The effect of errors is discussed in detail elsewhere (Fox, 1984).

C. Oligonucleotide families, signatures, and other matters

With the publication of the complete *E. coli* 16 S rRNA sequence (Brosius *et al.*, 1978) a much deeper appreciation of the relationship between catalogue and sequence became possible as is illustrated in Fig. 8. The oligomers of length six and longer are well distributed over the entire sequence. In fact, base-paired regions and single-stranded regions both seem to be well sampled by these oligomers. Seldom do the larger oligomers that comprise the catalogue include both interacting portions of a helical region. Paired positions evolve in a coordinated fashion and thus are redundant in their information content. That catalogues seldom detect both halves of these paired positions is a source of frustration to those who wish to use comparative data to search for these interactions. For phylogenetic studies it is a decided advantage and probably accounts for the fact that the cataloguing method works better than one intuitively expects. For example, the large oligomers of *E. coli* 16 S rRNA sample 595 of the 1542 positions (39%) in the molecule. A current secondary structure model (Woese *et al.*, 1983) which can be regarded as minimal (Fox, 1984) contains 428 bp. Thus, in reality, at least 53% of the most useful positions are sampled by the catalogue.

The S_{AB} analysis has one overriding virtue; it gives reasonable results! Nevertheless, it has been obvious since the first catalogues were compared that a method that ignored all the information in the non-identical oligomers could not be all good. In fact, some reflection shows that any two catalogues are only partially overlapping in that the actual positions sampled are not all homologous. An intercomparison between any two catalogues includes three major categories of oligomers: identities from homologous positions, non-identities from homologous positions, and non-identities from regions which are sampled in one catalogue but not the other. In addition, the real situation is fuzzy in that some oligomers in one catalogue may sample some but not all of the same positions as a non-identical oligomer in the other catalogue. If it were possible accurately to

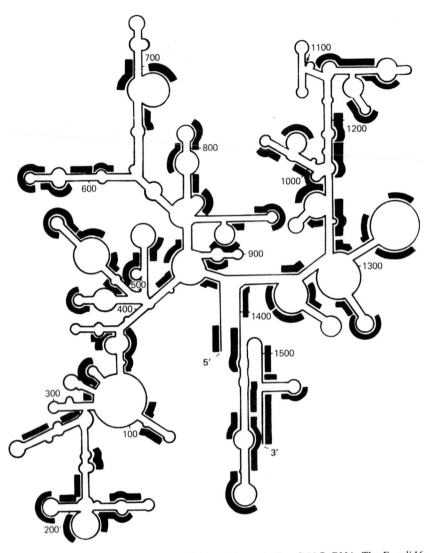

Fig. 8. Distribution of oligomers of length six and larger in *E. coli* 16 S rRNA. The *E. coli* 16 S rRNA is displayed schematically in a form which corresponds to a recent high-quality secondary structure model (Woese *et al.*, 1983). The 5' and 3' ends are indicated and every hundredth nucleotide is numbered. Parallel thin lines indicate stem regions which are thought to be base paired. The short, thick, dark segments show the location of the individual oligomers of six or more nucleotides that are produced by a complete ribonuclease T₁ digestion. These are well distributed throughout the molecule and do not show an obvious preference for either helical or single-stranded regions. It is apparent that it is a rare occurrence for two nucleotides which are thought to be base paired as part of a stem structure to both occur in oligomers of length six or larger.

assign oligomers to the three classes and to resolve the fuzziness we would have a very good method of estimating the true similarity between the underlying primary sequences. Unfortunately no general algorithm is available for this purpose.

It is possible nevertheless to identify many examples of non-identical oligomers that arise from homologous positions. Thus, for example, the oligomer CUCACCAAG which is present in most *Bacillus* strains is replaced by CUUACCAAG in *B. pasteurii*. An algorithm has been proposed (Fox *et al.*, 1977) for searching for such families of related oligonucleotides. A detailed study on such oligomer families in the genus *Bacillus* supported a branching order for the strains analysed that was essentially identical to that generated from the association coefficient. This was encouraging since the two approaches utilise an entirely different subset of the oligomers in the catalogues being compared. More recently, as sequence data has emerged it has been demonstrated that the approach used to identify such related oligomers is highly reliable. The problem is that the method is partially subjective and not readily amenable to complete automation. Thus as the number of catalogues has grown it has proven unfeasible to continue to generate these oligomer families. The oligomers in a few such families are sufficiently conserved that they can be tracked through the entire data set and work is now underway (Fox and Woese, unpublished results) to assemble this information. It is hoped that this data may be usable in correlation studies to locate tertiary interactions in the 16 S rRNA that exhibit patterns of covariation.

One can also look at the S_{AB} approach in a somewhat different way. The S_{AB} values are based entirely on oligomer identities which may not be equally significant. Certainly a few oligomer identities, especially among hexamers, are spurious. More importantly, some oligomers which are actually homologous reappear frequently in quite different phylogenetic groups while others are isolated and indeed quite characteristic of one or two specific groups. It is apparent then that these more informative oligomers should be given more weight in the calculation of the association coefficient than the others. A formal algorithm utilising this insight in a bootstrapping fashion may be feasible, though this has not yet been attempted. Efforts so far have involved the independent construction of an oligonucleotide signature (Woese *et al.*, 1980; Gibson *et al.*, 1979) of various groupings whose memberships may then reflect flaws in the association coefficient based dendrograms. These initial attempts at refinement of the dendrograms may offer the best hope of overcoming the related problems of rapid changes in evolutionary rate and drastic changes in 16 S rRNA base composition.

VII. Practical considerations

The previous sections have dealt with the theoretical background of sequence and data analysis and have given detailed protocols which begin with the isolation of 16 S rRNA and continue through the construction of a dendrogram which indicates the phylogenetic position of the strain. This section contains additional information and advice for the interested reader who may wish to begin work of this type. The major equipment needed includes a high-voltage electrophoresis apparatus consisting of two up-and-over tanks, a 5000-V power supply, a large incubator with glass windows which can be set to 70°C, a centrifuge with a swinging-bucket rotor to hold up to 80 1.5-ml reaction tubes, and a centrifuge that can be connected to a vacuum pump for rapid drying of ethanol-precipitated samples.

A single 16 S rRNA catalogue takes 4–5 weeks to complete, though much of this is waiting time for autoradiography. The operating cost excluding personnel is approximately 500–700 U.S. dollars, although this will be higher during the learning phase. The most expensive items are [γ-^{32}P]ATP (which can readily be made from inorganic phosphate (Johnson and Walseth, 1979)), T_4 polynucleotide kinase, cellulose–acetate strips, DEAE cellulose thin-layer foils, and X-ray films. Other items used in large quantities include gloves, Whatman 3MM paper, and reaction tubes. About 60 20 × 40 × 0.1-cm glass plates are needed as a support for the plastic DEAE cellulose foils and another 60 of size 20 × 40 × 0.2 cm are needed as radiation shields during autoradiography. The number of acetate strips and DEAE cellulose foils depend upon the number of spots found in the fingerprint and upon the number of samples requiring tertiary analysis. On the average 35 strips, foils, and X-ray films are necessary. Novices will find that it is far better to do extra analyses than to do too few and thus have to begin again. DEAE cellulose thin-layer glass plates are used only for the fingerprint and for the one-dimensional separation of oligonucleotides during tertiary analysis. In other situations the plastic-backed plates are used. Rather large amounts of ethanol (5 litres) and urea are needed. Each DEAE cellulose thin-layer foil is rinsed with about 100 ml of ethanol to remove urea and other salts and laboratory benches on which the transfer of oligonucleotides from the first to the second dimension is performed are thoroughly cleaned with ethanol. Ethanol can be used as 96% denatured solution and urea of p.a. grade was found to be sufficiently pure. The efficiency of the T_4 labelling reaction depends to some extent on the size of the oligonucleotides with the labelling better. Even oligonucleotides of comparable size which occur in one copy in the 16 S rRNA primary structure are not always labelled equally. This makes the determination of the number of comigrating oligonucleotides in a single spot on the fingerprint by Cerenkov counting unreliable.

The data handling and analysis package has been developed for a National Advanced Systems AS-9000N computer which emulates an IBM 370/3033 utilising WYLBUR operation under OS/MVS. Except for the dendrogram construction program, which is in FORTRAN, all the analysis programs are written in PL-1 and are readily transported to other mainframe computers which support PL-1. The main difficulty that will be encountered is with the data base structure. The current implementation is set up under the Indexed Sequential Access Method (ISAM) so that records can be retrieved either sequentially or by a four-digit key that has been assigned to each catalogue. This file structure is machine dependent and would have to be rewritten for implementation on a different hardware configuration. The programs in the analysis package are not calculationally intensive though many runs may be required. In an environment where computer time is expensive, data analysis may become a cost consideration.

On the whole, the single most difficult problem the novice will encounter is learning how to deduce reliably the sequences from the mobility-shift data. The ideal solution would be to spend 6 months with a group that is actively doing this type of work. If that is not possible the beginner would be well advised to begin working with small molecules such as 5 S rRNA, for which the spots are usually all resolved, and/or 16 S rRNAs, for which the appearance of the fingerprint and actual sequences of the individual oligomer are already known.

Acknowledgements

This work was supported by a NATO grant for international collaboration in research to E.S. and G.F., by National Aeronautics and Space Administration Grant NSG-7440 to G.F., and by a grant from the Deutsche Forschungsgemainschaft to W.L. and E.S.

References

Aaronson, R. P., Young, J. F., and Palese, P. (1982). *Nucleic Acids Res.* **10,** 237–246.
Anderberg, M. R. (1973). "Cluster Analysis for Applications." Academic Press, New York.
Balch, W. E., Fox, G. E., Magrum, L. J., Woese, C. R., and Wolfe, R. S. (1979). *Microbiol. Rev.* **43,** 260–296.
Bambara, R., Jay, E., and Wu, R. (1974). *Nucleic Acids Res.* **1,** 1503–1520.
Bonen, L., and Doolittle, W. F. (1975). *Proc. Natl. Acad. Sci. U.S.A.* **72,** 2310–2314.

Brosius, J., Palmer, M. L., Kennedy, P. S., and Noller, H. F. (1978). *Proc. Natl. Acad. Sci. U.S.A.* **75**, 4801–4805.

Brownlee, G. G. (1972). Determination of sequences in RNA. *In* "Laboratory Techniques" (T. S. Work and E. Work, Eds.). North-Holland/American Elsevier, Amsterdam and New York.

Brownlee, G. G., and Sanger, S. J. (1969). *Eur. J. Biochem.* **11**, 395–399.

Cameron, V., Soltis, D., and Uhlenbeck, O. C. (1979). *Nucleic Acids Res.* **5**, 825–833.

Chen, K. N. B. (1980). "Computer Simulation and Analysis of the 16S rRNA Sequence." M.S. Thesis, University of Houston Central Campus.

D'Alessio, J. M. (1982). *In* "Gel Electrophoresis of Nucleic Acids" (D. Rickwood and B. D. Hames, Eds.), p. 173. IRL Press, Oxford and Washington, D.C.

De Wachter, R., and Fiers, W. (1972). *Anal. Biochem.* **49**, 184–197.

De Wachter, R., and Fiers, W. (1982). *In* "Gel Electrophoresis of Nucleic Acids" (D. Rickwood and B. D. Hames, Eds.), p. 77. IRL Press, Oxford and Washington, D.C.

Diamond, A., and Dudock, B. (1983). *In* "Methods in Enzymology" (R. Wu, L. Grossman, and K. Moldave, Eds.), Vol. 100, p. 431. Academic Press, New York.

Domdey, H., and Gross, H. J. (1979). *Anal. Biochem.* **98**, 346–352.

Domdey, H., Jank, P., Saenger, H. L., and Gross, H. J. (1978). *Nucleic Acids Res.* **5**, 1221–1236.

Fowler, V., Ludwig, A., and Stackebrandt, E. (1985). *In* "The Use of Chemotaxonomic Methods for Bacteria." Academic Press, London.

Fox, G. E. (1985). *In* "The Bacteria" (I.C. Gunsalus, C. R. Woese, and R. S. Wolfe, Eds.) Vol. 8. Academic Press, New York.

Fox, G. E., Pechman, K. R., and Woese, C. R. (1977). *Int. J. Syst. Bact.* **27**, 44–57.

Fox, G. E., Stackebrandt, E., Hespell, R. B., Gibson, J., Maniloff, J., Dyer, T. A., Wolfe, R. S., Balch, W. E., Tanner, R. S., Magrum, L. J., Zablen, L. B., Blakemore, R., Gupta, R., Bonen, L., Lewis, B. J., Stahl, D. A., Luehrsen, K. R., Chen, K. N., and Woese, C. R. (1980). *Science* **209**, 457–463.

Gibson, J., Stackebrandt, E., Zablen, L. B., Gupta, R., and Woese, C. R. (1979). *Curr. Microbiol.* **3**, 59–64.

Gross, H. J., Domdey, H., Lossow, C., Jank, P., Raba, M., Alberty, H., and Sänger, H. L. (1978). *Nature (London)* **273**, 203–208.

Hassur, S. M., and Whitlock, Jr., H. W. (1974). *Anal. Biochem.* **59**, 162–166.

Hori, H., Itoh, T., and Osawa, S. (1982). *Zentralbl. Bakteriol., Parasitenkd., Infectionskr. Hyg., Abt. 1: Orig., Reihe C3*, 18–30.

Jay, E., Bambara, R., Padmanabhan, R., and Wu, R. (1974). *Nucleic Acids Res.* **1**, 331–353.

Johnson, J. L., and Harich, B. (1983). *Curr. Microbiol.* **9**, 111–120.

Johnson, R. A., and Walseth, T. F. (1979). *Adv. Cyclic Nucleotide Res.* **10**, 135–168.

Pechman, K. J., and Woese, C. R. (1972). *J. Mol. Evol.* **1**, 230–240.

Raba, M., Limburg, K., Burghagen, M., Katze, J. R., Simsek, M., Heckman, J. E., RajBhandary, U. L., and Gross, H. J. (1979). *Eur. J. Biochem.* **97**, 305–318.

Sanger, S. J., Brownlee, G. G., and Barrell, B. G. (1965). *J. Mol. Biol.* **13**, 373–398.

Silberklang, M., Gillum, A. M., and RajBhandary, U. L. (1977). *Nucleic Acids Res.* **4**, 4091–4108.

Silberklang, M., Gillum, A. M., and RajBhandary, U. L. (1979). *In* "Methods in Enzymology" (K. Moldave and L. Grossman, Eds.), Vol. 59G, pp. 58–109. Academic Press, New York.

Simsek, M., Ziegenmeyer, J., Heckman, J. E., and RajBhandary, U. L. (1973). *Proc. Natl. Acad. Sci. U.S.A.* **70**, 1041–1045.

Sobieski, J., Chen, K. N., Filiatreau, J., Pickett, M., and Fox, G. E. (1984). *Nucleic Acids Res.* **12**, 141–148.

Southern, E. M. (1974). *Anal. Biochem.* **62,** 317–318.

Stackebrandt, E., and Woese, C. R. (1981). *In* "Molecular and Cellular Aspects of Microbial Evolution" (M. J. Carlile, J. F. Collins, and B. E. B. Moseley, Eds.), pp. 1–32. Cambridge University Press, Cambridge.

Stackebrandt, E., Ludwig, W., Schleifer, K. H., and Gross, H. J. (1981). *J. Mol. Evol.* **17,** 227–236.

Stackebrandt, E., Seewaldt, E., Ludwig, W., Schleifer, K. H., and Huser, B. A. (1982). *Zentralbl. Bakteriol., Parasitenkd., Infektionskr. Hyg., Abt. 1: Orig., Reihe C3,* 90–100.

Uchida, T., Bonen, L., Schaup, H. W., Lewis, B. J., Zablen, L., and Woese, C. R. (1974). *J. Mol. Evol.* **3,** 63–77.

Wengler, G., Wengler, G., and Gross, H. J. (1979). *Nature (London)* **282,** 754–756.

Woese, C. R., Sogin, M. L. and Sutton, L. A. (1974). *J. Mol. Evol.* **3,** 293–299.

Woese, C. R., Fox, G. E., Zablen, L., Uchida, T., Bonen, L., Pechman, K., Lewis, B. J., and Stahl, D. (1975). *Nature (London)* **254,** 83–85.

Woese, C. R., Sogin, M., Stahl, D., Lewis, B. J., and Bonen, L. (1976). *J. Mol. Evol.* **7,** 197–213.

Woese, C. R., Maniloff, J., and Zablen, L. B. (1980). *Proc. Natl. Acad. Sci. U.S.A.* **77,** 494–498.

Woese, C. R., Gutell, R., Gupta, R., and Noller, H. F. (1983). *Microbiol. Rev.* **47,** 621–669.

Young, J. F., Taussig, R., Aaronson, R. P., and Palese, P. (1981). *In* "Replication of Negative Strand Viruses" (D. H. L. Bishop and R. W. Compans, Eds.), pp. 209–215. Elsevier/North Holland, Amsterdam and New York.

Zablen, L., Bonen, L., Meyer, R., and Woese, C. R. (1975). *J. Mol. Evol.* **4,** 347–358.

4

Analysis of Ribosomal Proteins by Two-Dimensional Gel Electrophoresis

AUGUST BÖCK

Institut für Genetik und Mikrobiologie, Universität München, Munich
Federal Republic of Germany

I. Introduction

Electrophoretic analysis of the protein complement of a microbial cell provides a rough measure of the genetic capacity and of the number and physicochemical properties of the gene products. Electropherograms of

METHODS IN MICROBIOLOGY
VOLUME 18

the cellular proteins thus deliver precise "fingerprints" of an organism and may be used for identification and classification purposes. One-dimensional separations have been applied extensively in this respect (for a review, see Kersters and De Ley, 1980); the results obtained closely paralleled those from DNA homology studies. Their application, however, is limited due to superposition effects and requires computerised comparison of the densitograms (Kersters and De Ley, 1975).

Theoretically, at least, two-dimensional separations of total cellular proteins (O'Farrell, 1975) are superior to one-dimensional ones, since they resolve individual gene products without extensive superposition. For practical application, however, they suffer from several drawbacks: (1) Cell proteins must be labelled with high specific radioactivity; this is achieved only for organisms cultivable on synthetic media. Moreover, protein patterns may be compared only when the labelling procedure can be carried out under identical nutritional conditions. (2) Because of the extreme sensitivity of resolution and due to the fact that many protein components apparently possess high evolutionary rates, the O'Farrell patterns differ greatly amongst strains of even the same species. As an example, patterns from B and K12 strains of *Escherichia coli* differ in the position of almost 20% of the proteins resolved (cf. Neidhardt *et al.*, 1983). (3) Quantitative comparison of O'Farrell patterns of microbial cells requires equipment ($X-Y$ densitometers with computerised treatment of the two-dimensional densitograms) which is not generally available.

A compromise, both from the methodological and from the informational points of view, is provided by the separation of ribosomal proteins. It offers a number of advantages: (1) It does not require radioactive labelling of proteins. (2) Because of their basic nature ribosomal proteins migrate differently from the bulk of cell proteins; therefore, no pre-fractionation is required. In fact, whole cells may be extracted with acetic acid and the extract directly employed for electrophoresis. (3) The high cellular content of ribosomal proteins allows the performance of a single analysis with only 10 A_{420} units of cells. (4) Ribosomal proteins constitute a group of cell components with a homogeneous "physiological" function. They possess evolutionary rates much lower than those for other proteins (Hori *et al.*, 1977). Because of many structural and functional constraints, their variability is more limited than that of other proteins. (5) Ribosomes from procaryotes are made up of a limited number of 50–60 proteins; this renders comparison—even visually—easier than that of O'Farrell gels.

Two principally different methods have been published for the electrophoretic analysis of ribosomal proteins. The first was devised by Kaltschmidt and Wittmann (1970); it has been used for disclosure of the protein components of bacterial ribosomes. The second was originally set up by Mets and Bogorad (1974) for separation of ribosomal proteins from *Chlamydomonas* and later from *E. coli* (Subramanian, 1974). The method described in this chapter (Geyl *et al.*, 1981) combines the first-dimension separation system of Mets and Bogorad (1974) with the second-dimension system of Kaltschmidt and Wittmann (1970). It offers several advantages over the original procedures: (1) Proteins can be kept in the sample solution; so electrophoresis can be repeated without any waste of material. (2) The first dimension run requires only 7 h as opposed to about 18 h necessary for the Kaltschmidt and Wittmann (1970) procedure. (3) No equilibration of the first-dimension gels with the second-dimension separation buffer is required; this keeps spots small and sharp. (4) The first-dimension gels can be removed from the glass tubes very easily due to the low acrylamide concentration.

II. Equipment

The first-dimension separation is carried out in glass tubes of 6 mm internal width and 110 mm length. For electrophoresis they are placed into any commercially available or machine shop-made disc electrophoresis apparatus. A general design of such an apparatus has been published in this series (Cooksey, 1971).

For the second-dimension the apparatus devised by Kaltschmidt and Wittmann (1970) is used. It was miniaturised to a chamber size of $10 \times 10 \times 0.3$ cm. Detailed construction information and exploded views are given by Kaltschmidt and Wittmann (1970).

III. Procedure

A. Isolation of ribosomal proteins

1. Extraction from whole cells

Cells from a 100-ml culture with an optical density of about 2 (420 nm) are collected by centrifugation, washed in 10 ml TMNSH buffer (20 mM Tris–HCl, 10 mM magnesium acetate, 30 mM ammonium chloride, and 6 mM 2-mercaptoethanol, pH 7.5), and resedimented. Five millilitres of 67% acetic acid containing 0.2 M magnesium acetate is added and the extraction mixture is stirred vigorously at 4°C for 45 min (Hardy *et al.*, 1969).

The precipitate which contains ribosomal RNA is removed by centrifugation (10 min at 10,000 g) and the supernatant containing the ribosomal proteins is dialysed for 60 min each against four changes of 2 litres of 1% acetic acid. The dialysate is frozen and lyophilised.

2. Extraction from isolated ribosomes

Cells are harvested, washed, and resedimented as described above. They are resuspended in 2 ml 10 mM Tris–HCl, pH 7.5, containing 10 mM magnesium acetate, 0.5 M ammonium chloride, 6 mM 2-mercaptoethanol (buffer A) and 10 μg desoxyribonuclease ml^{-1} (Hardy et al., 1969). The cells are broken by a method suitable for the organism studied. S30 extracts are prepared by two consecutive centrifugations, first for 10 min at 12,000 g and then for 30 min at 30,000 g. Five millilitres of the S30 extract are layered on top of 3 ml 30% sucrose solution (in buffer A) and centrifuged for 150 min at 47,000 rpm in a Beckman Ti 50 or Ti 75 rotor. The pellet which contains the ribosomes is suspended in 10 mM Tris–HCl, pH 7.5, containing 10 mM magnesium acetate and 6 mM 2-mercaptoethanol. Final A_{260} of the ribosome suspension should be below 250, ideally between 200 and 250. To this suspension, 2 volumes of glacial acetic acid and 43 mg magnesium acetate ml^{-1} of the total volume are added simultaneously. The further extraction procedure is identical to that described above for whole-cell extracts.

To avoid the time-consuming dialysis steps Barritault et al. (1976) have employed acetone precipitation for recovery of ribosomal proteins from the acetic acid extraction mixture. The ribosomal RNA-free supernatant is mixed with 5 volumes of acetone. After 5 min, the precipitated protein is collected by centrifugation (5 min at 7000 g); the protein sediment is suspended in a small aliquot of cold ($-20°$C) acetone, collected by filtration, and dried in vacuo for 5 min. The protein is then dissolved (at about 5 mg ml^{-1}) in 10 mM triethanolamine–HCl buffer (pH 7.5) containing 8 M urea (room temperature). The acetone precipitation and washing procedures described above are repeated and the filtrate obtained is finally dried in vacuo for 15 min. Acetone-prepared or lyophilised proteins can be stored in the dried state at $-20°$C for prolonged times.

B. Preparation of the first-dimension gel

The gel mixture is prepared by dissolving acrylamide (4 g); N,N'-methylene bisacrylamide (0.1 g); bis [2-hydroxyethyl]imino–tris[hydroxymethyl]methane (bis-tris) (1.19 g); urea (36 g); EDTA (ethylenedinitrilotet-raacetic acid) (0.2 g). The pH is adjusted to 5.0 with acetic acid and the volume is made up to 100 ml. The mixture should be kept refrigerated.

To 10 ml of the ice-cold gel mixture 35 μl of (N,N,N',N'-tetramethyl-enediamine) (TEMED) and 100 μl of 7% ammonium peroxydisulphate are added. (The peroxydisulphate solution should be stored refrigerated.) The mixture is poured into the gel tubes to a height of 10 cm. A 4 M urea solution is layered onto it (1–2 mm). Gelation will be completed within 20 min at room temperature.

C. Preparation of the sample

Ribosomal protein samples are dissolved at a concentration of 100–200 μg 100 μl^{-1} in the sample loading buffer, which contains urea (36 g); dithiothreitol (1 M, 1 ml); 10× concentrated upper electrode buffer (10 ml); H_2O to 100 ml.

To 20 ml of the sample loading buffer three drops of basic fuchsin (0.5 mg ml^{-1}) are added prior to use. Dissolved samples may be kept refrigerated in tightly sealed tubes (such as Eppendorf tubes) for several weeks without causing any adverse effects.

D. First-dimension buffers

Stock solutions of 10-fold concentration are prepared as follows. Upper electrode buffer: bis-tris (20.9 g); acetic acid (glacial) (45 ml); H_2O to 1.000 ml. The pH of the 1× working solution will be 4.0. Lower electrode buffer: potassium acetate (175.7 g); acetic acid (glacial) (49 ml); H_2O to 1.000 ml. The pH of the 1× working solution will be 5.0.

E. First-dimension electrophoresis

The gel tubes are placed in the electrophoresis chamber and the buffer reservoirs filled with cold (4°C) electrode buffer. Air bubbles are removed and the samples are underlaid on top of the gels. Usually 50–100 μl of samples are applied; up to 200 μl may be applied without impairing the resolution. Electrophoresis is carried out at 4°C and at 1 mA per gel for the first 30 min and then at 4 mA per gel until the dye migrates to the bottom. This usually takes 7 h. When more than 10 gels per chamber are analysed, a buffer change after about 3 h is necessary to avoid compression of slower migrating spots.

F. Preparation of the second-dimension gel

The second-dimension gel mixture is prepared by dissolving acrylamide (186 g); bis-acrylamide (4.8 g); urea (372 g); acetic acid (glacial) (54 ml); KOH (5 N, 10 ml); H_2O to 1.000 ml. The final pH will be 4.2; the gel mixture should be kept refrigerated.

The electrode buffer can be stored as a $10\times$ solution which contains glycine (140 g); acetic acid (glacial) (15 ml); H_2O to 1.000 ml. The final pH of the $1\times$ solution will be 4.0.

G. Second-dimension electrophoresis

All steps are carried out at room temperature. Per chamber of the second-dimension (which holds an assembly for five slabs; Kaltschmidt and Wittmann, 1970) about 350 ml of gel mixture is needed. The mixture is deaerated for 10 min, mixed with 2.1 ml TEMED, and kept on ice. One hundred fifty millilitres of the mixture is taken to seal the bottom of the slab grooves. For this purpose, 150 ml is mixed with 6 ml of 7% ammonium persulphate and poured into the bottom tray (Kaltschmidt and Wittmann, 1970) in which a sheet of cooking foil is stretched. This facilitates better detachment of the chamber assembly from the bottom gel after polymerisation. The assembled chambers (Kaltschmidt and Wittmann, 1970) are then placed into the tray to seal the bottom.

When the bottom gel has cooled down (after about 30 min), 7 ml of 7% ammonium persulphate solution is added to the remainder of the gel mixture and mixed and poured into the grooves of the chamber assembly. The first-dimension gels are then taken out from their tubes by carefully inserting a syringe needle and introducing 80% glycerol. These first-dimension gels are then immediately placed on top of the freshly poured slabs. A small spatula is inserted into the groove and run along underneath the first-dimension gel back and forth to eliminate air bubbles and to minimise any local heterogeneity of gel concentration which may have been introduced with the first-dimension gel. Gelation will be completed in 30 min. During this time add more gel mixture into each groove, if necessary.

After gelation the chamber assembly is detached from the bottom gel and placed into the buffer reservoir. Both reservoirs are filled with second dimension electrode buffer (stock solution diluted 10-fold with cold water). Electrophoresis is then run at 80 V in the cold (4°C) until the tracking dye, which was used in the first-dimension run and is at the bottom of the first-dimension gels, reaches the bottom. This usually takes 16–18 h.

IV. Staining procedure

After the run, gel slabs may be stained either with Coomassie brilliant blue or with amido black, following standard procedures. Gels are placed on stainless steel wire-net supports (Kaltschmidt and Wittmann, 1970) and immersed into the staining solution. Coomassie stain: 0.2% Coomassie brilliant blue, 30% methanol, and 10% acetic acid; destaining is with 30% methanol, 10% acetic acid. Amido black stain: 0.5% amido black, 5% acetic acid; after staining for 30 min the slabs are destained in 1% acetic acid.

The ribosomal protein patterns are then photographed in transmitted light.

V. Immunological analysis of two-dimensionally separated ribosomal proteins

A. Principle of the method

As exemplified below, ribosomal protein patterns may aid in the task of identifying an isolate and/or in tentatively classifying it relative to authentic standard organisms, provided that they are closely or moderately related. More quantitative information on the relationship is gained by combining two-dimensional separation with immunological comparison of the separated proteins. Figure 1 depicts the principle of the method originally developed by Towbin *et al.* (1979): Ribosomal proteins are separated by two-dimensional electrophoresis and then electrophoretically transferred onto nitrocellulose sheets. After staining, the nitrocellulose is saturated with bovine serum albumin and treated with antibodies directed against ribosomal proteins from homologous or heterologous organisms. After washing, antigen–antibody complexes on the nitrocellulose are detected by attachment of ^{125}I-iodinated A protein from *Staphylococcus aureus* and by subsequent autoradiography. The method described below is a modification of the procedures published by Towbin *et al.* (1979) and Howe and Hershey (1981).

B. Transfer of ribosomal proteins to nitrocellulose sheets

Ribosomal proteins are transferred to nitrocellulose sheets by electrophoresis in any commercially available equipment, such as the Biorad Transblot Chamber. Three gels can be processed simultaneously with this apparatus.

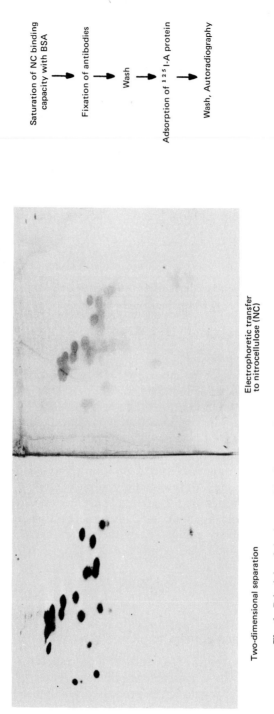

Two-dimensional separation

Electrophoretic transfer
to nitrocellulose (NC)

Saturation of NC binding
capacity with BSA

→ Fixation of antibodies

→ Wash

→ Adsorption of ^{125}I-A protein

→ Wash, Autoradiography

Fig. 1. Principle of the immunoblotting procedure (Burnette, 1981). NC, nitrocellulose; BSA, bovine serum albumin.

The gels are covered on one side by a nitrocellulose sheet (Schleicher and Schüll BA85, 0.45-μm pore size) and sandwiched between two sheets of Whatman 3MM filter paper. They are placed into the gel holder of the blotting chamber such that the nitrocellulose-covered side of the slab is directed towards the anode. The assembly is placed in the chamber filled with 6 M urea in 0.7% acetic acid. Care should be taken that no air bubbles are present; electrophoresis is carried out for 6 h at 60 V.

The nitrocellulose is then stained for 10 min with 0.1% amido black, 45% methanol, and 10% acetic acid, and destained for about 10 min with 90% methanol and 2% acetic acid. The nitrocellulose sheets are then rinsed with water and may be stored refrigerated and submersed in water for 1–2 days.

C. Fixation of antibodies

After staining, filters are incubated for 1 h at 37°C in 25 ml of 10 mM Tris–HCl, pH 7.4, containing 3% bovine serum albumin and 0.9% NaCl. They are then washed twice, 15 min each time, with 30 ml 10 mM Tris–HCl (pH 7.4), 0.5% Triton X-100, 0.2% sodium dodecylsulphate (SDS), and 0.9% NaCl.

Filters are then placed into a suitably diluted antibody solution (usually between 1 : 50 and 1 : 500) and incubated at room temperature with gentle agitation for 16 h. Antibodies are diluted in 10 mM Tris–HCl (pH 7.4), 0.5% Triton X-100, 0.2% SDS, 0.9% NaCl, 0.01% NaN$_3$, and 0.5% bovine serum albumin (buffer I). The filters are then washed five times (15 min each) in the same buffer but without bovine serum albumin (buffer II).

D. Attachment of *Staphylococcus aureus* protein A and autoradiography

Filters are placed in 20 ml buffer I containing 2.5 μCi ^{125}I-iodinated *Staphylococcus aureus* protein A and incubated with shaking for 3 h at room temperature. They are washed subsequently with five changes (30 min each) of buffer II and dried at room temperature. They are finally taped on a straight support, covered with a thin polyethylene cooking foil, and exposed to X-ray films.

VI. Examples of applications

A. Comparison of two-dimensional ribosomal protein patterns

The use of two-dimensional separation of ribosomal proteins for identification and classification purposes was first demonstrated by Wittmann and co-workers (Kaltschmidt *et al.*, 1970; Geisser *et al.*, 1973a,b) for members of the Enterobacteriaceae and Bacillaceae. It has been extended since to other groups of organisms (e.g., Adouette-Panview *et al.*, 1980; Cuny *et al.*, 1982; Madjar and Traut, 1980; Le Goff and Begueret, 1984) and shown to be especially useful in the analysis of organisms (such as methanogenic bacteria) for which classical identification criteria are not readily available (Douglas *et al.*, 1980; Schmid and Böck, 1982).

Figure 2 shows examples of two-dimensional ribosomal protein electropherograms from Gram-negative organisms. Figure 2,B and D show the patterns for *Salmonella typhimurium* and *Proteus mirabilis* when proteins from whole cell extracts were analysed. The comparison of Fig. 2D with the separation of ribosomal proteins isolated from purified ribosomes (Fig. 2C) proves that the relative spatial distribution is essentially identical; the only difference being that non-ribosomal proteins are migrating into the gel in the area of the origin of electrophoresis (upper left-hand corner) and that a few basic non-ribosomal proteins (see arrows in Fig. 2D) are migrating into the ribosomal protein area. It is evident, however, that the pattern is clear enough for separation of total cell extracts to be employed for identification purposes. What one needs then for identification is a "library" of ribosomal protein patterns from authentic organisms; as many as twenty preparations may be handled at the same time (four 2nd-dimension chamber assemblies) and the results—once the cells have been cultivated—are available within 4 days. One analysis only requires about 10 ml of culture of optical density 1.0 (420 nm).

For more quantitative purposes it is recommended that ribosomes are isolated by ultracentrifugation to avoid the overlap of nonribosomal proteins on electrophoresis (Figs. 2A,C,E). For classifying organisms the ratio of ribosomal proteins in the patterns of two organisms migrating into different positions relative to the total number of proteins has to be determined. This can be achieved in different ways: (1) Visual comparison can be made of the spatial distribution of protein spots. For example, it is immediately evident that the ribosomal protein pattern from *S. typhimurium* (Fig. 2B) resembles more that of *E. coli* (Fig. 2A) than that of *P. mirabilis* (Fig. 2C and D) or *Pseudomonas aeruginosa*. (Fig. 2E). (2) Measurements can be made for the R_f value of each protein which is highly reproducible in different runs (Kaltschmidt and Wittmann, 1970) (Fig. 1E). This may conveniently be done by photographing the protein pattern on a translucent support with a coordinate grid. (3) The five gels run in one and the same chamber assembly are usually completely superimposable in respect of their protein patterns. Differences in migration of some proteins can be monitored on the light box when two gels are placed on top of one another. (4) 1 : 1 mixtures of ribosomal proteins from two

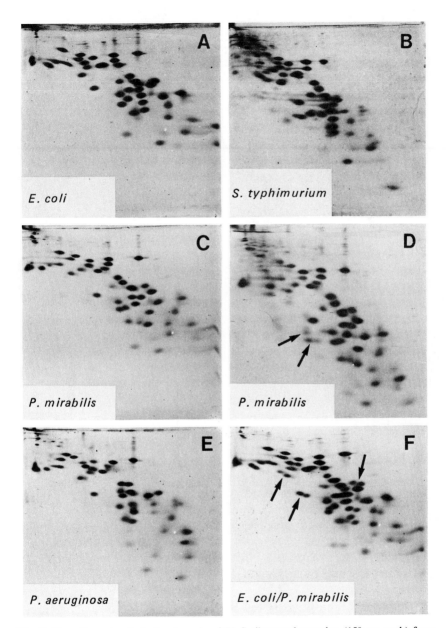

Fig. 2. Two-dimensional electrophoresis of 70 S ribosomal proteins (150 μg each) from (A) *E. coli*, (B) *S. typhimurium*, (C and D) *P. mirabilis*, (E) *P. aeruginosa*, and (F) a mixture (100 μg each) of *E. coli* and *P. mirabilis*. (B) and (D) show the separation of protein from a whole cell extract, the others of proteins from purified ribosomes.

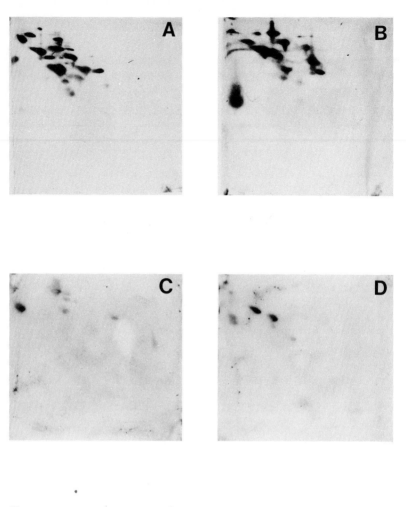

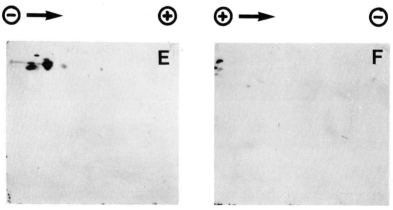

organisms can be separated to detect migration differences. Figure 2F shows an example of a combined electrophoresis of ribosomal proteins from *E. coli* and *P. mirabilis*. Some differently migrating protein pairs are indicated by arrows.

The number of electrophoretically differentiated proteins may be used to set up a difference matrix and to construct a phylogenetic tree of organisms. For treatment of the data the procedure of Hori and Osawa (1978) should be consulted. They quantified the relationship of 70 organisms belonging to the Enterobacteriaceae by co-elution of their ribosomal proteins relative to standard organisms from ion-exchange columns. Limitations of the method for classification purposes are also discussed by Hori and Osawa (1978).

B. Comparison by immunoblotting analysis

The immunoblotting technique originally devised by Burnette (1981) and modified for ribosomal protein comparison by Towbin *et al.* (1979) combines the resolution of two-dimensional separation with the power of immunological analysis. It provides more detailed information on relationships of two-dimensionally separated proteins. Questions may be answered such as whether proteins from different organisms migrating to the same position are indeed homologous, whether they have identical immunological determinants, or which of the proteins on an electropherogram are conserved relative to those from a reference organism.

Figure 3 shows an example. Here the cross-reaction of two-dimensionally separated archaebacterial ribosomal proteins (*Methanobacterium bryantii, Methanococcus vannielii, Halobacterium halobium*) was determined with antibodies directed against ribosomal subunits from *M. bryantii*. The three organisms are very distant phylogenetically (Fox *et al.*, 1977) and their two-dimensional ribosomal protein patterns do not bear any obvious resemblance (Schmid and Böck, 1982). Immunoblotting analysis, however, reveals that several proteins possess conserved immunological determinants, *Methanococcus* more than *Halobacterium*. Immunoblotting, therefore, gives a semiquantitative measure of the degree of relationship and also information on which of the proteins from the electropherogram are conserved.

◄ **Fig. 3.** Cross-reaction of two-dimensionally separated ribosomal proteins with antibodies raised against *Methanobacterium bryantii* ribosomes. (A) 80 μg 30 S proteins of *M. bryantii*, (B) 100 μg 50 S proteins of *M. bryantii*, (C) 80 μg 30 S proteins of *M. vannielii*, (D) 100 μg 50 S proteins of *M. vannielii*, (E and F) 150 μg 70 S proteins of *H. halobium* separated on gels with opposite polarisation to resolve acidic and basic proteins (G. Schmid and A. Böck, unpublished).

References

Adouette-Panvier, A., Davies, J. E., Gritz, L. R., and Littlewood, B. S. (1980). *Mol. Gen. Genet.* **179**, 273–282.

Barritault, D., Expert-Bezancon, A., Guerin, M. F., and Hayes, D. (1976). *Eur. J. Biochem.* **63**, 131–138.

Burnette, W. N. (1981). *Anal. Biochem.* **112**, 195–203.

Cooksey, K. E. (1971). *In* "Methods in Microbiology" (J. R. Norris and D. W. Ribbons, Eds.), Vol. *5B,* pp. 573–594. Academic Press, New York.

Cuny, M., Millet, M., and Hayes, D. H. (1982). *Biochimie* **64**, 1049–1058.

Douglas, C., Achatz, F., and Böck, A. (1980). *Zentralbl. Bakteriol., Parasitenkd., Infektionskr. Hyg., Abt. 1: Orig. Reihe C1,* 1–11.

Fox, G. E., Magrum, L. J., Balch, W. E., Wolfe, R. S., and Woese, C. R. (1977). *Proc. Natl. Acad. Sci. U.S.A.* **74**, 4537–4541.

Geisser, M., Tischendorf, G. W., and Stöffler, G. (1973a). *Mol. Gen. Genet.* **127**, 129–145.

Geisser, M., Tischendorf, G. W., Stöffler, G., and Wittmann, H. G. (1973b). *Mol. Gen. Genet.* **127**, 111–128.

Geyl, D., Böck, A., and Isono, K. (1981). *Mol. Gen. Genet.* **181**, 309–312.

Hardy, S. H. S., Kurland, C. G., Voynow, P., and Mora, G. (1969). *Biochemistry* **8**, 2897–2905.

Hori, H., and Osawa, S. (1978). *J. Bacteriol.* **133**, 1089–1095.

Hori, H., Higo, K., and Osawa, S. (1977). *In* "Molecular Evolution and Polymorphism" (M. Kimura, Ed.), Proceedings of the Second Taniguchi International Symposium on Biophysics, pp. 240–260. Mishima, Japan.

Howe, J. G., and Hershey, J. W. B. (1981). *J. Biol. Chem.* **256**, 12836–12839.

Kaltschmidt, E., and Wittmann, H. G. (1970). *Anal. Biochem.* **36**, 401–412.

Kaltschmidt, E., Stöffler, G., Dzionara, M., and Wittmann, H. G. (1970). *Mol. Gen. Genet.* **109**, 303–308.

Kersters, K., and De Ley, J. (1975). *J. Gen. Microbiol.* **87**, 333–342.

Kersters, K., and De Ley, J. (1980). *In* "Microbial Classification and Identification" (M. Goodfellow and R. G. Board, Eds.), Society for Applied Bacteriology Symposium Series. No. 8, pp. 273–297. Academic Press, New York.

Le Goff, V., and Begueret, J. (1984). *Mol. Gen. Genet.* **193**, 143–148.

Madjar, J. J., and Traut, R. R. (1980). *Mol. Gen Genet.* **179**, 89–101.

Mets, L. J., and Bogorad, L. (1974). *Anal. Biochem.* **57**, 200–210.

Neidhardt, F. C., Vaughn, V., Phillips, T. A., and Bloch, P. L. (1983). *Microbiol. Rev.* **47**, 231–284.

O'Farrell, P. (1975). *J. Biol. Chem.* **250**, 4007–4021.

Schmid, G., and Böck, A. (1982). *Zentralbl. Bakteriol., Parasitenkd., Infektionskr. Hyg., Abt. 1: Orig. C3,* 347–353.

Subramanian, A. R. (1974). *Eur. J. Biochem.* **45**, 541–546.

Towbin, H., Staehelin, T., and Gordon, J. (1979). *Proc. Natl. Acad. Sci. U.S.A.* **76**, 4350–4354.

5

Analysis of the Chemical Composition and Primary Structure of Murein

KARL HEINZ SCHLEIFER

Technische Universität München, Munich, Federal Republic of Germany

I. Introduction

Murein (peptidoglycan, mucopeptide) is the main cell wall polymer of eubacteria and is common to both Gram-negative and Gram-positive bacteria. There are only a few prokaryotic organisms, such as mycoplasmas and archaebacteria, which lack murein (Kandler, 1982; Schleifer and Stackebrandt, 1983). It is a heteropolymer consisting of polysaccharide strands (glycan) cross-linked through short peptides. The glycan moiety is made up of β-(1,4)-glycosidically linked N-acetylated residues of glucosamine and its 3-O-D-lactyl ether derivative, muramic acid. An oligopeptide (stem peptide) containing L- and D-amino acids is linked to the carboxyl group of muramic acid. Adjacent stem peptides may be cross-linked directly or via an interpeptide bridge.

123

METHODS IN MICROBIOLOGY
VOLUME 18

The glycan moiety is rather uniform and shows only a few variations such as O-acetylation or O-phosphorylation or the exceptional absence of peptide substituents. Some bacilli (Hayashi *et al.*, 1973) and rhodopseudomonads (Schmelzer *et al.*, 1982) lack, at least partially, *N*-acetyl substitution of glucosamine, whereas in *Micrococcus luteus* the amino group of muramic acid is not acetylated (Mirelman and Sharon, 1967). In certain mycobacteria and nocardiae, muramic acid is not present as an *N*-acetyl but as an *N*-glycolyl derivative (Lederer *et al.*, 1975; Uchida and Aida, 1977).

The peptide moiety of the murein shows, in contrast to the glycan part, a considerable variation. Extensive investigation of the chemical structure of murein has demonstrated the existence of almost 100 different variations of the peptide moiety (reviewed in Schleifer and Kandler, 1972; Schleifer and Stackebrandt, 1983; Schleifer and Seidl, 1984). The stem peptide always consists of alternating L- and D-amino acids. Even *meso*-diaminopimelic acid (*m*-A$_2$pm) follows this rule and is bound with its L-asymmetric centre in the stem peptide. The stem peptides can be cross-linked in two different ways (Fig. 1).

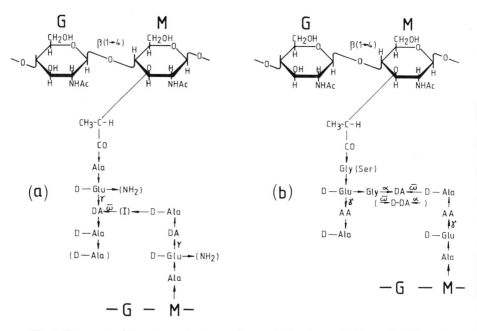

Fig. 1. Fragments of the primary structures of group A (a) and group B (b) mureins. Unusual abbreviations: AA, L-amino acid; DA, diamino acid; G, *N*-acetylglucosamine; I, interpeptide bridge; M, *N*-acetylmuramic acid; α, α-amino group of diamino acid; γ, γ-carboxyl group of D-glutamic acid; ω̄, distal amino group of diamino acid.

Depending on the mode of cross-linking two main groups of murein, named A and B, have been distinguished (Schleifer and Kandler, 1972). The cross-linkage of group A extends, as depicted in Fig. 1a, from the distal amino group of the diamino acid in position 3 of one stem peptide to the carboxyl group of D-alanine in position 4 of an adjacent stem peptide either directly or via an interpeptide bridge. The interpeptide bridge varies considerably in its amino acid composition and sequence.

Mureins of group B are characterised by a cross-linkage between the α-carboxyl group of D-glutamic acid in position 2 of one stem peptide and the C-terminal D-alanine in position 4 of another stem peptide (Fig. 1b). There is always an interpeptide bridge containing a diamino acid present since two carboxyl groups have to be cross-linked. Moreover, there is an interesting correlation between the configuration of the diamino acid present in the interpeptide bridge and its mode of linkage. D-Diamino acids are bound with their α-amino group to D-alanine whereas L-amino acids are linked with their distal amino group. The distribution of group B murein is quite restricted in comparison to that of group A. The former is only found in some coryneform and a few anaerobic bacteria (see Schleifer and Seidl, 1984).

Most mureins of Gram-negative bacteria are directly cross-linked and contain *meso*-diaminopimelic acid (m-A$_2$pm) as the diamino acid in position 3 of the stem peptide. There are only a few exceptions known (reviewed in Schleifer and Seidl, 1984) in which m-A$_2$pm is replaced by ornithine (*Spirochaeta stenostrepta*, *Treponema pallidum*), m-lanthionine (some fusobacteria), or lysine (*Bacteroides asaccharolyticus*).

Detailed information on the chemical composition and structure of murein, the different murein types, and their taxonomic implications is presented in the review by Schleifer and Kandler (1972). The more recently described murein types have been summarised by Schleifer and Seidl (1984).

II. Description of analytical methods

A. Preparation of murein

The different structure of cell walls of Gram-positive and Gram-negative bacteria necessitates discrete methods for preparing cell walls. The cell walls of Gram-positive bacteria reveal in profile one thick and more or less homogeneous layer, whereas Gram-negative bacteria have thinner, but distinctly layered cell walls with an outer membrane resembling the cytoplasmic membrane in profile (Bayer, 1974). The polymers found in the cell walls of Gram-positive and Gram-negative bacteria are chemically quite different. The former contain as major components murein (usually ≥30% of the dry weight of walls) with covalently attached protein, polysaccharide and/or teichoic acid, or teichuronic acid. Cell walls of Gram-negative bacteria, on the other hand, are composed mainly of lipopolysaccharide, phospholipid, protein, lipoprotein, and relatively little murein (usually less than 10% of the dry weight of walls). Lipoprotein is the only cell wall polymer of Gram-negative bacteria that is covalently linked to murein.

1. Preparation of cell walls and murein from Gram-positive bacteria

Cells are harvested, suspended in water, and boiled for 10 min to inactivate autolytic enzymes. Cells which contain large amounts of lipids (e.g., Actinomycetales) are difficult to disintegrate because of clumping, and should be defatted by stirring with chloroform–methanol (2 : 1, v/v) overnight at room temperature (Minnikin et al., 1977) or by a modified Bligh and Dyer procedure (Card, 1973). In the latter case cells are suspended in aqueous 0.3% NaCl (10 ml) and added to methanol (100 ml), and the suspension is refluxed over a boiling water bath for 5 min. The suspension is then cooled and chloroform (50 ml) and 0.3% NaCl (30 ml) are added to give final concentrations of chloroform–methanol–water of 1 : 2 : 0.8 (by volume). The suspension is stirred at room temperature for at least 2 h and is then filtered and collected in a separating funnel. Chloroform and 0.3% NaCl are added to give final concentrations of chloroform–methanol–water of 2 : 2 : 1.8 (by volume) and the phases are allowed to separate. The aqueous phase containing the cells is centrifuged and washed with methanol.

Cell walls are usually prepared from disrupted cells by exposing them to mechanical forces, ultrasonic energy, or shearing stress. The standard method for breaking open bacterial cells is shaking them with small glass or plastic beads (diameter 0.1–0.3 mm) in specially designed cell mills (e.g., Vibrogen cell mill, Bühler, Tübingen, Federal Republic of Germany). The use of styrene divinylbenzene copolymer (Shockman et al., 1967) generates less heat during disruption and no pH change is observed in contrast to some samples of glass beads which may liberate considerable amounts of alkali. The wet cell paste is mixed with the beads to a stiff mash and rapidly shaken with continuous cooling in the cell mill for 20–30 min. Foaming can be prevented by addition of one or two drops of octanol. The disruption of the cells can be followed microscopically by Gram-staining or by phase-contrast microscopy. Broken cells appear grey in the phase-contrast microscope and can easily be distinguished from the dark intact cells. The beads are removed by filtering through a coarse-grade sintered-glass filter and rinsing with water or buffer. The glass beads are then carefully cleaned by washing them successively with concentrated hydrochloric acid, hot distilled water, acetone–ethanol (1 : 1, v/v), and distilled water. The glass beads are also dried before being used again. Other, less frequently applied methods for preparation of cell walls include grinding the cells with Al_2O_3, passage of a suspension through slits or orifices (e.g., Hughes- and French-press) or treatment with ultrasonic waves. The suspensions of broken cells are centrifuged at 30,000–40,000 g for 20 min. The sediment consists of crude cell walls, still contaminated by membrane fragments and cytoplasmic contents. Crude cell walls are carefully suspended in 0.05 M phosphate buffer (pH 7.8). Trypsin (0.5 mg ml⁻¹) and 1 ml toluene, to prevent microbial growth, are added and the suspension is incubated at 37°C for about 12 h. The mixture is then centrifuged and washed at least four times with distilled water. The pellet consists of purified cell walls (trypsin-treated cell walls) containing, besides murein, cell wall polysaccharides and/or teichoic acids.

Acid hydrolysates of trypsin-treated cell walls of most Gram-positive bacteria reveal only the amino acids of murein and no or only traces of other amino acids (contaminants). Rare cases, e.g., strains of *Streptococcus pyogenes,* need an additional treatment with other proteases since their cell walls contain trypsin-resistant protein(s). Traces of DNA and RNA can easily be removed from cell walls in incubation with DNase and/ or RNase before or after treatment with trypsin. Crude cell walls can also be purified by extracting them twice in a 2% solution of sodium dodecylsulphate (SDS) in the cold (Conover *et al.,* 1966). Treatment with anionic detergent removes membranous material that adheres firmly to the walls. The detergent is removed by repeated washing with buffer containing 1 M NaCl and water.

Murein can be prepared from purified cell walls by removing non-murein material with different extraction procedures (Schleifer, 1975). In the case of Gram-positive bacteria most of the cell wall polymers are covalently linked to murein and rather harsh extraction procedures have to be applied to remove them. Cell wall polysaccharides can be extracted with hot formamide (150–180°C, 30 min) but O- and N-formyl groups are introduced into the murein (Perkins, 1965). Cell wall teichoic acids can be extracted with hot (60–90°C, 10–30 min) or cold (4°C, 24–28 h) 10% (w/v) aqueous trichloroacetic acid. However, a partial acid degradation of the murein can occur. Cell wall teichoic acids can also be extracted with dilute alkali (0.1 N NaOH, 20–37°C, 16–24 h). Using this procedure glycine-containing interpeptide bridges can be partially hydrolysed (Archibald *et al.*, 1969) and the reducing termini of the glycan strands can also be degraded. The mildest procedure for extracting most of the polysaccharides and teichoic acid is treatment with 70% (w/w) hydrogen fluoride at 2°C for 2–3 h (Glaser and Burger, 1964; Lipkin *et al.*, 1969; Fiedler *et al.*, 1981). The HF is evaporated *in vacuo*, the dried material is then mixed with dilute NaOH to neutralise residual HF and centrifuged, and the sediment is repeatedly washed with distilled water.

Crude and even trypsin-treated cell walls of Gram-positive bacteria are often contaminated with lipoteichoic acids. They can be extracted with 70% (w/v) aqueous phenol at pH 4.7 with stirring at 65°C for 1 h (Fischer *et al.*, 1983). After cooling, the mixture is centrifuged and the insoluble residue (lipoteichoic acid-free cell walls) is washed several times with distilled water.

2. Preparation of murein from Gram-negative bacteria

Weidel *et al.* (1960) were the first to isolate the so-called rigid or R layer from a Gram-negative bacterium. Cells of *Escherichia coli* B are disrupted by repeated treatment with 0.03 M NaOH and subsequent suspension in 0.4% SDS. The cell debris is then extracted several times with 90% (w/v) aqueous phenol and repeatedly washed to remove residual lipopolysaccharide. The insoluble residue is suspended again in 0.4% SDS and shaken with glass beads; after separation of the glass beads the suspension is centrifuged and washed. The sediment is the rigid layer that consists of murein and covalently attached lipoprotein. The latter can be removed by digestion with either trypsin or pepsin. A simpler procedure for the preparation of murein from Gram-negative bacteria (Braun and Sieglin, 1970; Braun *et al.*, 1970; Goodwin and Shedlarski, 1975) is as follows: Freshly grown or frozen cells, preferentially harvested in the logarithmic growth phase, are mechanically opened by shaking with glass beads or disrupted by several freeze–thaw cycles. The glass beads are removed on a coarse sintered filter and washed with ice-cold water. The suspension is adjusted to a final EDTA concentration of 0.01 M by adding 0.1 M EDTA, pH 7.4. The mixture is centrifuged at 22,000 g at 2°C for 1.5 h. The pellet is resuspended in ice-cold water and again centrifuged. The washing procedure is repeated once.

The sediment consisting of crude cell wall and residual cytoplasmic membrane is suspended in water (10–15 g wet wt. ml^{-1}) and added dropwise with stirring into a boiling solution of 4% SDS (Braun and Sieglin, 1970; Braun *et al.*, 1970) or 4% SDS–1 M NaCl solution (Goodwin and

Shedlarski, 1975) at a ratio of 2–3 ml g wet wt.$^{-1}$ of crude cell walls. The temperature of the mixture should not fall below 90°C while adding the crude walls. The suspension is stirred for 2 h while it cools and then kept overnight at room temperature. The cell walls are centrifuged at 78,000 g at 20°C for 20 min and washed at least three times with distilled water. In many cases the SDS treatment has to be repeated to obtain a clean preparation of the rigid layer (Braun et al., 1970). The pellet obtained by centrifugation should consist of translucent material. White opaque material has to be retreated with SDS. The clear pellet has to be washed several times with water to remove detergent. The sediment obtained from the final centrifugation is designated as rigid layer or partially purified murein.

The rigid layer is suspended and incubated with trypsin (0.01 M Tris–HCl, pH 7.8, 37°C, 4–12 h), pronase P (0.01 M Tris–HCl, pH 7.4, 37°C, 2 h), or papain (0.01 M Tris–HCl, pH 6.4, 37°C, 4 h) to yield a murein free of protein. The only non-murein amino acids which can remain are lysine and, usually to a minor extent, arginine. This indicates that the two amino acids constitute the covalent link between the murein and the lipoprotein. Murein prepared as described above is chemically pure but may still contain traces of endotoxin as indicated by its high pyrogenic activity (Rotta and Schleifer, 1974). Treatment with hot formamide (180°C, 30 min) or 0.25 M NaOH at 56°C for 60 min destroys residual endotoxin activity. Murein of cyanobacteria can be prepared in the same way as that of Gram-negative bacteria (Golecki, 1977). The extraction with 4% SDS can also be replaced by treatment with 2% SDS containing 0.7 M 2-mercaptoethanol at 60°C for 30 min followed by a second treatment at 100°C for 5 min (Rosenbusch, 1974).

A detailed description of the preparation of a clean peptidoglycan of the unicellular cyanobacterium Synechocystis is given by Jürgens et al. (1983). Sodium dodecylsulphate-insoluble fractions of the cell wall (rigid layer) are suspended in 0.02 M Tris buffer (pH 7.4) and treated with pronase and trypsin at 37°C to remove residual protein. The protease-treated cell walls are extracted twice with boiling 4% SDS, and the detergent is removed by washing with distilled water. Covalently bound polysaccharides are extracted with ice-cold 48% hydrofluoric acid (HF). Lyophilised detergent-extracted and protease-treated cell walls (30–50 mg) are suspended in 2 ml of ice-cold 48% HF and kept at 0°C for 48 h. HF is removed by evaporation in vacuo. The lyophilised residue is suspended in ice-cold distilled water and neutralised with 0.1 M LiOH. After several washes with distilled water, the HF-treated walls are collected by centrifugation (175,000 g, 4°C, 2 h). The pellet consists of clean murein.

3. Rapid preparation of murein for routine studies

The preparation of cell walls by mechanical disruption, sonication, or grinding with aluminum oxide is rather tedious and time consuming. Therefore, several extraction procedures have been tried to prepare relatively pure cell walls from intact cells without disintegrating the cells by mechanical or other physical forces. Extracting cells with diluted NaOH (Kandler *et al.*, 1958; Boone and Pine, 1968; Richter, 1977) has the disadvantage that it hydrolyses some peptide linkages and can lead to a dissolution of cells (Archibald *et al.*, 1969, 1970). Moreover, the extent of purification obtained with the alkali extraction procedure is not sufficient with some organisms (O'Donnell *et al.*, 1982).

In our hands (Schleifer and Kandler, 1972) a modification of the method of Park and Hancock (1960) turned out best for a rapid preparation of relatively pure murein. An overnight culture (30–50 ml) is harvested and resuspended in 10% (w/v) aqueous trichloroacetic acid. The suspension should have at 1 : 10 dilution an optical density of about 1.0 at 650 nm. The suspension is kept in a boiling water bath for 20 min, then cooled and centrifuged. The pellet is carefully rinsed with distilled water and resuspended in 0.1 M phosphate or Tris–HCl buffer, pH 7.8–8.0. Trypsin (2 mg 10 ml buffer^{-1}) is added and the mixture is incubated at 37°C on a shaker for about 2 h. The suspension is centrifuged and the sediment washed twice with distilled water. The washed pellet can be hydrolysed with 4–6 M HCl to determine the amino acid composition of the murein. The murein preparation obtained by the rapid procedure may still contain minor contaminating amino acids.

B. Determination of murein constituents

Acid hydrolysates of purified cell walls are prepared to determine their amino acid and amino sugar composition. Hydrolysis with 4 M HCl (0.1 ml mg cell wall^{-1}) at 100°C for 16 h is usually sufficient to break down the murein completely to free amino acids and amino sugars. In a few cases rather acid-stable peptides can be formed that are not completely split to the corresponding acids under normal hydrolysis conditions. Prolonged acid hydrolysis (6 M HCl, 100°C, 18 h) is necessary to degrade peptides consisting of aspartic acid or threonine linked with their carboxyl group to the distal amino group of lysine or ornithine. HCl is removed by drying in a stream of warm air. The dry residue is dissolved in double distilled water and insoluble, black degradation products can be removed by filtration.

*1. Qualitative determination of the amino sugar and
amino acid composition*

Two-dimensional descending paper or ascending thin-layer cellulose chromatography is often sufficient to determine the qualitative amino acid and amino sugar composition of the purified cell wall. Extraction of trypsin-treated cell walls with hot formamide, trichloroacetic acid, or 70% HF is often necessary to remove amino sugar containing cell wall polysaccharides. The most suitable method for the qualitative amino acid and amino sugar determination is two-dimensional descending paper chromatography. Schleicher and Schüll 2043b Mgl paper or a comparable chromatography paper should be used. To shorten the running time only a quarter (29 × 30 cm) of the original sheet is used. A hydrolysate of 1–3 mg of a purified cell wall preparation is applied as a single spot in one corner of the chromatogram. The solvent isopropanol (2-propanol)–acetic acid–water (75 : 10 : 15, by volume) is used for the first and α-picoline (2-methylpyridine)–25% NH_4OH–water (70 : 2 : 28, by volume) for the second direction. Chromatograms are sprayed with ninhydrin (5 g of ninhydrin is dissolved in 70 ml acetic acid plus 930 ml *n*-butanol) and heated at 100°C for 5–10 min. Amino acids and amino sugars give violet to purple spots with the exception of aspartic acid which appears steel blue at the beginning and turns to purple within 10–15 min. Figure 2 shows a two-dimensional paper chromatogram of a total acid hydrolysate of a trypsin-treated cell wall. The mobilities of various amino acids and amino sugars after two-dimensional descending paper chromatography in the solvent systems isopropanol and α-picoline are listed in Table I. An unequivocal determination of the diamino acids requires an additional one-dimensional paper chromatography run using the solvent methanol–pyridine–water–12 *M* HCl (32 : 4 : 7 : 1, by volume) or methanol–pyridine–water–formic acid (80 : 10 : 19 : 1, by volume). The R_{Lys} and R_{Orn} values of various diamino acids are given in Table II. Homoserine and threonine cannot be separated using the solvent systems mentioned above. However, the two amino acids do not occur in the same murein. Nevertheless, homoserine can be identified by its, at least partial, conversion to homoserine lactone. It stains with ninhydrin yellow, turning slowly brown and then purple.

2. Quantitative determination of murein constituents

Quantitative determination of all the amino acids and amino sugars of murein is usually carried out with an automated amino acid analyser (Schleifer and Kandler, 1972). A cell wall hydrolysate (0.1–0.3 mg) is

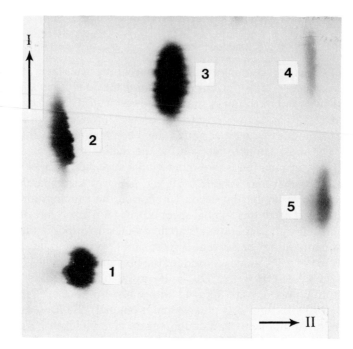

Fig. 2. Two-dimensional chromatogram of a total hydrolysate (4 N HCl, 100°C, 16 h) of cell walls of *Micrococcus kristinae*. I, isopropanol–acetic acid–water (75 : 10 : 15, by volume); II, α-picoline–25% NH₄OH–water (70 : 2 : 28, by volume). 1, Lys; 2, Glu; 3, Ala; 4, muramic acid; 5, glucosamine.

applied and eluted with buffer A (0.2 M sodium citrate buffer, pH 3.22) followed by buffer B (0.2 M sodium citrate, pH 4.25) and buffer C (0.05 M sodium citrate, pH 6.28 + 1 M NaCl). The total time for the elution of all amino compounds of murein including ammonia is 165 min. The temperature of the column is increased from 37 to 55°C. after the start of elution. It is also possible to apply a shorter elution program (Jürgens *et al.,* 1983) or to separate the amino acids and amino sugars by gas–liquid chromatography (Husek and Macek, 1975; O'Donnell *et al.,* 1982). The correction for destruction of amino acids and, in particular, amino sugars during acid hydrolysis of cell walls is based on the analysis of a calibration mixture that is hydrolysed in the same way as the cell walls.

The only amino acids which cannot be separated by the amino acid analyser are homoserine and muramic acid. To obtain a quantitative estimation of these two amino compounds they have to be isolated from one-

TABLE I

Mobilities of various amino acids and amino sugars after two-dimensional separation by paper chromatography using the isopropanol (I) and α-picoline (II) solvent systems

Amino acid or amino sugar	R_{Ala} values	
	Solvent system	
	I	II
Alanine	1.0	1.0
Aspartic acid	0.55	0.28
Diaminobutyric acid	0.28	0.58
Diaminopimelic acid	0.11	0.14
Galactosamine	0.57	1.68
Glucosamine	0.60	1.80
Glutamic acid	0.81	0.29
Glycine	0.60	0.66
Homoserine	0.83	1.01
Threo-3-hydroxyglutamic acid	0.43	0.27
Lanthionine	0.12	0.25
Lysine	0.30	0.42
Mannosamine	0.71	1.80
Muramic acid	1.01	1.70
Ornithine	0.27	0.41
Serine	0.59	0.81
Threonine	0.84	1.02

TABLE II

Mobilities of diamino acids on paper chromatogram[a]

Diamino acid	R_{Lys}	R_{Orn}
meso-3-Hydroxydiaminopimelic acid	0.30	0.37
Lanthionine	0.38	0.46
meso- and DD-Diaminopimelic acid	0.48	0.59
LL-Diaminopimelic acid	0.64	0.75
Threo-β-hydroxyornithine	0.70	0.83
2,4-Diaminobutyric acid	0.73	0.88
δ-Hydroxylysine	0.81	0.98
Ornithine	0.83	1.0
Lysine	1.0	1.20

[a] Paper: Whatman No. 1 paper; solvent system: methanol–water–pyridine–12 M HCl (32:7:4:1, by volume). Running time: 9 h at 27–28°C. After staining with ninhydrin, diaminopimelic acid and meso-3-hydroxydiaminopimelic acid show a characteristic colour change from purple to yellow.

dimensional paper chromatograms. Homoserine is present in acid hydrolysates partly as the lactone which should be converted to the open chain
form by heating at 100°C with 10 M ammonia for 5 min. Muramic acid and
alanine show almost identical mobilities in the isopropanol solvent system
whereas in the α-picoline solvent system homoserine and alanine behave
similarly. The compounds are eluted and can be separated and quantified
using an amino acid analyser (Schleifer, 1970).

Total and individual amino sugars can also be determined by colorimetric methods. A modified Morgan–Elson reaction (Ghuysen *et al.*, 1966) is
used for the estimation of total amino sugars. Cell wall hydrolysates (4 M
HCl, 100°C, 4 h) containing 10–100 nmol of total amino sugars are neutralised with 3 M NaOH. The amino sugars are N-acetylated by adding 10 μl
saturated NaHCO$_3$ and 10 μl 5% (v/v) acetic anhydride to 30 μl solution.
The mixture is incubated at room temperature for 10 min. Excess acetic
anhydride is destroyed by heating in boiling water for exactly 3 min and
then immersing in cold water. Fifty microlitres of 5% (w/v) potassium
tetraborate is added. Samples are mixed and heated in boiling water for 7
min, cooled, and mixed with 600 μl of diluted Morgan–Elson reagent (16 g
p-dimethylaminobenzaldehyde dissolved in acetic acid to a final volume
of 95 ml and 5 ml concentrated HCl is added; 2 volumes of the mixture are
diluted with 5 volumes of acetic acid). The colour is developed at 37°C in 20
min and measured at 585 nm. If the heating times are strictly followed,
both glucosamine and muramic acid have a molar extinction coefficient of
11,500.

Glucosamine and muramic acid can be separated before estimation as
described by Hughes (1968). A sample containing 0.2–1.0 μmol of each
amino sugar is evaporated to dryness *in vacuo* over solid NaOH. A suspension of charcoal (125 mg in 2.5 ml) is added and shaken at room
temperature for 10 min. The suspension is washed with small portions of
water on a sintered glass filter. Ten millilitres, containing glucosamine, is
collected. Muramic acid is eluted with 10% (v/v) ethanol. The fractions
are evaporated and dissolved in water.

Muramic acid can also be determined enzymatically (Tipper, 1968). A
cell wall hydrolysate containing about 50 nmol of muramic acid is neutralised and incubated in 0.04 M sodium phosphate, pH 12.5, at 37°C for 1 h.
Alkali treatment eliminates the D-lactyl residue of muramic acid. D-Lactate is quantitatively converted to pyruvate and NADH in the presence of
D-lactate dehydrogenase (Boehringer Mannheim) and NAD.

In most cases amino sugar constituents of the murein are released with
4 M HCl at 100°C within 3–4 h. Prolonged acid hydrolysis yields lower
amounts of amino sugars since they are partly destroyed under these
conditions. Murein hydrolysates of some bacteria such as certain bacilli

(Araki *et al.*, 1971; Hayashi *et al.*, 1973) or rhodopseudomonads (Schmelzer *et al.*, 1982) reveal only very low amounts of glucosamine and muramic acid under normal hydrolysis conditions. Instead, the disaccharide GlcNH$_2\beta$(1,4)Mur could be identified as a main component of the hydrolysate (Araki *et al.*, 1972; Schmelzer *et al.*, 1982). The lack of N-acetyl substitution of glucosamine or of muramic acid (in the murein of *Micrococcus luteus*, Mirelman and Sharon, 1967) renders the glycosidic linkage rather acid resistant. Occurrences of non-acetylated glucosamine residues also causes resistance to lysozyme (Hayashi *et al.*, 1973). N-Acetylation of intact cell walls with acetic anhydride yields normal values of glucosamine and muramic acid after acid hydrolysis. Moreover, after N-acetylation the murein of *Bacillus cereus* and *Rhodopseudomonas viridis* becomes lysozyme sensitive.

Amide ammonia is estimated after hydrolysis of cell walls with 4 M HCl at 100°C for 4 h on an amino acid analyser or colorimetrically (Ghuysen *et al.*, 1967). Two hundred microlitres of a sample containing 5–20 nmol of ammonia is placed in a glass vial (5 ml capacity) and 80 μl of 4 M NaOH is added. The vial is provided with a glass rod passing through a stopper and extending about half-way down. Five microlitres of 2 M H$_2$SO$_4$ is placed on the end of the rod, 200 μl of saturated K$_2$CO$_3$ is added to the hydrolysate, and the vial is immediately stoppered. The vial is slowly rotated for 1 h, then the rod is washed in a tube with ammonia-free water, and 4 μl of 2 M NaOH is added to the washings. The ammonia content of the solution is determined colorimetrically (Ternberg and Hershey, 1960).

The O-acetyl content of murein can be measured by direct reaction of cell walls with alkaline hydroxylamine and estimation of acethydroxamate formation, with ethylacetate as standard (Hestrin, 1949). Acethydroxamate is identified by paper chromatography in isopropanol–acetic acid–water (75 : 10 : 15, by volume) visualised with 0.37 M FeCl$_3$ · 6H$_2$O dissolved in 0.1 M HCl (Schleifer and Steber, 1974). O-Acetyl groups can also be released by incubating purified cell walls in 0.05 M NaOH at room temperature for 4 h. The suspension is centrifuged and the acetic acid content of the supernatant is determined enzymatically with acetate kinase (Rose, 1953; Bergmeyer and Möllering, 1974) or by gas-liquid chromatography (Fromme and Beilharz, 1978).

The murein of some mycobacteria and nocardiae contains N-glycolylated instead of N-acetylated muramic acid (Adam *et al.*, 1969; Azuma *et al.*, 1970; Guinand *et al.*, 1970; Vilkas *et al.*, 1970; Bordet *et al.*, 1972). Later studies showed that some other coryneform bacteria and Actinomycetales also possess glycolyl residues in their murein (Uchida and Aida, 1977, 1979; Kawamoto *et al.*, 1981). Uchida and Aida (1977) introduced a simple method for the identification of glycolylated murein: Bacterial cells

(10 mg dry wt.) or cell walls are hydrolysed (0.1 ml 4 M HCl, 100°C, 20 h) and the hydrolysate is applied to a microcolumn (0.5 × 0.05 cm) filled with Dowex 50W × 8 (H$^+$). The column is eluted twice with 0.5 ml distilled water. The eluate is applied on another microcolumn filled with Dowex 1 × 8 (acetate) and washed with 1.0 ml of H_2O and 1.0 ml of 0.5 M HCl. Glycolic acid is eluted twice with 1 ml of 0.5 M HCl. A total of 0.1 ml of the solution is mixed with 2 ml of a solution of 0.02%, 2,7-dihydroxynaphthaline in conc. H_2SO_4 in a capped tube, and immediately heated in a boiling water bath for 10 min, then cooled in running water and diluted with 1.9 ml of 2 M H_2SO_4. A purple-red colour is developed if glycolic acid is present. The absorption of the solution is measured at 530 nm.

3. Determination of the configuration of amino acids

The configuration of the amino acids can be determined by stereospecific enzyme reactions or physical methods. The amino acid should be purified prior to analysis to exclude any interference from other murein constituents.

A combination of D-amino acid oxidase and L-lactate dehydrogenase is used in an optical test for the estimation of D-alanine, D-serine, D-threonine, or D-diaminobutyric acid (Schleifer and Kandler, 1967; Larsen *et al.*, 1971; Fiedler *et al.*, 1973). Various L-amino acids can be determined by L-amino acid decarboxylases (Work, 1971; Niebler *et al.*, 1969; Kandler and König, 1978). For the estimation of L-lysine, L-glutamic acid, and L-ornithine the corresponding decarboxylases (Sigma, St. Louis) are applied (Kandler and König, 1978). L-Glutamic acid can also be determined with L-glutamate dehydrogenase (Schleifer and Kandler, 1967).

meso-Diaminopimelic acid is quantitatively converted to L-lysine by diaminopimelate decarboxylase (Work, 1963). Diaminopimelate epimerase catalyses the interconversion of *meso*- and LL-A$_2$pm (White *et al.*, 1969). The DD isomer is not attacked by either enzyme. Both enzymes are present in a crude cell extract of a wild-type *E. coli*. Such a crude enzyme preparation converts both *meso* and LL isomers to L-lysine. A partially purified A$_2$pm-carboxylase attacks only the *meso* isomer (White and Kelly, 1965). Therefore, all three isomers can be identified by a combination of paper chromatography and enzymatic reactions. LL-A$_2$pm can be separated from the other two isomers by paper chromatography (Table II). *meso*- and DD-A$_2$pm can be distinguished using the crude enzyme preparation from *E. coli* or a purified diaminopimelate decarboxylase. The *meso* isomer is quantitatively converted to lysine.

Another possibility for determining the configuration of most amino acids is to measure the optical rotatory dispersion of their 2,4-dinitrophenyl (DNP) derivatives (Bricas *et al.*, 1967; Kandler *et al.*, 1968a). The DNP-amino acid is purified by paper or thin-layer chromatography, eluted, and dried. The sample is dissolved in acetic acid. The extinction at 412 nm should be about 1.0. The solution is measured in a spectropolarimeter (380–500 nm) and compared with the authentic DNP L- and/or D-amino acids. The different stereoisomers give a positive or negative Cotton effect. The configurations of the following amino acids have been determined by this method: alanine, lysine, ornithine, diaminopimelic acid, glutamic acid, and aspartic acid.

Di-DNP derivatives of m-A$_2$pm can be distinguished from the di-DNP derivatives of LL and DD isomers (Bricas *et al.*, 1967; Perkins, 1969). The di-DNP derivative of m-A$_2$pm travels slower than that of the other two isomers on silica gel (Kieselgel G) thin-layer plates using benzyl alcohol–chloroform–methanol–water–15 M ammonia (30 : 30 : 30 : 6 : 2, by volume) as solvent system.

Another method for determining the configuration of amino acids is their coupling with an L-amino acid N-carboxyanhydride (Manning and Moore, 1968). The resulting pairs of LL- and LD-dipeptides are separable on an amino acid analyser. Even 0.1% of an isomer can be detected.

C. Determination of murein type

The primary structure (amino acid sequence) of the murein, i.e., the murein type, can be determined by use of cell wall lytic enzymes or by chemical methods. The primary structures of the mureins of *E. coli* (Weidel and Pelzer, 1964) and *Staphylococcus aureus* (Ghuysen and Strominger, 1963a,b) were established by analysing fragments (muropeptides) obtained from the degradation of bacterial cell walls with cell wall lytic enzymes. After the basic structure of the murein was known, it was possible to introduce a rather simple chemical method to elucidate the amino acid sequence of different mureins (Schleifer and Kandler, 1967, 1972; Schleifer and Stackebrandt, 1983; Schleifer and Seidl, 1984). The main steps of this method are summarised in Table III.

In most cases the first three steps are sufficient to establish the murein type. Only mureins with complicated or new structures need additional data (steps 4 and 5) to establish their primary structure.

1. Determination of the amino and carboxyl terminal amino acids in intact cell walls

The cross-linkage of the stem peptides is never complete. Therefore, the determination of the major NH$_2$-terminal amino acid of the undegraded

TABLE III

Chemical method for the determination of the murein type

---

1. Quantitative determination of the amino acid and amino sugar composition of purified cell walls. Determination of the configuration of the amino acids when an interpeptide bridge is present
2. Determination of the amino and carboxyl terminal amino acids in intact cell walls
3. Isolation and identification of oligopeptides from partial acid hydrolysates of purified cell walls
4. Isolation and identification of nucleotide-activated murein precursors
5. Isolation and identification of muropeptides from lysates of cell walls

---

cell wall is a good indication for the NH_2-terminus of the interpeptide bridge. The NH_2-terminal amino acids are determined by dinitrophenylation. Undegraded cell walls are dinitrophenylated by using the method of Takebe (1965): 20 mg of purified cell walls is suspended in 5 ml distilled water, 5 ml of 2.5% (w/v) $NaHCO_3$, and 20 ml of a freshly prepared 1.5% ethanolic solution of 1-fluoro-2,4-dinitrobenzene (FDNB). The suspension is shaken in the dark at 37°C for 12–14 h and then centrifuged at 30,000 g for 20 min. The sediment is washed with ether, 96% ethanol, 50% ethanol, and distilled water successively. Dinitrophenylated cell walls are hydrolysed (6 M HCl, 100°C, 6 h) (Schleifer et al., 1969) and dinitrophenylated (DNP) amino acids are separated and identified by one-dimensional thin-layer chromatography on silica gel in the following solvent systems: benzene–acetic acid (8:2, v/v), chloroform–methanol–acetic acid (95:5:1, by volume) or chloroform–methanol–acetic acid–water (65:25:13:8, by volume). The latter solvent system is especially useful for the identification of mono-DNP-diamino acids. Mono- and di-DNP-diamino acids are prepared as described by Hammes et al. (1973): 80 mg of diamino acid is dissolved in 20 ml of 2.5% (w/v) of $NaHCO_3$, 40 ml of ethanol and 0.073 ml of FDNB are added, and the solution is incubated at 50°C for 30 min. Ethanol is removed in a stream of warm air, and unreacted FDNB is extracted with ether. The aqueous phase is acidified with 2 ml of 4 M HCl. Di-DNP-diamino acids are extracted with ethyl acetate, and mono-DNP-diamino acids with a mixture of ethyl acetate and n-butanol (1:1, v/v). Mono-DNP-diamino acids can also be identified on chromatograms by their reaction with ninhydrin (brown colour).

The extent of cross-linkage of the murein can be established by quantitative determination of DNP-amino acids either photometrically at 360 nm (Rao and Sober, 1954; Takebe, 1965) or by comparison of the amino acid content of hydrolysates of untreated and dinitrophenylated cell walls (Schleifer et al., 1968).

COOH-terminal amino acids are determined following hydrazinolysis. Anhydrous hydrazine is prepared by the method of Fraenkel-Conrat and Tsung (1967). Lyophilised cell walls (3 mg) are suspended with 0.3 ml of anhydrous hydrazine in a tube, which is immediately sealed and incubated at 80°C for 48 h. The excess reagent is removed *in vacuo* over conc. H_2SO_4. The dried sample is dissolved in 0.3 ml of distilled water containing 0.1 μmol norleucine as internal standard. Redistilled benzaldehyde (0.05 ml) is added, and the tube is thoroughly mixed for 10 s at 5 min intervals for 1 h. The emulsion is centrifuged at 12,000 g for 5 min, and the clear supernatant is treated again with benzaldehyde and extracted twice with 0.3 ml ether. The aqueous solution is finally dried *in vacuo*. The amount of COOH-terminal amino acid is determined by the amino acid analyser or by photometric measurement of the DNP derivative. It was established that the hydrazinolysis of cell walls is not as effective as that of peptides. Cell walls of *Micrococcus luteus,* in which all of the glycine residues are COOH terminal, showed yields of 65–68% of glycine, whereas soluble peptides gave yields of 80–85% of the theoretical amount of the COOH-terminal amino acid (Hammes *et al.,* 1973). These figures should be used as correction factors.

2. Isolation and identification of oligopeptides from partial acid hydrolysates of purified cell walls

The most important step for the determination of the murein type is the isolation and identification of oligopeptides after partial acid hydrolysis of clean cell wall preparations (Schleifer and Kandler, 1967, 1972). The condition for hydrolysis is usually 4 M HCl (0.1 ml mg cell wall^{-1}) at 100°C for 45–60 min. Using this condition mainly dipeptides, amino acids, and amino sugars will be released. For the isolation of tripeptides and acid-labile peptides such as muramylalanine or muramylglycine, a shorter hydrolysis time (10–15 min) is applied. Descending paper chromatography is used for the separation of peptides and amino compounds. There are two approaches for the isolation of peptides. In the first a partial acid hydrolysate of about 40–50 mg cell walls is spotted on a streak of 40–45 cm on paper (Schleicher-Schüll 2043b MgI) and separated using the solvent isopropanol–acetic acid–water (75 : 10 : 15, by volume, for 3 × 24 h). After drying the chromatogram, longitudinal strips from both margins and the middle are cut out and stained with ninhydrin to determine the location of peptides and other amino compounds. The non-stained peptide-containing bands are cut out, eluted with water, and separated again by one-dimensional paper chromatography using the solvent α-picoline–25%

NH$_4$OH–water (70 : 2 : 28, by volume) for 3 × 24–30 h. Repeated one-dimensional paper chromatography in both solvent systems is usually necessary to obtain pure peptides.

In the second approach the partial acid hydrolysate is separated by two-dimensional descending paper chromatography (~5 mg cell wall hydrolysate per chromatogram). One chromatogram is stained with ninhydrin to determine the location of peptides. The corresponding areas are cut out from the other unsprayed parallel chromatograms and eluted with water. A more exact alignment of the peptides can also be obtained by staining adjacent amino acids with ninhydrin. Peptides can also be isolated from fluorescamine-stained chromatograms (Undenfriend *et al.,* 1972; Kusser and Fiedler, 1983). A typical picture of a two-dimensional paper chromatogram of a partial acid hydrolysate of a cell wall is shown in Fig. 3.

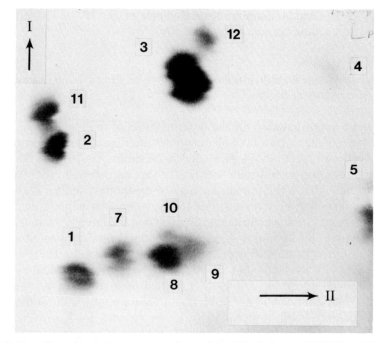

Fig. 3. Two-dimensional chromatogram of a partial acid hydrolysate (4 *N* HCl, 100°C, 2 h) of cell walls of *Micrococcus kristinae* (murein type: Lys-Ala$_3$). I, isopropanol–acetic acid–water (75 : 10 : 15, by volume); II, α-picoline–25% NH$_4$OH–water (70 : 2 : 28, by volume). 1, Lys; 2, Glu; 3, Ala; 4, muramic acid; 5, glucosamine; 7, Lys-D-Ala; 8, N^6-Ala-Lys; 9, N^6-Ala$_2$-Lys; 10, N^6-Ala-Lys-D-Ala; 11, Ala-D-Glu; 12, Ala-Ala and D-Ala-Ala.

The mobilities of various peptides after two-dimensional paper chromatography in the isopropanol and α-picoline solvent systems are listed in Table IV. The R_{Ala} values can vary slightly depending on the conditions used for chromatography (solvent composition, temperature, quantity of hydrolysate, etc.). Therefore, it is advisable to standardise the system under defined chromatography conditions by using authentic amino acids and peptides or partial acid hydrolysates of cell walls with known murein types. Some of the peptides show characteristic colours after spraying with ninhydrin and heating at 100°C. Peptides with NH_2-terminal glycine residues appear yellow at the beginning whereas those with NH_2-terminal serine develop an orange colour during the first minutes after heating. However, within 10–15 min all peptides give the usual purple to violet colour.

Diaminopimelic acid-containing peptides are separated by paper chromatography using the solvent methanol–pyridine–formic acid–water (80 : 10 : 1 : 19, by volume; Rhuland et al., 1955) or by preparative electrophoresis (Bogdanovsky et al., 1971) using Whatman 3MM paper (2 h, 60 V cm^{-1}) and buffers pH 1.9, 2% formic acid or pH 4.1, pyridine–acetic acid–water (6 : 23 : 976, by volume). Peptides containing hydroxyglutamic acid and glutamic acid are also separated by high-voltage electrophoresis under the following conditions: 58 V cm^{-1}, 2–4 h, formic acid–acetic acid–water (5 : 15 : 80, by volume), pH 1.9, Whatman 3MM paper (Schleifer et al., 1967; Cziharz et al., 1971).

The isolated peptides are identified by determining the quantitative amino acid composition and the NH_2-terminal amino acid by dinitrophenylation (Ghuysen et al., 1968). The dried peptide ($\sim$0.1 μM) is dissolved in 0.1 ml ethanol–triethylamine–water (8 : 1 : 1, by volume), 15 μl of FDNB (130 μl FDNB, 10 ml ethanol) is added, and the mixture is incubated at 60°C for 30 min. The procedure is repeated if the peptide contains a diamino acid. The dinitrophenylated peptides are dried in a stream of warm air and then hydrolysed (6 M HCl, 100°C, 6 h). Mono-DNP-amino acids and di-DNP-diamino acids are extracted with ether and mono-DNP-diamino acids with water-saturated butanol. The DNP-amino acids are separated and identified by thin-layer chromatography on silica gel as described above.

DNP derivatives containing a dinitrophenylated amino group adjacent to a carboxyl group are photochemically unstable (Russell, 1963). Photolysis of dinitrophenylated peptides containing NH_2-terminal aspartic acid or glutamic acid is used to determine whether the dicarboxylic amino acids are bound with their α- or distal carboxyl groups (Perkins, 1967; Niebler et al., 1969; Bogdanovsky et al., 1971). The purified DNP-peptides are dissolved in 1% (w/v) NaHCO$_3$ and the absorption at 360 nm is

TABLE IV

Mobilities of various peptides after two-dimensional separation by paper chromatography using the isopropanol (I) and α-picoline (II) solvent systems

Peptide	R_{Ala} values Solvent system		Colour[a]	Peptide	R_{Ala} values Solvent system		Colour[a]
	I	II			I	II	
D-Ala-Ala	1.17	1.20		Lys-D-Ala-Glu	0.37	0.27	
D-Ala-D-Ala or Ala-Ala	1.26	1.25		Lys-Lys	0.19	0.34	
Ala-Ala-Ala	1.30	1.32		Lys-Lys-D-Ala	0.20	0.49	
D-Ala-Gly	0.93	0.91		Mur-Ala	1.20	1.60	
D-Ala-Gly-Gly	0.78	0.83		Mur-Gly	0.98	1.39	
Ala-Thr	1.05	1.27		Mur-Ala-D-Glu	1.08	0.81	
Ala-D-Glu or D-Ala-Glu	0.95	0.23		Mur-GlcNH$_2$	0.30	1.36	
D-Ala-D-Asp	0.95	0.23		N^2-Gly-Lys	0.27	0.53	Brownish
D-Ala-D-Glu	1.05	0.29		N^2-D-Ala-D-Lys	0.39	0.60	
Ala-γ-D-Glu-Lys	0.19	0.32		N^6-Gly-Lys	0.22	0.62	Brownish
D-Ala-γ-Glu-Gly	0.67	0.31		N^6-Ser-Lys	0.24	0.77	Brownish
D-Ala-γ-Glu-Ala	0.92	0.41		N^6-Ala-Lys	0.40	0.87	
D-Ala-Ala-D-Glu	1.05	0.54		N^6-Thr-Lys	0.32	0.91	
D-Ala-Dpm	0.28	0.34		N^6-Gly-Lys-D-Ala	0.33	0.76	Yellow

Compound			Colour
D-Asp-Ala	0.55	0.33	Steel blue
D-Asp-Lys-Lys	0.17	0.63	
Dpm-D-Ala	0.31	0.40	
γ-D-Glu-Lys	0.20	0.23	
γ-Glu-Gly	0.56	0.27	
γ-Glu-Ala	0.92	0.40	
γ-D-Glu-Gly	0.77	0.20	
D-Glu-Dpm-D-Ala	0.15	0.08	
Gly-Gly	0.61	0.59	Yellow
Gly-Gly-Gly	0.51	0.62	Yellow
Gly-Gly-Gly-Gly	0.39	0.64	Yellow
Gly-Ser	0.50	0.74	Yellow
Gly-Ala	0.90	0.85	Yellow
Gly-D-Glu	0.72	0.17	Yellow
Gly-Gly-Ala	0.71	0.83	Yellow
Gly-α-Hyg-Gly	0.35	0.15	Yellow
Gly-γ-Hyg-Hsr	0.36	0.20	Yellow
Gly-LL-, Dpm-D-Ala	0.21	0.14	Brownish
Lys-D-Ala	0.39	0.60	
Lys-D-Ala-D-Ala	0.52	0.87	

Compound			Colour
N^6-Ser-Lys-D-Ala	0.33	0.85	
N^6-Ala-Lys-D-Ala	0.47	0.95	
N^6-γ-D-Glu-D-Lys	0.36	0.27	
ε-(Aminosuccinyl)lysine	0.32	0.77	
N^6-(Ala-Ala)-Lys	0.46	1.0	
N^6-(D-Ala-D-Asp)-Lys	0.39	0.97	
N^6-(Ala-Thr)-Lys	0.40	1.02	
N^6-(D-Asp-Ala)-Lys	0.17	0.60	
N^6-D-Ala-D-Orn	0.37	0.58	Brownish
N^5-Gly-Orn	0.20	0.58	Brownish
N^5-Ser-Orn	0.22	0.75	Yellow
N^5-Gly, N^2-D-Ala-D-Orn	0.29	0.72	
N^5-Ala-Orn	0.40	0.87	
Orn-D-Ala	0.37	0.58	
Ser-Ala	0.93	1.01	Yellow
Ser-D-Glu	0.70	0.23	Yellow
Ser-Gly	0.66	0.80	Yellow
Ser-Ser	0.55	0.90	Yellow
Thr-Ala	1.16	1.36	Yellow

[a] Most spots give the usual violet colour with ninhydrin; only the unusual colours are specified.

adjusted to 1.0. The solution is placed in a quartz cuvette and subjected to UV light (273 nm) for 20–30 min. The absorption of the solution is recorded during photolysis at 348 and 284 nm. A decrease of the absorbance at 348 nm and a simultaneous increase at 284 nm indicate photochemical degradation of the compound. Dinitrophenylated peptides containing an NH$_2$-terminal glutamic acid or aspartic acid bound with their distal carboxyl groups are photochemically degraded, whereas corresponding peptides with dicarboxylic amino acids linked with their α-carboxyl group are photochemically stable. Photolysis of dinitrophenylated derivatives is also useful in distinguishing peptides containing COOH-terminal diamino acids. COOH-terminal diamino acids bound with their distal amino group are photolabile in contrast to those linked with their α-amino group.

3. Isolation and identification of nucleotide-activated murein precursors

Nucleotide-activated precursors are extracted with 25% (w/v) trichloroacetic acid at 4°C for 30 min (Chatterjee and Perkins, 1966) from non-inhibited cells in the stationary growth phase (Hammes *et al.*, 1973) or from cells grown in the presence of D-cycloserine (0.1 mg ml^{-1}; Roze and Strominger, 1966; Cziharz *et al.*, 1971). Trichloroacetic acid is extracted with ether, and the aqueous phase is neutralised and concentrated *in vacuo* at 20°C and applied to a column (80 × 2.5 cm) of Sephadex G-25 following the procedure of Rosenthal and Sharon (1964). Elution is carried out with distilled water at 4°C and a flow rate of 20 ml h^{-1}. Fractions absorbing at 260 nm and containing bound *N*-acetylamino sugar (Ghuysen *et al.*, 1967) are pooled, concentrated, and separated by paper chromatography on Whatman 3MM using the solvent isobutyric acid–25%NH$_4$OH–water (250 : 5 : 145, by volume). The nucleotide-activated precursors are detected on chromatograms by UV absorption. The UV-absorbing areas are eluted with water and rechromatographed in the solvent ethanol–1 *M* ammonium acetate (7 : 3, by vol.). The amino acid sequence of the precursors is elucidated by quantitative determination of muramic acid and amino acids of a total acid hydrolysate and by isolation and identification of peptides from a partial acid hydrolysate (4 *M* HCl, 100°C, 5–15 min). The amino acid sequence of nucleotide-activated precursors is the basis for that of the stem peptide of the murein. In the presence of D-cyloserine mainly nucleotide-activated muramyltripeptide is accumulated whereas a variety of precursors can be extracted from non-inhibited cells.

4. Isolation and identification of muropeptides

Cell wall-lytic enzymes are used to digest cell walls and to isolate fragments of murein (muropeptides). Hen egg white lysozyme is a cheap and readily available muramidase for the preparation of muropeptides. Purified cell walls (2 g) are suspended in 200 ml of 0.1 M ammonium acetate buffer (pH 6.2), and 200 mg of lysozyme is added. The suspension is incubated at 37°C for 20 min and then centrifuged at 40,000 g for 30 min. The clear supernatant (lysate) is concentrated under reduced pressure to a final volume of about 10 ml and separated by gel filtration on two columns (80 × 2.5 cm); Sephadex G-50 (fine) and G-25 (fine), connected in series with a flow rate of 20 ml h^{-1} at 20°C (Hammes et al., 1973). For each separation only 1.0 ml of the concentrated lysate should be applied and eluted with distilled water.

The low-molecular-weight fraction can be further separated by preparative paper chromatography on Whatman 3MM paper using the solvent n-butanol–acetic acid–water (62 : 25 : 25, by volume). The primary structure of the muropeptides can be established by applying the previously described methods. By using this procedure muropeptides have been isolated from cell walls of staphylococci (Schleifer et al., 1969) and of Lactobacillus cellobiosus (Hammes et al., 1973).

There are also other cell wall lytic enzymes now commercially available that can be used for the degradation of cell walls and isolation of muropeptides, e.g., lysostaphin (Sigma, St. Louis; Iversen and Grov, 1971), mutanolysin (Sigma, St. Louis; Yokogawa et al., 1974), Kyowa lytic No. 2 enzyme (Kyowa Fermentation Industry, Tokyo, Japan; source and mode of action very similar to those of Flavobacterium L-11 enzyme; Kotani et al., 1975), or achromopeptidase (Wako Pure Chemical Industries, Ltd., Japan, Masaki et al., 1981; Ezaki and Suzuki, 1982). A detailed description of the use of cell wall-lytic enzymes to elucidate the primary structure of murein has been published by Ghuysen (1968).

D. Rapid methods for the determination of known murein types

1. Chemical method

Known murein types can be easily recognised by using the rapid screening method of Schleifer and Kandler (1972). Cell walls prepared by the hot trichloroacetic acid method as described above (Section A,3). A sample of the cell wall preparation can be hydrolysed with 6 M HCl at 105°C for 6 h to determine the quantitative amino acid composition. In most cases, however, it is sufficient to prepare a partial acid hydrolysate (4 M HCl, 100°C, 45–60 min) and to subject it to two-dimensional paper chromatography (Schleicher and Schüll 2043b Mgl, 29 × 30 cm) using the isopropanol and α-picoline solvent systems. Each direction is run twice for 18–20 h. The type of murein can be recognised from the characteristic fingerprint of its partial acid hydrolysate. It is advisable for comparison to run in parallel a two-dimensional chromatogram of a partial acid hydrolysate of purified cell walls with a known murein type.

2. Serological detection of murein types

It is known that proteins substituted with oligopeptides revealing terminal amino groups yield antisera that will react with the NH_2-terminal part of the peptide determinants (Schechter et al., 1970). Since all mureins studied so far are not completely cross-linked, some of the interpeptide bridges always show unsubstituted amino groups. Oligopeptides with structural similarity to the interpeptide bridges can be covalently linked to denatured human serum albumin and antisera can be obtained from rabbits immunised with the synthetic peptidyl-protein antigens (Seidl and Schleifer, 1978a). Antisera to pentaglycyl-substituted albumin react with mureins prepared from staphylococci but not with other mureins lacking oligoglycine interpeptide bridges. On the other hand, antisera with trialanyl-substituted albumin react with mureins of certain micrococci and streptococci that contain oligo-alanine interpeptide bridges but not with other mureins (Schleifer and Seidl, 1978a). These specific antisera can be used for the serological separation of staphylococci from micrococci (Seidl and Schleifer, 1978b). Bacterial cells are extracted with 10% (w/v) trichloroacetic acid at 100°C for 30 min to remove non-murein cell wall components. Antisera to pentaglycyl-substituted albumin specifically agglutinate extracted cells of staphylococci (Seidl and Schleifer, 1978b). The test is highly specific and very sensitive. Only 10 ml of a liquid culture grown overnight is necessary, and 10^6–10^7 cells ml^{-1} can be detected. Using antisera against synthetic peptidyl-protein antigens may be a general method for the serological differentiation of bacteria that differ in the NH_2-terminal amino acid of their mureins.

III. Taxonomic significance of chemical composition or type of murein

The murein structure fulfills, at least among the Gram-positive bacteria, all of the prerequisites of a valuable taxonomic marker (Schleifer and Kandler, 1972):

1. Murein is widely distributed and is a typical constituent of the cell wall of almost all eubacteria.

2. The involvement of many genes in the biosynthesis of murein guarantees a certain stability and makes a convergent formation of identical variations rather unlikely. No single-step mutations are known so far that lead to an altered murein structure (Schleifer et al., 1976).

3. Considerable mutational changes of the murein structure have been achieved during the course of evolution. This diversification of the chemical structure is obviously compatible with survival since many different murein types are still conserved especially amongst Gram-positive bacteria.

4. Endogenous and exogenous factors do not usually affect the composition and structure of the murein (Schleifer et al., 1976). However, the amino acid composition of the growth medium can influence the composition of the interpeptide bridges of staphylococcal mureins (Schleifer et al., 1969, 1976). Under unusual conditions, two genetically different strains can even show the same interpeptide bridge; e.g., a strain of S. epidermidis grown in a glycine-enriched medium reveals interpeptide bridges in its murein similar to those of a strain of S. aureus grown in a serine-enriched medium. However, the genetic differences become obvious when the same strains are cultivated in a balanced medium containing sufficient glycine. Phenotypic variations seem to be an exception rather than a rule, since studies with other bacteria have shown that growth in various media did not change their murein structure (Schleifer et al., 1976).

A. Chemical composition of the murein

Simple qualitative amino acid analysis of the cell wall is often sufficient to provide useful chemotaxonomic information, especially in cases in which the chemical structure of the murein is quite distinct. There are various examples, such as the separation of Enterococcus faecalis from other enterococci (Kandler et al., 1968b; Kilpper-Bälz et al., 1982; Schleifer and Kilpper-Bälz, 1984) or of staphylococci from micrococci (Schleifer and Kloos, 1976), distinguishing certain propionibacteria (Schleifer and Kandler, 1972) or streptococci of the S. mutans group (Schleifer et al., 1984), and the differentiation of oerskoviae (Seidl et al., 1980), certain micrococci (Schleifer et al., 1981), or bifidobacteria (Lauer and Kandler, 1983). A simple and rapid procedure is the prerequisite for using the qualitative amino acid composition of cell walls as a chemotaxonomic criterion in the routine classification of certain Gram-positive bacteria. Cell walls are prepared by the rapid method (Section A,3), hydrolysed with 6 M HCl at 105°C for 6 h, and analysed by two-dimensional paper chromatography using the isopropanol and α-picoline solvent systems. It is advisable to prepare two chromatograms with different concentrations of the hydrolysate to distinguish minor contaminating amino acids from the murein amino acids.

Presence of glycolyl residues in cell walls of certain coryneform bacteria may also be a useful chemotaxonomic character for their identification (Uchida and Aida, 1979).

B. Primary structure of murein (murein type)

Qualitative, or even quantitative, amino acid determination of the cell wall is in many cases not sufficient to characterise unequivocally the murein type. There are some instances known in which the amino acid composition of the cell wall is identical but the murein type is quite different, e.g., *Micrococcus luteus* and *Bifidobacterium breve, Arthrobacter aurescens* and *A. citreus, Lactobacillus viridescens* and *Leuconostoc cremoris,* or *Clostridium leptum* and *Staphylococcus epidermidis* (Schleifer and Kandler, 1972; Weiss *et al.,* 1981; Schleifer and Seidl, 1984). Moreover, the allocation of mureins to the cross-linkage groups A or B is possible only after establishing the amino acid sequence of the murein or at least after the detection of characteristic oligopeptides. The rapid screening method proposed by Schleifer and Kandler (1972) is usually sufficient to recognise known murein types (see Section D,1).

C. Taxonomic implication of murein structure

Studies on murein structure have had considerable impact on the taxonomy of many Gram-positive bacteria, whereas with most Gram-negative bacteria the murein is chemically uniform and therefore not useful as a chemotaxonomic marker. Among the Gram-negative bacteria there are only a few examples, such as the spirochaetes (Schleifer and Joseph, 1973; Umemoto *et al.,* 1981) and fusobacteria (Kato *et al.,* 1981; Vasstrand *et al.,* 1982), that can be distinguished on the basis of murein structure. The value of murein structure as a chemotaxonomic criterion has been discussed in detail (Schleifer and Kandler, 1972; Schleifer and Seidl, 1984). The correct determination of murein type is helpful in the differentiation of genera, species, or subspecies and in the recognition of misclassified strains. False identifications of mureins have been repeatedly reported in the literature. The majority of these mistakes are based on incorrect or insufficient amino acid determinations (Kandler and Schleifer, 1980; Schleifer and Stackebrandt, 1983). Studies on the murein structure correlate well with those of other useful chemotaxonomic markers such as lipid composition (Minnikin and Goodfellow, 1980) and DNA base composition (Schleifer, 1973). Occurrence of the same murein type in different bacteria does not necessarily mean that they are closely related; e.g., *E. coli* and *Bacillus megaterium* have the same murein type. On the other hand, closely related eubacteria usually possess the same murein type whereas eubacteria with different murein types are normally not closely related. However, a few closely related anaerobic cocci differ in their murein structure (Huss *et al.*, 1984).

The presence of murein is one of the key characters for separating eubacteria from archaebacteria (Kandler, 1982). The absence of typical murein constituents, muramic acid, and D-amino acid in cell wall or even whole cell hydrolysates of an organism is a good indication for its allocation to archaebacteria. There is only one exception of the rule, namely, cell wall-less eubacteria such as mycoplasmas or L forms.

IV. Biological activities of murein

In addition to providing rigidity to the bacterial cell, mureins of different bacteria have been shown to exhibit various biological activities (Heymer, 1975; Heymer and Rietschel, 1977; Stewart-Tull, 1980; Schleifer, 1984). The most important biological properties of murein are listed in Table V.

A. *In vivo* activity of murein

Purified cell walls or murein preparations are poorly immunogenic for rabbits (Abdulla and Schwab, 1965; Nguyen-Dang *et al.*, 1976). However, antisera prepared against various Gram-positive bacteria often contain antibodies with specificity to murein (Krause, 1975). In particular, antisera against *Streptococcus* group A-variant are rich in murein antibodies (Karakawa *et al.*, 1968). There are at least five antigenic determinants in the murein. Specific antibodies are formed against the glycan moiety (Schleifer and Krause, 1971b; Wikler, 1975), the pentapeptide (Schleifer and Krause, 1971a) and tetrapeptide subunits (Seidl and Schleifer, unpublished results), and the NH_2-terminus (Seidl and Schleifer, 1978a) and COOH-terminus (Helgeland *et al.*, 1973; Ranu, 1975; Seidl and Schleifer, 1977) of the interpeptide bridges. Murein antibodies are not only found in hyperimmune sera but also in normal animal and human sera (Heymer *et al.*, 1976; Seidl *et al.*, 1983).

The active component of bacterial cell walls responsible for immunopotentiating effects (adjuvant activity) is murein (Holton and Schwab, 1966;

TABLE V
Biological activity of murein

In vivo activity		In vitro activity	
Phenomenon observed	Test animal	Phenomenon observed	Source of cells or serum
Antigenicity	Man, rabbit	Complement activation	
Adjuvant activity	Guinea pig, rabbit	by classical pathway	Man
		by alternative	Man, mouse
Pyrogenicity	Rabbit	pathway	
Local Shwartzman reaction	Rabbit	Release of histamine from mast cells	Rat
Non-specific resistance to bacterial infections	Mouse	Polyclonal B-cell activation (mitogenic effect)	Man, mouse, rabbit, rat
Introduction of inflammation	Guinea pig, rat	Effects on macrophages	Guinea pig, man, mouse,
Increase of capillary permeability	Guinea pig, rat	(secretion of pyrogen, interleukin	rabbit
Tumoricidal activity	Guinea pig, mouse	I, prostaglandin; stimulation of	
Lesions in internal organs (heart, liver)	Mouse, rabbit	migration and chemotaxis; increase	
Sleep-promoting effect	Cat, rabbit, rat	of superoxide production)	
		Effects on granulocytes (inhibition of migration; phagocytosis and chemotaxis)	Guinea pig, man
		Activation of synovial cells	Man
		Activation of bone marrow myelopoietic precursor cells	Rat

Stewart-Tull, 1980). The minimal chemical structure of murein required for adjuvant activity is N-acetylmuramyl-L-alanyl-D-isoglutamine (Ellouz *et al.*, 1974). However, depending on the dose and time of injection, murein fragments may show not only immunopotentiating but also immunosuppressive (tolerance) effects (Stewart-Tull, 1980).

Some of the biological activities of murein, such as pyrogenicity, local Shwartzman reaction, or non-specific resistance to bacterial infections, are similar to those of endotoxin (Rotta, 1975; Heymer and Rietschel,

1977). Purified bacterial cell walls or murein preparations solubilised by ultrasonic treatment are pyrogenic (Atkins and Morse, 1967; Rotta, 1969; Rotta and Schleifer, 1974; Rotta et al., 1979). The property to induce fever is lost after degrading murein with cell wall lytic enzymes such as lysozyme (Rotta, 1975). Repeated injections of murein preparations result in the induction of tolerance to the fever effect (Rotta, 1975). The pyrogenic activity of murein is different from that of endotoxin (Heymer and Rietschel, 1977). Moreover, murein is less pyrogenic than endotoxin. The smallest murein dose required to induce fever in rabbits is 50 μg kg body weight^{-1} in contrast to 0.008 μg of endotoxin kg^{-1}.

Murein also elicits local but not generalized Shwartzman reaction in rabbits and guinea pigs (Rotta, 1975). Non-specific resistance to bacterial infections is another property that murein shares with endotoxin. Mice injected intraperitoneally with 250 μg of streptococcal murein become resistant to a subsequent challenge with virulent group A streptococci (Rotta, 1975).

Cell walls of bacteria can persist within macrophages for weeks or months (Ginsberg and Sela, 1976). The inadequate clearance of cell walls is probably responsible for the induction of inflammation. An acute, intense inflammatory reaction is induced by injection of a high dose of purified murein, whereas cell walls containing murein and other cell wall polymers such as polysaccharides and/or teichoic acid cause a long-lasting chronic inflammation (Heymer et al., 1982). However, it is important to stress that murein is the active principle for this reaction (Heymer, 1975; Cromartie et al., 1977). The induction of acute inflammation of the skin is closely linked to the capability of murein to increase capillary permeability (Ohta, 1981).

Studies on the anti-tumor activity of murein are inconsistent and results must be interpreted with great caution. Intraveneous injections of bacterial cell walls cause degenerative changes in intestinal organs, particularly in the heart. Intraventricular infusion of muramyl peptides can mimic the sleep-promoting effects of sleep factor and induces slow-wave sleep in cats, rabbits, and rats (Krueger et al., 1982a,b).

B. *In vitro* activity of murein

Murein is able to activate the immune, complement, and clotting systems. At least some of the endotoxin-like activities of murein may be explained by the ability to activate complement. It can be activated by both the classic (Petersen and Rosenthal, 1982) and the alternative (Greenblatt et al., 1978; Petersen et al., 1978) pathway. Murein interacts with different cells and induces the release of various mediators such as 5-hydroxytryptamine (Ryc and Rotta, 1975), histamine (Kimural *et al.*, 1981), pyrogen (Oken *et al.*, 1979), interleukin factor I, prostaglandin (Ginsberg *et al.*, 1977; Windle *et al.*, 1983), and superoxide (Kaku *et al.*, 1983). Moreover, murein exerts a pronounced mitogenic effect on B lymphocytes (Dziarski, 1982) and stimulates migration and chemotaxis of macrophages (Ogawa *et al.*, 1982, 1983) and granulocytes (Osada *et al.*, 1982). However, the results on the effects on granulocytes are somewhat conflicting. There are also reports that murein inhibits phagocytosis and chemotaxis of granulocytes (Musher *et al.*, 1981). Murein also exerts other *in vitro* effects such as the activation of synovial cells (Hamilton *et al.*, 1982) and of myelopoietic precursor cells in rat bone marrow (Monner *et al.*, 1981).

In conclusion, it should be stressed that murein not only is an exoskeleton responsible for maintaining the shape of cells, but also exerts biological activity. However, some of these properties have to be viewed with great caution, in particular when conflicting results have been reported. Some of these discrepancies may be explained by differences in the chemical structure and physical form of the preparation (Schleifer, 1975) or in the test system.

References

Abdulla, E. H., and Schwab, J. H. (1965). *Proc. Soc. Exp. Biol. Med.* **118,** 359–362.

Adam, A., Petit, J. F., Wietzerbin-Falzspan, J., Sinay, P., Thomas, D. W., and Lederer, E. (1969). *FEBS Lett.* **4,** 87–92.

Araki, Y., Nakatani, T., Hayashi, H., and Ito, E. (1971). *Biochem. Biophys. Res. Commun.* **42,** 691–697.

Araki, Y., Nakatani, T., Nakayama, K., and Ito, E. (1972). *J. Biol. Chem.* **247,** 6312–6322.

Archibald, A. R., Coapes, H. E., and Stafford, G. H. (1969). *Biochem. J.* **113,** 899–900.

Archibald, A. R., Baddiley, J., and Goundry, J. (197). *Biochem. J.* **116,** 313–315.

Atkins, E., and Morse, S. I. (1967). *Yale J. Biol. Med.* **39,** 297–309.

Azuma, I., Thomas, D. W., Adam, A., Ghuysen, J. M., Bonaly, R., Petit, J. F., and Lederer, E. (1970). *Biochim. Biophys. Acta* **208,** 444–451.

Bayer, M. E. (1974). *Ann. N. Y. Acad. Sci.* **235,** 6–28.

Bergmeyer, H. U., and Möllering, H. (1974). *In* "Methoden der enzymatischen Analyse" (H. U. Bergmeyer, Ed.), pp. 1566–1583. Verlag Chemie, Weinheim.

Bogdanovsky, D., Interschick-Niebler, E., Schleifer, K. H., Fiedler, F., and Kandler, O. (1971). *Eur. J. Biochem.* **22,** 173–178.

Boone, C. J., and Pine, L. (1968). *Appl. Microbiol.* **16,** 279–284.

Bordet, C., Karahjoli, M., Gateau, O., and Michel, G. (1972). *Int. J. Syst. Bacteriol.* **22,** 251–259.

Braun, V., and Sieglin, U. (1970). *Eur. J. Biochem.* **13,** 336–347.

Braun, V., Rehn, K., and Wolff, H. (1970). *Biochemistry* **9,** 5041–5049.

Bricas, E., Ghuysen, J. M., and Dezélée, Ph. (1967). *Biochemistry* **6,** 2598–2607.

Card, G. L. (1973). *J. Bacteriol.* **114,** 1125–1137.

Chatterjee, A. N., and Perkins, H. R. (1966). *Biochem. Biophys. Res. Commun.* **24,** 489–494.

Conover, M. H., Thompson, J. S., and Shockman, G. D. (1966). *Biochem. Biophys. Res. Commun.* **23,** 713–719.

Cromartie, W. J., Craddock, J. G., Schwab, J. H., Anderle, S. K., and Yang, Ch. H. (1977). *J. Exp. Med.* **146,** 1585–1602.

Cziharz, B., Schleifer, K. H., and Kandler, O. (1971). *Biochemistry* **10,** 3574–3578.

Dziarski, R. (1982). *Infect. Immun.* **35,** 507–514.

Ellouz, F., Adam, A., Ciorbaru, R., and Lederer, E. (1974). *Biochem. Biophys. Res. Commun.* **59,** 1317–1325.

Ezaki, T., and Suzuki, S. (1982). *J. Clin. Microbiol.* **16,** 844–846.

Fiedler, F., Schleifer, K. H., and Kandler, O. (1973). *J. Bacteriol.* **113,** 8–17.

Fiedler, F., Schäffler, M. J., and Stackebrandt, E. (1981). *Arch. Microbiol.* **129,** 85–93.

Fischer, W., Koch, H. U., and Haas, R. (1983). *Eur. J. Biochem.* **133,** 523–530.

Fraenkel-Conrat, H., and Tsung, C. H. (1967). *In* "Methods in Enzymology" (S. P. Colowick and N. O. Kaplan, Eds.), Vol. 11, pp. 151–155. Academic Press, New York.

Fromme, I., and Beilharz, H. (1978). *Anal. Biochem.* **84,** 347–353.

Ghuysen, J. M. (1968). *Bacteriol. Rev.* **32,** 425–464.

Ghuysen, J. M., and Strominger, J. L. (1963a). *Biochemistry* **2,** 1110–1119.

Ghuysen, J. M., and Strominger, J. L. (1963b). *Biochemistry* **2,** 1119–1125.

Ghuysen, J. M., Tipper, D. J., and Strominger, J. L. (1966). *In* "Methods in Enzymology" (S. P. Colowick and N. O. Kaplan, Eds.), Vol. 8, pp. 685–699. Academic Press, New York.

Ghuysen, J. M., Bricas, E., Ley-Bouille, M., Lache, M., and Shockman, G. D. (1967). *Biochemistry* **6,** 2607–2619.

Ghuysen, J. M., Bricas, E., Lache, M., and Ley-Bouille, M. (1968) *Biochemistry* **7,** 1450–1460.

Ginsberg, I., and Sela, M. (1976). *Crit. Rev. Microbiol.* **4,** 249–332.

Ginsberg, I., Zor, U., and Floman, Y. (1977). *In* "Experimental Models of Chronic Inflammatory Diseases" (L. E. Glynn and H. D. Schlumberger, Eds.), pp. 256–299. Springer-Verlag, Berlin and New York.

Glaser, L., and Burger, M. (1964). *J. Biol. Chem.* **239,** 3187–3191.

Golecki, J. R. (1977). *Arch. Microbiol.* **114,** 35–41.

Goodwin, S. D., and Shedlarski, J. C. (1975). *Arch. Biochem. Biophys.* **170,** 23–36.

Greenblatt, J., Boackle, R. J., and Schwab, J. H. (1978). *Infect. Immun.* **19,** 296–303.

Guinand, M., Vacheron, M. J., and Michel, G. (1970). *FEBS Lett.* **6,** 37–39.

Hamilton, J. A., Zabriskie, J. B., Lachman, L. B., and Chen, Y.-S. (1982). *J. Exp. Med.* **155,** 1702–1718.

Hammes, W., Schleifer, K. H., and Kandler, O. (1973). *J. Bacteriol.* **116,** 1029–1053.

Hayashi, H., Araki, Y., and Ito, E. (1973). *J. Bacteriol.* **113** 592–598.

Helgeland, S., Grov, A., and Schleifer, K. H. (1973). *Acta Pathol. Microbiol. Scand., Sect. B* **81,** 413–418.

Hestrin, S. (1949). *J. Biol. Chem.* **180,** 249–261.

Heymer, B. (1975). *Klin. Wochenschr.* **53,** 49–57.

Heymer, B., and Rietschel, E. T. (1977). *In* "Microbiology" (D. Schlessinger, Ed.), pp. 344–349. Am. Soc. Microbiol., Washington, D. C.

Heymer, B., Schleifer, K. H., Read, S., Zabriskie, J. B., and Krause, R. M. (1976). *J. Immunol.* **117,** 23–26.

Heymer, B., Spanel, R., and Haferkamp, O. (1982). *Curr. Top. Pathol.* **71,** 123–152.

Holton, J. B., and Schwab, J. H. (1966). *J. Immunol.* **96,** 134–138.

Hughes, R. C. (1968). *Biochem. J.* **106**, 41–59.

Husek, P., and Macek, K. (1975). *J. Chromatogr.* **113**, 139–230.

Huss, V. A. R., Festl, H., and Schleifer, K. H. (1984). *Int. J. Syst. Bacteriol.* **35**.

Iversen, O.-J., and Grov, A. (1971). *Eur. J. Biochem.* **38**, 293–300.

Jürgens, U. J., Drews, G., and Weckesser, J. (1983). *J. Bacteriol.* **154**, 471–478.

Kaku, M., Yagawa, K., Nagao, S., and Tanaka, A. (1983). *Infect. Immun.* **39**, 559–564.

Kandler, O. (1982). *Zentralbl. Bakteriol., Parasitenk., Infektionskr., Hyg., Abt. 1 Orig.* **C3**, 149–160.

Kandler, O., and König, H. (1978). *Arch. Microbiol.* **118**, 141–152.

Kandler, O., and Schleifer, K. H. (1980). *Prog. Bot.* **42**, 234–252.

Kandler, O., Hund, A., and Zehender, C. (1958). *Arch. Mikrobiol.* **30**, 355–362.

Kandler, O., Koch, D., and Schleifer, K. H. (1968a). *Arch. Mikrobiol.* **61**, 181–186.

Kandler, O., Schleifer, K. H., and Dandl, R. (1968b). *J. Bacteriol.* **96**, 1935–1939.

Karakawa, W. W., Braun, D. G., Lackland, H., and Krause, R. M. (1968). *J. Exp. Med.* **128**, 325–329.

Kato, K., Umemoto, T., Fukuhara, H., Sagawa, H., and Kotani, K. (1981). *FEMS Microbiol. Lett.* **10**, 81–85.

Kawamoto, I., Oka, T., and Nara, T. (1981). *J. Bacteriol* **146**, 527–534.

Kilpper-Bälz, R., Fischer, G., and Schleifer, K. H. (1982). *Curr. Microbiol.* **7**, 245–250.

Kimural, Y., Norose, Y., Kato, T., Furuya, M., Hida, M., and Okabe, T. (1981). *In* "Basic Concepts of Streptococci and Streptococcal Diseases" (S. E. Holm and P. Christensen, Eds.), pp. 99–100. Reedbooks, Chertsey, England.

Kotani, S., Narita, T., Stewart-Tull, D. E. S., Shimono, T., Watanabe, Y., Kato, K., and Iwata, S. (1975). *Biken J.* **18**, 77–92.

Kotta, J., and Schleifer, K. H. (1974). *J. Hyg. Epidemiol. Microbiol. Immunol.* **10**, 50–59.

Krause, R. M. (1975). *Z. Immunitätsforsch.* **149**, 136–150.

Krueger, J. M., Pappenheimer, J. R., and Karnovsky, M. L. (1982a). *J. Biol. Chem.* **257**, 1664–1669.

Krueger, J. M., Pappenheimer, J. R., and Karnovsky, M. L. (1982b). *Proc. Natl. Acad. Sci. U.S.A.* **79**, 6102–6106.

Kusser, W., and Fiedler, F. (1983). *FEMS Microbiol. Lett.* **20**, 391–394.

Larsen, D. H., Snetsinger, D. C., and Waibel, P. E. (1971). *Anal. Biochem.* **39**, 395–401.

Lauer, E., and Kandler, O. (1983). *Syst. Appl. Microbiol.* **4**, 42–64.

Lederer, E., Adam, A., Ciorbaru, R., Petit, J.-F., and Wietzerbin, J. (1975). *Mol. Cell. Biochem.* **7**, 87–104.

Lipkin, D., Philips, B. E., and Abrell, J. W. (1969). *J. Org. Chem.* **34**, 1539–1547.

Manning, J. M., and Moore, S. (1968). *J. Biol. Chem.* **243**, 5591–5597.

Masaki, T., Tanabe, M., Nakamura, K., and Soejima, M. (1981). *Biochem. Biophys. Acta* **660**, 44–55.

Minnikin, D., and Goodfellow, W. (1980). *In* "Microbiological Classification and Identification" (M. Goodfellow and R. G. Board, Eds.), pp. 189–256. Academic Press, New York.

Minnikin, D. E., Patel, P. V., Alshamaony, and Goodfellow, M. (1977). *Int. J. Syst. Bacteriol.* **27**, 104–117.

Mirelman, D., and Sharon, N. (1967). *J. Biol. Chem.* **242**, 3414–3427.

Monner, D. A., Gmeiner, J., and Mühlradt, P. F. (1981). *Infect. Immun.* **31**, 957–964.

Musher, D. M., Verbrugh, H. A., and Verhoef, J. (1981). *J. Immunol.* **127**, 84–88.

Niebler, E., Schleifer, K. H., and Kandler, O. (1969). *Biochem. Biophys. Res. Commun.* **34**, 560–568.

Nguyen-Dang, H., Nauciel, C., and Wermuth, C.-G. (1976). *Eur. J. Biochem.* **66**, 79–84.

O'Donnell, A. G., Minnikin, D. E., Goodfellow, M., and Parlett, J. H. (1982). *FEMS Microbiol. Lett.* **15**, 75(E)–78(E).
Ogawa, T., Kotani, S., Fukuda, K., Tsukamoto, Y., Mori, M., Kusumoto, S., and Shiba, T. (1982). *Infect. Immun.* **38**, 817–824.
Ogawa, T., Kotani, S., Kusumoto, S., and Shiba, T. (1983). *Infect. Immun.* **39**, 449–451.
Ohta, M. (1981). *Nippon Ika Daigaku Zasshi* **48**, 402–409.
Oken, M. M., Peterson, P. K., and Wilkinson, B. J. (1979). *Clin. Res.* **27**, 352 A.
Osada, Y., Otani, T., Sato, M., Une, T., Matsumoto, K., and Ogawa, H. (1982). *Infect. Immun.* **38**, 848–854.
Park, J. T., and Hancock, R. (1960). *J. Gen. Microbiol.* **22**, 249–258.
Perkins, H. R. (1965). *Biochem. J.* **95**, 876–882.
Perkins, H. R. (1967). *Biochem. J.* **102**, 29c–32c.
Perkins, H. R. (1969). *Biochem. J.* **115**, 797–805.
Petersen, B. H., and Rosenthal, R. S. (1982). *Infect. Immun.* **35**, 442–448.
Peterson, P. K., Wilkinson, B. J., Kim, Y., and Schmeling, D. (1978). *J. Clin. Invest.* **61**, 597–609.
Ranu, R. S. (1975). *Med. Microbiol. Immunol.* **161**, 53–61.
Rao, K. R., and Sober, H. A. (1954). *J. Am. Chem. Soc.* **76**, 1328–1331.
Rhuland, L. E., Work, E., Denman, R. F., and Hoare, D. A. (1955). *J. Am. Chem. Soc.* **77**, 4844–4846.
Richter, G. (1977). *Veröff. Inst. Meeresforsch. Bremerhaven* **16**, 125–138.
Rose, J. A. (1955). *In* "Methods in Enzymology" (S. P. Colowick and N. O. Kaplan, Eds.), Vol. 1, pp. 591–594. Academic Press, New York.
Rosenbusch, J. P. (1974). *J. Biol. Chem.* **249**, 8019–8029.
Rosenthal, S., and Sharon, N. (1964). *Biochim. Biophys. Acta* **83**, 378–380.
Rotta, J. (1969). *Curr. Top. Microbiol. Immunol.* **48**, 63–101.
Rotta, J. (1975). *Z. Immunitätsforsch.* **149**, 230–244.
Rotta, J., and Schleifer, K. H. (1974). *J. Hyg. Epidemiol. Microbiol. Immunol.* **18**, 50–59.
Rotta, J., Rye, M., Masek, K., and Zaoral, M. (1979). *Exp. Cell Biol.* **47**, 258–268.
Roze, U., and Strominger, J. L. (1966). *Mol. Pharmacol.* **2**, 92–94.
Russel, D. W. (1963). *Biochem. J.* **87**, 1–4.
Ryc, M., and Rotta, J. (1975). *Z. Immunitätsforsch.* **149**, 265–272.
Schechter, B., Schechter, I., and Sela, M. (1970). *J. Biol. Chem.* **245**, 1438–1447.
Schleifer, K. H. (1970). *Arch. Mikrobiol.* **71**, 271–282.
Schleifer, K. H. (1973). *In* "Staphylococci and Staphylococcal Infections. Recent Progress" (J. Jeljaszewicz, Ed.), pp. 13–23. Karger, Basel.
Schleifer, K. H. (1975). *Z. Immunitätsforsch.* **149**, 104–117.
Schleifer, K. H. (1984). *In* "Staphylococci and Staphylococcal Infections" (C. S. F. Easmon and C. Adlam, Eds.), Vol. 2, pp. 385–428. Academic Press, New York.
Schleifer, K. H., and Joseph, R. (1973). *FEBS Lett.* **36**, 83–86.
Schleifer, K. H., and Kandler, O. (1967). *Arch. Mikrobiol.* **57**, 335–364.
Schleifer, K. H., and Kandler, O. (1972). *Bacteriol. Rev.* **36**, 407–477.
Schleifer, K. H., and Kilpper-Bälz, R. (1984). *Int. J. Syst. Bacteriol.* **35**, 31–34.
Schleifer, K. H., and Kloos, W. E. (1976). *Zentralbl. Bakteriol. Parasitenkd., Infektionkr., Hyg. Abt. 1: Suppl.* **5**, 3–9.
Schleifer, K. H., and Krause, R. M. (1971a). *J. Biol. Chem.* **246**, 986–993.
Schleifer, K. H., and Krause, R. M. (1971b). *Eur. J. Biochem.* **19**, 471.
Schleifer, K. H., and P. H. Seidl (1984). *In* "Chemotaxonomic Methods for Bacteria" (M. Goodfellow and D. Minnikin, Eds.), pp. 201–219. Academic Press, New York.

Schleifer, K. H., and Stackebrandt, E. (1983). *Annu. Rev. Microbiol.* **37**, 143–187.
Schleifer, K. H., and Steber, J. (1974). *Arch. Mikrobiol.* **98**, 251–270.
Schleifer, K. H., Plapp, R., and Kandler, O. (1967). *Biochem. Biophys. Res. Comm.* **28**, 566–570.
Schleifer, K. H., Ried, M., and Kandler, O. (1968). *Arch. Mikrobiol.* **62**, 198–208.
Schleifer, K. H., Huss, L., and Kandler. O. (1969). *Arch. Mikrobiol.* **68**, 387–404.
Schleifer, K. H., Hammes, W. P., and Kandler, O. (1976). *Adv. Microb. Physiol.* **13**, 245–292.
Schleifer, K. H., Kloos, W. E., and Kocur, M. (1981). *In* "The Prokaryotes" (M. P. Starr, H. Stolp, H. G. Trüper, A. Balows, and H. G. Schlegel, Eds.), pp. 1539–1547. Springer-Verlag, Berlin and New York.
Schleifer, K. H., Kraus, J., Kilpper-Bälz, R., and Gehring, F. (1984). *J. Dent. Res.* **63**, 1047–1050.
Schmelzer, E., Weckesser, J., Warth, R., and Mayer, M. (1982). *J. Bacteriol.* **149**, 151–155.
Seidl, P. H., and Schleifer, K. H. (1977). *In* "Microbiology" (D. Schlessinger, Ed.), pp. 339–343. Am. Soc. Microbiol., Washington, D. C.
Seidl, P. H., and Schleifer, K. H. (1978a). *Arch. Microbiol.* **118**, 185–192.
Seidl, P. H., and Schleifer, K. H. (1978b). *Appl. Environ. Microbiol.* **35**, 479–482.
Seidl, P. H., Faller, A. H., Loider, R., and Schleifer, K. H. (1980). *Arch. Microbiol.* **127**, 173–178.
Seidl, P. H., Franken, N., and Schleifer, K. H. (1983). *In* "The Target of Penicillin" (R. Hakenbeck, J. V. Höltje, and H. Labischinski, Eds.), pp. 209–304. de Gruyter, Berlin.
Shockman, G. D., Thompson, J. S., and Connover, M. J. (1967). *Biochemistry* **6**, 1054–1065.
Stewart-Tull, D. E. S. (1980). *Annu. Rev. Microbiol.* **34**, 311–340.
Takebe, I. (1965). *Biochim. Biophys. Acta* **101**, 124–126.
Ternberg, J. L., and Hershey, F. B. (1960). *J. Lab. Clin. Med.* **56**, 766–776.
Tipper, D. J. (1968). *Biochemistry* **7**, 1441–1449.
Uchida, K., and Aida, K. (1977). *J. Gen. Appl. Microbiol.* **23**, 249–260.
Uchida, K., and Aida, K. (1979). *J. Gen. Appl. Microbiol.* **25**, 169–183.
Umemoto, T., Ota, T., Sagawa, H., Kato, K., Takada, H., Tsujimoto, M., Kawasaki, A., Ogawa, T., Harada, K., and Kotani, S. (1981). *Infect. Immun.* **31**, 767–774.
Undenfriend, S., Stein, S., Böhlen, P., Dairman, W., Leimgruber, W., and Weigele, M. (1972). *Science* **178**, 871–872.
Vasstrand, E. N., Jensen, H. B., Miron, T., and Hofstad, T. (1982). *Infect. Immun.* **36**, 114–122.
Vilkas, E., Massot, J. C., and Zissmann, E. (1970). *FEBS Lett.* **7**, 77–79.
Weidel, W., and Pelzer, H. (1964). *Adv. Enzymol.* **26**, 193–232.
Weidel, W., Frank, H., and Martin, H. H. (1960). *J. Gen. Microbiol.* **22**, 158–166.
Weiss, N., Schleifer, K. H., and Kandler, O. (1981). *Rev. Inst. Pasteur Lyon* **14**, 3–12.
White, P. J., and Kelly, B. (1965). *Biochem. J.* **96**, 75–84.
White, P. J., de Jeune, B., and Work, E. (1969). *Biochem. J.* **113**, 589–601.
Wikler, M. (1975). *Z. Immunitätsforsch.* **149**, 193–200.
Windle, P. E., Murphy, P. A., and Cooperman, S. (1983). *Infect. Immun.* **39**, 1142–1146.
Work, E. (1963). *In* "Methods in Enzymology" (S. P. Colowick and N. O. Kaplan, Eds.), Vol. 6, pp. 624–634. Academic Press, New York.
Work, E. (1971). *In* "Methods of Microbiology" (J. R. Norris and D. W. Ribbons, Eds.), Vol. 5a, pp. 361–418. Academic Press, New York.
Yokogawa, K., Kawata, S., Nishimura, S., Ideda, Y., and Yushimura, Y. (1974). *Antimicrob. Agents Chemother.* **6**, 156–165.

6

Analysis of Lipopolysaccharides of Gram-Negative Bacteria

H. MAYER,* R. N. THARANATHAN,*,[1] AND J. WECKESSER†

*Max-Planck-Institut für Immunbiologie, Freiburg i.Br.
Federal Republic of Germany*

*† Institut für Biologie II, Mikrobiologie, der Universität, Freiburg i.Br.
Federal Republic of Germany*

I. Introduction

A. General description of lipopolysaccharides

Lipopolysaccharides (LPS) are characteristic components of the cell wall of all Gram-negative bacteria (Westphal *et al.*, 1981; Lüderitz *et al.*, 1978; Rietschel *et al.*, 1983a) and of at least some cyanobacteria (Weckesser *et al.*, 1979). They are localised in the outer layer of the outer membrane (Nikaido and Nakae, 1979; Funahara and Nikaido, 1980; Lugtenberg and

[1] On leave from Discipline of Biochemistry, C.F.T.R.I., Mysore, India.

METHODS IN MICROBIOLOGY
VOLUME 18

van Alphen, 1983), and are, in noncapsulated strains (Ørskov *et al.*, 1977; Wilkinson, 1977), exposed on the cell surface. They contribute to the integrity of the outer membrane and, as has been shown for enteric bacteria, protect the cell against the action of bile salts and (lipophilic) antibiotics (Schlecht and Westphal, 1970; Lugtenberg and van Alphen, 1983).

Since lipopolysaccharides confer antigenic properties on the cell, they have been termed "O antigens." As main antigens of invading infecting strains, lipopolysaccharides are involved in various host–parasite interactions (Lindberg, 1977; Rietschel *et al.*, 1982b). They seem to protect Gram-negative bacteria from phagocytosis and lysis. Due to their exposed position they are frequently part of the receptor complexes for bacteriophages and colicines (Schmidt and Lüderitz, 1969; Lindberg, 1973, 1977).

Isolated lipopolysaccharide, upon injection into higher organisms, exhibits a variety of biological ("endotoxic") activities, such as the induction of fever, changes in the white blood cell count, diarrhoea, or even shock and death (Galanos *et al.*, 1977a; Rietschel *et al.*, 1982b). Since the toxic activities are due to substances released from disintegrating bacteria, and not to an excreted exotoxin, lipopolysaccharide has been termed an endotoxin. Thus, the terms lipopolysaccharide, O antigen, and endotoxin are used synonymously and describe either the chemical structure or discrete biological functions of the same substance (Westphal *et al.*, 1977; Rietschel *et al.*, 1982a).

Distinct architectural elements are common to lipopolysaccharides of various Gram-negative bacteria despite their extensive structural diversity. It has been shown by Westphal and Lüderitz and their associates (Lüderitz *et al.*, 1966) that lipopolysaccharides are built up of three distinct structural regions (Fig. 1): Region I represents the O-specific chains;

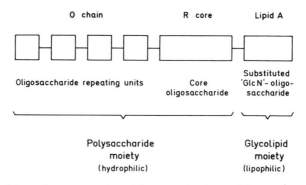

Fig. 1. Schematic representation of the general makeup of lipopolysaccharides.

region II, the core oligosaccharide, and region III, the lipid A moiety. Each region is under separate genetic control (Mäkelä and Stocker, 1984; Stocker and Mäkelä, 1978), shows different but characteristic constituents, and exhibits distinct biological activities (Weinbaum et al., 1971). The three structural regions are of different taxonomic relevance and, hence, will be described separately. Special emphasis will be placed on the analysis of nonenterobacterial species, since earlier reviews deal largely with Enterobacteriaceae (Nikaido, 1970; Rietschel and Lüderitz, 1980) and data on the former became available only more recently.

1. Region I: O-specific chains

In most cases O-specific chains are built up of repeating units of oligosaccharides which exhibit a strain-specific structural diversity. The sugar constituents, their sequence, and their mode of linkage determine the serological O specificity of respective strains, and are thus of considerable taxonomic and diagnostic importance (Lüderitz et al., 1966, 1971; Nikaido, 1970). They are the main determinants for the classification of the serotypes of Salmonella species as used in the Kauffmann–White scheme (Kauffmann, 1966). Similar classifications are also available for Escherichia, Citrobacter, Proteus, Xanthomonas, and others. Grouping of bacteria based on bacterial agglutination and cross-absorption was possible on the basis of shared O-antigenic factors (groups A, B, C, . . . , of Salmonella) (Kauffmann, 1966).

In many O groups specific sugar components of unusual structure are found as immunodominant sugars. They often occur in terminal positions, e.g., the 3,6-dideoxyhexoses: paratose (in group A of the Kauffmann–White scheme), abequose (in groups B and C_2), and tyvelose (in group D_2). These sugars occur also in Yersinia pseudotuberculosis (Samuelsson et al., 1974). In other bacterial genera and families O-methylated, O-acetylated, or otherwise substituted sugars or unusual amino sugars or uronic acids have been found and they may also play a role as immunodominant constituent. Table I gives a list of sugar components so far reported from lipopolysaccharides.

In a number of pathogenic and non-pathogenic bacteria, e.g., Neisseria, Haemophilus, Pasteurella, phototropic bacteria of the Rhodospirillaceae family, and Nitrobacteriaceae (Mäkelä and Stocker, 1984; Mayer et al., 1983a), region I is either totally missing or is represented by a few sugar residues only or by very short chains (Weckesser et al., 1983). This fact may be illustrated by three phylogenetically related species of the Rhodospirillaceae family (Fig. 2). All three species share the same lipid A

TABLE I

Sugar and non-sugar constituents identified in lipopolysaccharides[a]

Neutral sugars	Amino sugars
Pentose (5)	4-Amino-4-deoxypentose (1)
4-Deoxypentose (1)	2-Amino-2-deoxyhexose (3)
Pentulose (1)	2-Amino-2,6-dideoxyhexose (4)
Hexose (3)	3-Amino-3,6-dideoxyhexose (2)
6-Deoxyhexose (7)	4-Amino-4,6-dideoxyhexose (3)
3,6-Dideoxyhexose (5)	2,3-Diamino-2,3-dideoxyhexose (1)
Hexulose (1)	2,4-Diamino-2,4,6-trideoxyhexose (2)
6-Deoxyheptose (1)	2-Amino-2-deoxyheptose (1)
Heptulose (1)	
	Acidic sugars
Branched sugars	Hexuronic acid (2)
3-*C*-Methyl-6-deoxyhexose (2)	2-Amino-2-deoxyhexuronic acid (3)
3-*C*-Hydroxymethylpentose (1)	3-Deoxyoctulosonic acid (1)
4-*C*-Hydroxyethyl-3,6-dideoxyhexose (1)	3-*O*-Lactyl-6-deoxyhexose (2)
	4-*O*-Lactylhexose (1)
O-Methyl sugars (total 30)	2,3-Diamino-2,3-dideoxyhexuronic acid (3)
O-Methylpentose	2,3(1-Acetyl-2-methyl-2-imidazolino-5,4)-hexuronic acid (1)
O-Methylhexose	
O-Methylheptose	
O-Methyl-6-deoxyhexose	Non-sugar constituents
O-Methyl-2-amino-2-deoxyhexose	2-Dihydroxybutanoic acid
	Glycine
	Alanine
	Threonine
	Lysine
	Pyruvic acid
	Ethanolamine

[a] Fatty acids are not included; with modifications according to Lüderitz et al. (1983). For further details see Kenne and Lindberg (1983), Schramek et al. (1983), Gorshkova et al. (1983), Kotelko (1985), Salimath et al. (1984). Numbers in parentheses are the number of isomers identified.

backbone; two of them, *Rhodospirillum tenue* and *Rhodocyclus purpureus,* have the same lipid A composition (Mayer and Weckesser, 1984). Only in the latter species, however, are typical O-specific chains present, whereas in lipopolysaccharide of *R. tenue* strains only very short O chains are found (Weckesser et al., 1977). In most strains of *Rhodopseudomonas gelatinosa* the O-specific region is completely missing. These strains do possess strain-specific haptenic polysaccharides not bound to the LPS core (Weckesser et al., 1975a,b). The questions regarding the anchoring in the outer membrane of these O chains and whether *R. gelatinosa,* being phylogenetically an old species, is lacking the O-translocase system—which ligates O chain and R core—are yet to be answered.

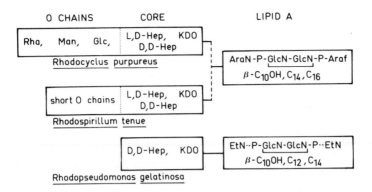

Fig. 2. The three regions and their characteristic constituents of the lipopolysaccharides from *Rhodocyclus purpureus, Rhodospirillum tenue,* and *Rhodopseudomonas gelatinosa.* A distinction between O chains and core is hypothetical for *Rhodocyclus purpureus.* In *R. tenue* lipid A an additional, unsubstituted glucosamine is linked at C-4 of the glucosamine backbone. (See Fig. 7 for abbreviations).

The diversity of O chains in Enterobacteriaceae may have developed during evolution to allow enteric bacteria to escape the host's immune system by again and again developing new specificities on their cell surface (Nikaido, 1970; Rietschel and Lüderitz, 1980). However, this cannot be the only mechanism used by bacteria to survive in animal hosts as is shown by *Neisseria* and by *Y. pseudotuberculosis* ssp. *pestis,* which both lack O chains in their lipopolysaccharides (Mäkelä and Stocker, 1984).

2. Region II: R-core

The finding that mutants of *Salmonella* showing rough (R) colony morphology have defects in the biosynthesis of their O antigens has greatly facilitated work on the chemical structure and biosynthesis of lipopolysaccharides. Depending on the biosynthetic step which is blocked a variety of incomplete lipopolysaccharide types were characterised (Lüderitz et al., 1966, 1971). According to the monosaccharide pattern of core oligosaccharides the designations Ra (for the complete *Salmonella* core), Rb, Rc, Rd, and Re for the various incomplete core structures (Fig. 3) have been introduced. These different mutants could be characterised also by using a set of R-specific phages (Schmidt and Lüderitz, 1969; Schmidt et al., 1969a) and new mutants could be isolated by their distinct phage sensitivity pattern (Lindberg, 1977). All *Salmonella* serotypes studied so far share the same core structure; deviations in complete core structures are, however, found amongst other Enterobacteriaceae. From

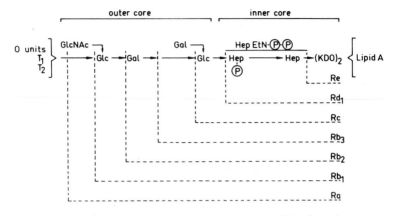

Fig. 3. The incomplete core structures (Ra–Re chemotypes) of the lipopolysaccharides of *Salmonella* R mutants; according to Lüderitz *et al.* (1971, 1982).

Escherichia coli serotypes five different complete R-core structures have been reported so far (Schmidt *et al.,* 1969b; Schmidt and Lüderitz, 1969; Mayer and Schmidt, 1973) and designated as R1–R4 and K-12 core types (Fig. 4). They all differ in the outer (hexose) region, whereas the inner region, composed of characteristic and LPS-specific components, L-glycero-D-mannoheptose, and 2-keto-3-deoxy-D-mannooctonate (KDO or dOclA), seems to be the same, as judged from ^{1}H NMR analysis (Jansson *et al.,* 1981). The different *E. coli* R core types occur also in *Shigella* (Mayer and Schmidt, 1973; Gamian and Romanowska, 1982); so far no *Shigella*-specific core types have been observed. The core structures can be differentiated by R-specific phages and in some cases by precipitation with lectins (Ahamed *et al.,* 1980). Core types with R1 and R4 structures (Feige *et al.,* 1977) are partly substituted by the enterobacterial common antigen (ECA). Sera prepared against these mutants therefore produce both R and ECA antibodies (Mayer and Schmidt, 1979). Other R mutant types, deeper R mutants, and S forms of enterobacterial strains possess ECA in the non-immunogenic haptenic form; i.e., it is not bound to the core. Although ECA is localised in the outer membrane, it is not readily available for antibodies in S forms (Acker *et al.,* 1982) and does not therefore interfere with the serological analysis by bacterial agglutination, unless R mutants are investigated (Marx *et al.,* 1977).

Galacturonic acid is the terminal sugar of the core in *Proteus mirabilis* (Kotelko, 1984; Radziejewska-Lebrecht, 1983). Some *P. mirabilis* R mutants possess in their LPS galactosamine and the rare D-glycero-D-manno-heptose in addition to the usual L-glycero-D-mannoheptose, indicating the

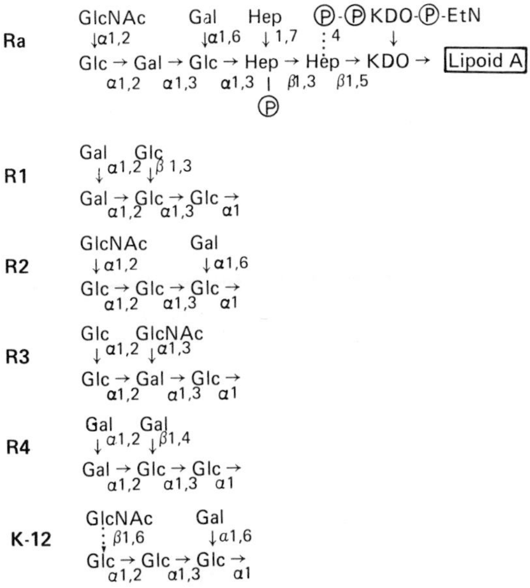

```
          GlcNAc     Gal   Hep   Ⓟ-ⓅKDO-Ⓟ-EtN
Ra        ↓α1,2           ↓α1,6  ↓1,7  : 4      ↓
          Glc → Gal → Glc → Hep → Hèp → KDO →  [Lipoid A]
            α1,2   α1,3   α1,3 |  β1,3   β1,5
                              Ⓟ
```

```
          Gal   Glc
R1        ↓α1,2 ↓β 1,3
          Gal → Glc → Glc →
            α1,2   α1,3   α1
```

```
          GlcNAc     Gal
R2        ↓α1,2           ↓α1,6
          Glc → Glc → Glc →
            α1,2   α1,3   α1
```

```
          Glc   GlcNAc
R3        ↓α1,2 ↓α1,3
          Glc → Gal → Glc →
            α1,2   α1,3   α1
```

```
          Gal   Gal
R4        ↓α1,2 ↓β1,4
          Gal → Glc → Glc →
            α1,2   α1,3   α1
```

```
          GlcNAc     Gal
K-12      : β1,6           ↓α1,6
          Glc → Glc → Glc →
            α1,2   α1,3   α1
```

Fig. 4. The hexose region of the different complete enterobacterial R core types; according to Rietschel and Lüderitz (1980).

existence of different R core types in *Proteus* (Kotelko *et al.*, 1977). The R-specific phages reacting with *Salmonella* and *Escherichia* R mutants do not attack *Proteus* R mutants, thus indicating a rather unique structure of the *Proteus* core. Like *Proteus* species, *Yersinia* has a core with L- and D-glycero-D-mannoheptose (Tomshich *et al.*, 1983).

Only a few data on R cores are available from non-enteric bacteria. Two reports demonstrate the presence of galactosamine, alanine, and rhamnose in addition to glucose, heptose, and KDO in two different strains of *Pseudomonas aeruginosa* (Wilkinson, *et al.*, 1973; Rowe and Meadow, 1983). Alanine is substituting the amino group of galactosamine (*N*-alaninylgalactosamine). The complete R core of *Coxiella burnetii* (Phase II) consists of only two sugars, D-mannose and D-glycero-D-mannoheptose occurring in equimolar amounts (Schramek and Mayer, 1982).

Some species were shown to lack heptose and/or KDO (Rietschel and Lüderitz, 1980). Heptose is missing in *Xanthomonas, Myxobacteria, Moraxella, Brahamella,* and in some Rhodospirillaceae species [*Rhodopseudomonas sphaeroides, Rhodopseudomonas capsulata, Rhodopseudomonas viridis,* and *Rhodomicrobium vannielii* (Drews *et al.*, 1978)]. The Rhodospirillaceae species belong to two related genealogical clusters of species according to their 16 S rRNA homology (Stackebrandt and Woese, 1981; Gibson *et al.*, 1979). KDO was reported to be lacking in lipopolysaccharides of *Vibrio cholerae, Leptotrichia,* and *Bacteroides.* Species of the latter genus lack also heptose(s). The recent identification of traces of KDO in a *Vibrio* strain (Brade *et al.*, 1983; Banoub *et al.*, 1983) indicates that substituted or exclusively chain-linked KDO may escape detection under usual conditions of analysis.

In cases in which R mutants are not easily available, R core material can be obtained from S forms by separating "degraded polysaccharide" on a Sephadex G-50 column (see Fig. 9). This method has been used for analysis of the complete core from *P. mirabilis* (Radziejewska-Lebrecht, 1983; Kotelko, 1985), where R mutants were difficult to isolate because of the swarming property of this species.

3. Region III: Lipid A

The third structural region of lipopolysaccharide, the so-called lipid A, is by far the most conservative part of the molecule. It is responsible for the endotoxic properties of lipopolysaccharides. The complete structure of the *Salmonella* lipid A has quite recently been elucidated (Rietschel *et al.*, 1984) (Fig. 5), although its basic structure, the so-called lipid A backbone, has been known for many years (Gmeiner *et al.*, 1969; Hase and Rietschel, 1976b, 1977). The R core region is linked via a KDO residue to lipid A. This ketosidic linkage is acid labile and can be split by 1 M acetic acid (100°C). The liberated lipid A has been termed "free lipid A" to discriminate it from bound lipid A that occurs in intact lipopolysaccharide. The complete structure of *Salmonella* "free lipid A," as shown in Fig. 5, consists of a β-1,6-linked D-glucosamine disaccharide carrying two phosphate groups. One is linked to the glycosidic hydroxyl group of the reducing glucosamine (GlcN I), and the other in an ester linkage to the 4' position of the non-reducing glucosamine (GlcN II). The backbone is substituted by 4 mol of (R)-3-hydroxy fatty acids; two of them are amide linked and two of them ester linked to positions 3 and 3'. These hydroxy fatty acids can carry additional ester-linked non-hydroxylated fatty acids, thus giving rise to acyloxyacyl residues. The precise location of the fatty acids in lipid A has been elucidated recently by two new experimental approaches: liberation of amide-linked double ester according to Wollenweber *et al.* (1983), making use of the Kraska methylation technique (Kraska *et al.*, 1976), and diazomethane methylation, operating at neutral pH (Ohno *et al.*, 1979), which allows the localisation of unsubstituted OH groups in lipid A. The presently accepted structure of "free lipid A" of

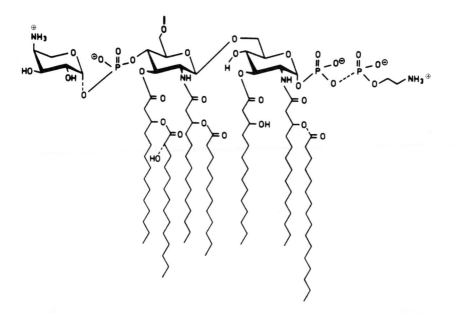

Fig. 5. Proposed chemical structure of *Salmonella minnesota* lipid A; according to Rietschel *et al.* (1984).

Salmonella (Fig. 5) has been strongly supported by fast atom bombardment mass spectrometry (FAB-MS) (Takayama *et al.*, 1984) and by laser desorption mass spectrometry (LDMS) (Seydel *et al.*, 1984). The OH group in position 4 of GlcN I is unsubstituted and that at the 6' position of the GlcN II carries the KDO (Imoto *et al.*, 1983; Lüderitz *et al.*, 1983). Both positions can expectedly be methylated in "free lipid A." The substitution of 3-OH fatty acids by additional fatty acids may vary from species to species and can be modified by changing the cultivation temperature. Especially, the fatty acid linked to the amino group of GlcN II can be substituted either by 12:0 or 14:0 (growth at 37°C) or by Δ⁹ 16:1 (palmitoleic acid, growth at 12°C), as was demonstrated with an Re mutant of *P. mirabilis* (Seydel *et al.*, 1984; Wollenweber *et al.*, 1983).

The two phosphate groups attached to the lipid A backbone are often substituted by polar head groups, such as 4-amino-4-deoxy-L-arabinose, ethanolamine, phosphorylethanolamine, glucosamine with free amino group, etc. The nature of the substituents is often characteristic for a given species, but the degree of substitution may vary between strains and depend on culture conditions, thus creating microheterogeneity in lipopolysaccharide (see below) (Rietschel and Lüderitz, 1980; Lehmann *et al.*, 1978).

The backbone structure of lipid A, i.e., the diphosphorylated glucosamine disaccharide, is very likely the most conservative partial structure
of lipopolysaccharide. It is common to all Enterobacteriaceae, Pseudomonadaceae, *Fusobacterium nucleatum,* and *Selenomonas ruminantium*
(Rietschel and Lüderitz, 1980), but also to some, but not all, Rhodospirillaceae (Mayer and Weckesser, 1984).

In *R. viridis* (Roppel *et al.,* 1975), *Pseudomonas diminuta* (Wilkinson
and Taylor, 1978), Nitrobacteriaceae (Mayer *et al.,* 1983), and in a
chloridazone-degrading species (Weisshaar and Lingens, 1983), as well as
in a few related species (Mayer *et al.,* 1984), non-phosphorylated 2,3-
diamino-2,3-dideoxy-D-glucose forms the backbone (as a monomeric unit)
of lipid A. All these species are phylogenetically related as revealed by
16 S rRNA homology studies (Stackebrandt and Woese, 1981; Seewaldt *et
al.,* 1982). This lipid A backbone sugar carries also amide-linked (and in
some species also ester-linked) 3-hydroxy fatty acids, and is substituted
by the polysaccharide moiety via KDO at position 6 of the diaminohexose
(Weisshaar and Lingens, 1983).

The correspondence of lipid A types with the 16 S rRNA phylogenetic
scheme of Rhodospirillaceae is shown in Fig. 6. Four different lipid A
types have been characterised so far (Mayer and Weckesser, 1984;
Tharanathan *et al.,* 1978b; Salimath *et al.,* 1983) and designated as Rhodospirillaceae types I–IV (Table II). Lipid A's of members of groups I–III
share with *Salmonella* the β-1,6-linked glucosamine disaccharide as lipid
A backbone sugar. 2,3-Diamino-2,3-dideoxy-D-glucose is the backbone
sugar in group IV. The backbone sugar is phosphorylated in groups I and
II but not in groups III and IV. According to the different substituents of
the lipid A backbone (ethanolamine in *R. gelatinosa,* 4-amino-L-arabinose
and D-arabinofuranose in *R. tenue*) and the nature of the heptose(s) in the
R core group I can be subdivided into Ia and Ib (Table II). Lipid A of
group II contains the rare 3-oxomyristic acid (3-oxo-14:0) as amide-
linked fatty acid. The phosphate-free lipid A of *R. vannielii* (group III) is
(partly) substituted at the 4' position of GlcN II by residues of D-manno-
pyranose. Whether other mannose-containing lipid A's reported from
Chromatiaceae share a similar structure has not yet been revealed
(Hurlbert *et al.,* 1976, 1978).

The structural similarities existing between *Salmonella* lipid A and
those of Rhodospirillaceae of groups I and II are also documented by the
mutual serological cross-reaction of lipid A's with lipid A antisera (Galanos *et al.,* 1977b; Strittmatter *et al.,* 1983).

Of all lipopolysaccharides of Rhodospirillaceae only that of *R. gelatinosa* shows high lethal toxicity (in mice) and high pyrogenicity (in
rabbits) (Galanos *et al.,* 1977b). Low lethal toxicity (about 1% of the
Salmonella and *R. gelatinosa* value) has been found for LPS of *R. tenue*

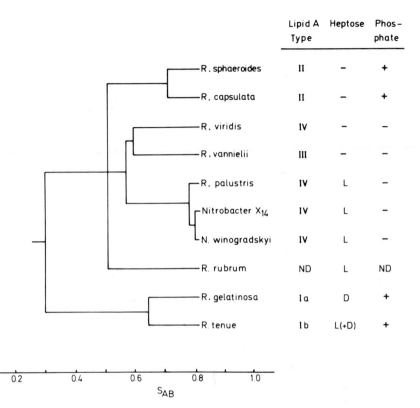

	Lipid A Type	Heptose	Phos-phate
R. sphaeroides	II	–	+
R. capsulata	II	–	+
R. viridis	IV	–	–
R. vannielii	III	–	–
R. palustris	IV	L	–
Nitrobacter X₁₄	IV	L	–
N. winogradskyi	IV	L	–
R. rubrum	ND	L	ND
R. gelatinosa	I a	D	+
R. tenue	I b	L(+D)	+

Fig. 6. Section of the dendrogram showing the phylogenetic relationship amongst species of the Rhodospirillaceae and Nitrobacteraceae families, based on their 16 S rRNA cataloguing; according to Seewaldt *et al.* (1982) and Mayer (1984).

(group Ib) and for *R. vannielii* (group III) (Holst *et al.*, 1983a). All other species have lipopolysaccharides of negligible toxicity.

The taxonomic value of the different fatty acid substituents in lipid A has been summarised by Rietschel *et al.* (1983a), Wollenweber *et al.* (1984), and Rietschel and Lüderitz (1980), and shows essentially the following.

Amide-linked fatty acids are exclusively (R)-3-OH fatty acids (in a few species additionally also 3- or 4-oxo fatty acids). The chain length may vary (generally between 10:0 and 16:0) depending on the species or other respective taxonomic units. Thus, 3-OH-14:0 is found in all Enterobacteriaceae, 3-OH-12:0 is characteristic for Pseudomonadaceae, and 3-OH-16:0 for several Rhizobiaceae and Rhodospirillaceae group III. A combination of distinct amide- and ester-linked fatty acids seems to be

TABLE II

Different lipid A types in Rhodospirillaceae

Species	Lipid A type	GlcN $\xrightarrow{\beta}$ GlcN	2,3-DAGlc	4-N-Ara	EtN	Neutral sugars	Amide-linked fatty acids	P
Rhodopseudomonas gelatinosa	Ia	+			+		3-OH-10:0	++
Rhodospirillum tenue	Ib	+		+		Ara	3-OH-10:0	+
Rhodocyclus purpureus	Ib	ND		+		Ara	3-OH-10:0	+
Rhodopseudomonas sphaeroides	II	+					3-OH-14:0 3-Oxo-14:0	++
Rhodopseudomonas capsulata	II	+					(3-OH-14:0) 3-Oxo-14:0	++
Rhodopseudomonas blastica	II	ND					3-OH-14:0 3-Oxo-14:0	ND
Rhodomicrobium vannielii	III	+				Man	3-OH-16:0	—
Rhodopseudomonas acidophila	III	ND				Man	3-OH-16:0	—
Rhodopseudomonas viridis	IV		+				3-OH-14:0	—
Rhodopseudomonas sulfoviridis	IV		+				3-OH-14:0	—
Rhodopseudomonas palustris	IV		+				3-OH-14:0	—

[a] GlcN, D-glucosamine; 2,3-DAGlc, 2,3-diamino-2,3-dideoxy-D-glucose; 4-N-Ara, 4-amino-L-arabinose; Ara, D-arabinofuranose; Man, D-mannose; EtN, ethanolamine; P, phosphorus content: +, 0.9–1.2%; ++, 2–4.5%; ND, not determined; —, not present in all strains of this species. (Mayer and Weckesser, 1984; Tharanathan *et al.*, 1985).

characteristic for a number of bacterial families or other major phylogenetic clusters of species and can thus be considered as a useful taxonomic
marker (Wollenweber *et al.*, 1984). 3-Hydroxy fatty acids do not exclusively occur in lipid A; they have been reported also as constituents of
ornithine lipids, where they also occur as amide-linked acylated fatty
acids (Madhavan *et al.*, 1981). The taxonomic value of lipid A spectra
may be illustrated by two examples (see Fig. 6): (1) lipopolysaccharides of
the two related species *R. tenue* and *R. gelatinosa* (Weckesser *et al.*,
1977, 1979) have very similar fatty acid patterns, which differ, however,
in the chain length of one ester-linked fatty acid (3-OH-10:0,-12:0, and
14:0 in *R. gelatinosa;* 3-OH-10:0, 12:0, and 16:0 in *R. tenue*); (2) amide-
linked 3-oxo-14:0 occurs only in the three closely related species *R.
sphaeroides* (Salimath *et al.*, 1983), *R. blastica,* and *R. capsulata* (lipid A
type II of the Rhodospirillaceae family, Table II) (Omar *et al.*, 1983).

B. Heterogeneity of lipopolysaccharides

Like polysaccharides in general, lipopolysaccharides show considerable
heterogeneity (Nowotny, 1971). Lipopolysaccharide of a distinct strain is
in reality a family of closely related molecules that share important characteristic structural features. The heterogeneity is caused by the mode of
biosynthesis which has been studied in detail for *Salmonella typhimurium* (Mäkelä and Stocker, 1984; Stocker and Mäkelä, 1978). Although
other routes of lipopolysaccharide biosynthesis have been suggested for
E. coli 09 and *Salmonella montevideo* (Weisgerber and Jann, 1982; Jann
et al., 1982), the biosynthetic steps leading to LPS assembly in *S. typhimurium* is taken here as an example.

The lipid A moiety (lipid A precursor) is biosynthesised first; then the
backbone structure may be substituted on both phosphate groups by additional constituents (Osborn, 1979; Osborn and Rothfield, 1971; Nikaido,
1973). Only in rare cases are these substitutions complete. Furthermore,
the chain length and the relative amount of amide- and ester-linked fatty
acids may vary. As pointed out above, the spectrum of ester-linked non-
hydroxylated fatty acids is dependent on growth temperature (Rottem *et
al.*, 1978, Wollenweber *et al.*, 1983; Wartenberg *et al.*, 1983). In general,
the degree of unsaturation of fatty acids increases on lowering the growth
temperature, and this is often accompanied by a decrease in the mean
chain length of the saturated fatty acids. The temperature-dependent incorporation of fatty acids allows the bacteria to maintain their membrane
fluidity also at lower temperatures (Sinensky, 1974; van Alphen *et al.*,
1979). In following steps of LPS biosynthesis the core region is built up by
sequential addition of single sugar units. Sugars present in branching
positions are often present in non-stoichiometric ratios, thus creating het-

erogeneity also in the core region (Ng *et al.*, 1976). In Enterbacteriaceae, and probably in a number of non-enteric strains as well (*Pseudomonas, Chromatium, Vibrio*) (Wilkinson *et al.*, 1973; Hurbert and Hurlbert, 1977), O-repeating units are assembled on a carrier lipid, the bactoprenol (Osborn and Rothfield, 1971). The degree of polymerisation varies highly (n in Fig. 7: 0–40) (Mäkelä and Stocker, 1984).

In a final step the polymerised chains are attached to the core region by the function of the O-translocase system. A complete core is needed for O-chain attachment, but by no means are all complete cores substituted in S forms. The amount of unsubstituted core stubs may exceed 50% (Palva and Mäkelä, 1980; Goldman and Leive, 1980). After the assembly of the O chains, additional modifications (e.g., acetylation, methylation, glycosylation, or attachment of phosphorylethanolamine) can take place. The different points of observed heterogeneity in LPS of *S. typhimurium* are depicted in Fig. 7 (Palva and Mäkelä, 1980).

In general, the different species present in a lipopolysaccharide preparation are not separable due to formation of stable co-micelles. Only in a few cases could two different LPS specimens be obtained after preparative ultracentrifugation for separation: two different LPS fractions were extracted from *P. mirabilis* S forms (Gmeiner, 1975; Kotelko, 1985). The pellet and the supernatant fraction contained 56 and 27% lipid A, respectively.

The observation that a whole palette of LPS species (obtained from one extraction) can be demonstrated by polyacrylamide gel electrophoresis in the presence of sodium dodecylsulphate (SDS–PAGE) was an important step in recognising the heterogeneity of LPS preparations (Jann *et al.*, 1975; Hurlbert and Hurlbert, 1977; Palva and Mäkelä, 1980; Goldman and Leive, 1980). A series of bands (in many cases doublets of bands) is usually obtained (Fig. 8). It is suggested that they correspond to LPS molecules with different numbers of O-repeating units. The fastest moving band is LPS with a complete but unsustituted core; the following ones contain the core with one, two, three, etc., repeating units attached. This interpretation was suggested by results obtained with genetically defined S, R, and especially SR forms. The latter mutant type (Naide *et al.*, 1965) has LPS with only one repeating unit attached due to a defect of the O-chain polymerase. The question of why in many cases doublets of bands are observed is not yet answered. It may be connected with additional heterogeneity in the core or lipid A region, such as the amount of phosphate-linked substituents or phosphate (Vaara *et al.*, 1981). A partial transition from S to R forms has been observed with *Yersinia enterocolitica* when grown at 40°C and not at 20°C or lower (Acker *et al.*, 1980; Wartenberg *et al.*, 1983). An opposite temperature dependence was ob-

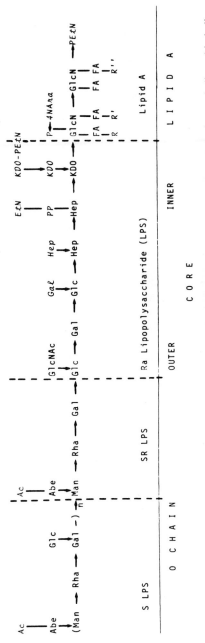

Fig. 7. Heterogeneity in LPS of *Salmonella typhimurium*. The points of observed or suspected heterogeneity are indicated in italics. Abbreviations: Ac-abe, 2-O-acetyl abequose; Man, D-mannose; Rha, L-rhamnose; Gal, D-galactose; Glc, D-glucose; GlcNAc, N-acetyl-D-glucosamine; Hep, L-glycero-D-mannoheptose; EtN, ethanolamine; PEtN, phosphorylethanolamine; PPEtN, pyrophosphorylethanolamine; KDO, 2-keto-3-deoxy-D-manno-octonate; 4NAra, 4-amino-L-arabinose; FAR, FAR′, FAR″, 3-hydroxy fatty acid with substituted 3-OH group; modified according to Mäkelä and Stocker (1984).

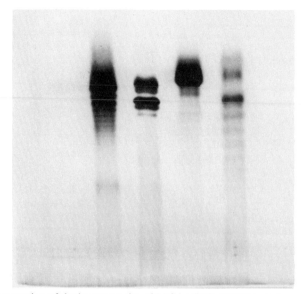

Fig. 8. Demonstration of the heterogeneity of LPS by SDS–PAGE technique, detection by periodic acid–silver staining (Tsai and Frasch, 1982).Left to right: lane 1, *Salmonella montevideo;* lane 2, *Aeromonas shigelloides;* lane 3, *Proteus rettgeri;* lane 4, *Shigella sonnei.*

served with *Salmonella anatum* (McConnell and Wright, 1979); here, the S- to R-transition was observed when grown at 20°C or below.

C. Value of lipopolysaccharide analyses for taxonomic questions

Knowledge of the detailed structure of lipopolysaccharides can give valuable information on the taxonomic position of a strain or its phylogenetic relatedness. It should be emphasised that even the detection of a few distinct LPS constituents, e.g., core-specific sugars, amide-linked fatty acids, or the nature of an unusual backbone sugar, may provide useful information. With lipopolysaccharides we are dealing with a "product of enormous complexity," as Nikaido (1970) pointed out, "and consequently we can determine the relationship between species by looking at the structure of the polysaccharide without going into the trouble of determining the amino acid sequences of enzymes." LPS is the "product of the collaboration of several dozens of enzymes," whereas some other taxonomic markers are determined by single genes or at the most by a small number.

This review is restricted to methods of chemical constituent and structure analysis contributing to taxonomic questions. For the large number of biological methods describing the endotoxic or other biological activities such as lethal toxicity, pyrogenicity, B-cell mitogenicity, complement inactivation, etc., of lipopolysaccharides, the reader is referred to "Microbial Toxins," Vol. 5 (Kadis *et al.*, 1971) or to "Handbook of Endotoxin," Vol. 2 (Hinshaw, 1984).

II. Extraction and Purification Methods

A. Isolation procedures

Only a few selected extraction methods are described here; for more detailed information and other methods the reader is referred to Nowotny (1971) and Wilkinson (1977).

1. Extraction with trichloroacetic acid

Extraction of cells with trichloroacetic acid (TCA) represents historically the first (Boivin and Mesrobeanu, 1933) and one of the most universal methods for obtaining endotoxin (Westphal *et al.*, 1977). The extracted material, commonly referred to as Boivin antigen, is a highly immunogenic LPS–protein–phospholipid complex.

Cells, suspended in five times their weight of ice-cold water, are treated with an equal volume of 0.5 N trichloroacetic acid at 4°C for 3 h. The suspension (at room temperature) is centrifuged, the supernatant neutralised with dilute sodium hydroxide, and the complex precipitated with 2 volumes of ethanol. The precipitate is taken up in water, dialysed, centrifuged to remove cell debris, and lyophilised.

On treatment with hot liquid phenol, protein-free LPS preparations are obtained, which contain full O antigenicity, but are poor immunogens (Staub, 1965).

2. Extraction with ethylene glycol

Diethylene glycol selectively extracts free polysaccharides and/or the "whole O-antigen complex" from cells (Morgan, 1965). The extraction conditions are very mild and the risk of degradation or modification of the extracted material is low. Acetone-dried cells are extracted with 10 parts of freshly distilled diethylene glycol by shaking for several hours each day (for 3–4 days) at room temperature. The extract is filtered, dialysed, and concentrated to a small volume. Four extractions are usually made. The

O-antigen complex is finally precipitated by the addition of acetone at −10°C. Further purification of the product is obtained by fractionation of a 1% aqueous solution with ammonium sulphate. The material precipitating between 35–40% ammonium sulphate saturation is recovered, dialysed, and lyophilised. Associated phospholipids are removed by extraction with ethyl ether–ethanol (1 : 1) containing 0.5% of 10 N HCl at −5°C. The procedure is lengthy and the yield is low.

3. Extraction with hot phenol–water

Although the two extraction methods described above are useful, the isolation procedure worked out by Westphal et al. (1952; Westphal and Jann, 1965) is the most widely used today. The material extracted is lipopolysaccharide which is only slightly contaminated with protein (Galanos et al., 1969). It still contains, however, glycans and the enterobacterial common antigen (ECA) (Männel and Mayer, 1978). The method works well for extraction of S- and R-type lipopolysaccharides, although it is often not complete for R mutants.

Dried cells suspended in hot water are treated (1 : 1) with hot (65°C) aqueous phenol (single phase) for 20 min with repeated shaking or in a rotatory evaporator, followed by cooling and centrifugation. The upper water layers of two extractions usually contain all the lipopolysaccharide (and RNA and glycans as well). It is siphoned off and exhaustively dialysed and then lyophilised. With some non-Enterobacteriaceae strains part or even the total LPS is extracted into the phenol layer (Weckesser et al., 1977). The lyophilised product of the aqueous phase is further purified by dispersing it in water (3% solution) and subjecting it to high-speed ultracentrifugation (105,000 g, 4 h, three times). The final pellet is taken up in a small amount of water (turbid solution), centrifuged (2000 g, 10 min) to remove cell debris, and then lyophilised. The resulting LPS may still contain small amounts of RNA. In the latter case it is advisable to incubate it with ribonuclease and to remove the split products of RNA by dialysis and to recover LPS by lyophilisation.

Further purification of LPS from the contaminating RNA and acidic K polysaccharides can be obtained by Cetavlon (cetyltrimethylammonium bromide) fractionation of an aqueous solution of LPS, followed by centrifugation and precipitation of the purified LPS with ethanol (Westphal and Jann, 1965). Cetavlon should be used only in small amounts since it tends to form micelles and can then react with LPS by forming co-micelles (Kennedy et al., 1983).

It has been claimed recently (Okuda et al., 1975; Tsang et al., 1974) that the hot phenol–water extraction may give rise to non-specific degradation of LPS and may lead to a loss of O-chain material (up to 5%) and of ester-linked fatty acids (up to 11% of the total amount). It has been assumed that such degradative losses may be one of the reasons for the observed heterogeneity of various LPS (and lipid A) preparations.

4. Extraction with dimethyl sulphoxide (DMSO)

At higher temperatures (>60°C) DMSO extracts quantitatively the LPS together with a large amount of protein (Adams, 1967). From this extract, LPS, free of protein, can be obtained by O-acetylation and subsequent extraction with chloroform. Finally, the product is deacetylated with sodium methoxide (2% in acetone at room temperature). It is probable that part of the ester-linked fatty acids are removed from lipid A during the deacylation step.

5. Extraction with phenol–chloroform–petroleum ether (PCP extraction)

As mentioned above, the phenol–water method often does not extract quantitatively the more hydrophobic lipopolysaccharides present in R mutants. An extraction method has been devised (Galanos et al., 1969) yielding R-type LPS of high purity. Dried bacteria are homogenised (for 2 min with cooling) with an extraction mixture consisting of liquid phenol (90 g solid phenol plus 11 ml water)–chloroform–light petroleum ether (2 : 5 : 8; v/v/v). The homogenate is then centrifuged (5000 g, 15 min) and the supernatant containing the LPS is filtered. The extraction is repeated twice. Chloroform and petroleum ether are removed by evaporation from the combined supernatants. Water is then added dropwise to the remaining phenol to precipitate LPS. The precipitate is centrifuged off, washed with petroleum ether followed by acetone to remove residual phenol, and lyophilised. Usually a further purification by ultracentrifugation (105,000 g, 4 h) is made.

   This method, commonly referred to as the PCP method, is superior to other methods because of the high purity of the LPS obtained and the amount of LPS recovered. In contrast to lipopolysaccharide extracted by phenol–water the PCP-extracted LPS is free of the enterobacterial common antigen (ECA). Extraction of whole cells with hot phenol–water followed by PCP extraction of the dried extract obtained from the aqueous phase is often used to obtain clean LPS also from S-type bacteria (Galanos et al., 1979).

6. *Other methods of LPS extraction*

A procedure has been described recently (Darveau and Hancock, 1983), whereby both S- and R-type LPS are quantitatively obtained in a high degree of purity.

The method includes digestion of the whole cells in Tris–HCl buffer (10 mM, pH 8) with DNase and RNase followed by the addition of SDS (2%) and salts: EDTA (0.1 M) and Tris–HCl (10 mM) and adjusting the pH to 9.5. The suspension is centrifuged (50,000 g, 30 min) to remove the peptidoglycan and the supernatant incubated twice at 37°C with pronase and centrifuged at 0°C. LPS is precipitated from the supernatant by the addition of 2 volumes of ethanol.

Although the method appears to be more laborious and requires incubation with enzymes, it is reported to provide material from both S and R strains which is free from contaminants.

Selective extraction of LPS can also be achieved by exposure of bacterial cells to EDTA (ethylenediamine tetraacetic acid, disodium salt, 5 mM solution) at 37°C for about 2 min (Leive, 1965). The reaction is terminated by adding $MgCl_2$ (20 mM solution) followed by a brief centrifugation (10,000 g, 10 min). The supernatant is extensively dialysed and then lyophilised.

The method extracts about 50% of the total LPS. However, it appears to be very specific since the extraction of several other cell wall components is apparently negligible.

B. Electrodialysis

Isolated lipopolysaccharides from both S and R forms are usually associated with a number of mono- and divalent cations, such as Na^+, K^+, Mg^{2+}, Ca^{2+}, and some polyamines, e.g., putrescine, spermidine, spermine, and cadaverine. These cations have strong influence on the solubility and aggregation of LPS. Electrodialysis of LPS was introduced by Galanos and Lüderitz (1975). Thereby, acidic LPS is obtained which can be converted to defined salt forms by neutralisation with respective bases. Triethylamine salt form of LPS is highly water soluble [solutions up to 200 mg ml^{-1} water could be made with *Salmonella* LPS (Galanos and Lüderitz, 1975)]. This salt form is not sedimentable in the ultracentrifuge and shows modified biological properties (increased lethal toxicity in rat, but decreased lethal toxicity for mice) (Lüderitz *et al.*, 1978). Better solubility is due to replacement of cross-linking polyvalent cations by the monovalent triethylamine. Unneutralised electrodialysed LPS preparations (acidic LPS) are not stable for long periods of time, deteriorating on storing because of autolysis.

III. Analytical methods

A. Identification and quantitative determination of constituents

1. Sugars

For structural analysis of complex carbohydrates, the knowledge of the chemical nature of the constitutent monosaccharides is a necessary prerequisite (Dutton, 1973) and their quantitative liberation and characterisation is generally the first analytical step.

Because of the presence of different types of glycosidic linkages in most heteropolysaccharides and of constituents having different stability towards acid, a single hydrolysis procedure will often not be sufficient to cleave every linkage and give each component in quantitative yield.

Commonly employed conditions for the liberation of neutral sugars from lipopolysaccharides are N HCl or 2 N H$_2$SO$_4$, 100°C, 1–4 h; or 0.1 N HCl, 100°C, 48 h (in a sealed tube and with occasional shaking) (Schmidt et al., 1970). Dideoxyhexoses, found frequently in enterobacterial lipopolysaccharides (Jann and Westphal, 1975) are particularly acid labile and require milder conditions of hydrolysis (Westphal and Lüderitz, 1960). After hydrochloric acid hydrolysis, the excess of HCl can easily be removed by evaporation over NaOH pellets in a vacuum desiccator. Neutralisation of H$_2$SO$_4$ hydrolysates can be done either with a solution of Ba(OH)$_2$ or with solid BaCO$_3$.

Hydrolysis with trifluoroacetic acid (2 M, 100°C, 2–3 h) has occasionally been used to cleave the glycosidic linkages of all cell wall components and of various glycoconjugates (Albersheim et al., 1967). This acid, like HCl, is volatile and can be removed without a neutralisation step. Glycuronic acids, reported in many instances as constituents of LPS (Kotelko, 1985), need strong conditions for their complete liberation. It is known that HCl rather than sulphuric acid causes considerable degradation of uronic acids (decarboxylation of uronic acids to pentoses). It is therefore advisable to use N H$_2$SO$_4$ (100°C, 4 h) for hydrolysis.

The liberated sugars may be separated and identified by chromatographic methods: paper chromatography (PC), thin-layer chromatography (TLC), or—after derivatisation—gas–liquid chromatography (GLC). Other separation techniques are high-pressure liquid chromatography (HPLC), which is especially useful for separating oligosaccharides, and paper electrophoretic techniques (Keleti and Lederer, 1974). A number of solvent systems are suggested for the efficient separation of most of the components on PC, TLC, and HPLC (see Hais and Macek, 1958, 1960; Lederer and Lederer, 1957; Randerath, 1962; Stahl, 1962).

Specific staining methods are available (Hais and Macek, 1958) to detect or identify the components separated on PC and TLC and after paper electrophoresis. Silver nitrate followed by NaOH, according to Trevelyan *et al.* (1950), is a very sensitive reagent, staining almost all sugars and even sugar alcohols; anilinium hydrogen phthalate stains reducing sugars often with a characteristic colour (Partridge, 1949); ninhydrin spray or dip techniques are available for amino sugars and for amino acids as well (Hais and Macek, 1960; Oden and Hofsten, 1954). Naphthoresorcinol has been recommended for uronic acids (Partridge, 1948). Periodate oxidation followed by thiobarbituric acid spray stains sugars forming malondialdehyde (or substituted malondialdehyde) upon periodate oxidation (2- and 3-deoxy sugars, 2-keto-3-deoxyoctonate (KDO), and *N*-acetylneuraminic acid (NANA, etc.) (Warren, 1959).

For GLC one has to convert the sugars into volatile derivatives such as acetates, methyl ethers, and trimethylsilyl ethers (TMS) before separation and analysis (Dutton, 1973). To avoid multiple peaks in GLC due to the formation of pairs of anomeric α- and β-pyranosyl and -furanosyl derivative, it is advantageous in most cases to convert the reducing sugar units into alditols or aldononitriles. Alditols and aldononitriles can be per-*O*-acetylated and appear as single peaks per individual sugar residue (Sawardeker *et al.*, 1965). In some cases, reduction to alditols is not advisable, e.g., when alditols are LPS constituents (Gmeiner *et al.*, 1977) or when two sugars, e.g., heptoses, are present in LPS which form the same alditol acetate derivative. Here it is preferable to make the aldononitrile derivatives. Reduction is performed at room temperature with sodium borohydride (or sodium borodeuteride in ^2H_2O, to have a marker for mass spectrometric identification) for about 12 h with a 4–10 times excess of the hydride. After reduction the excess of borohydride is destroyed by adding 1% acetic acid or a few microlitres of acetone (or by treatment with Dowex 50 × 8, H^+ form) and evaporation to dryness. The dry residue is repeatedly co-evaporated with methanol (5 × 1 ml) to remove borate as its volatile trimethylborate. The resulting dry mass is O-acetylated with acetic anhydride–pyridine (1:1, v/v; 100°C, for 30 min), followed by removal of excess reagent by co-evaporation with toluene. The acetylated derivatives (alditol acetates) are taken up in chloroform and subjected to GLC (Wong *et al.*, 1980; Lönngren and Svensson, 1974). A variety of stationary phases is used for GLC analysis. Commonly used phases are 3% ECNSS-M on Gas Chrom Q (100–120 mesh) at 175°C, 3% OV-225 on Gas Chrom Q (100–120 mesh) at 200°C, or capillary columns coated with either CP-Sil 5 or SE 52.

Per-*O*-acetyl aldononitriles are prepared (Lance and Jones, 1967) by adding pyridine containing hydroxylamine hydrochloride to the dried hydrolysate. The solution is heated in a sealed tube at 90°C for 30 min. Acetic anhydride is then added and the mixture again heated (90°C, 30 min). The resulting aldononitrile derivatives are directly injected into the gas chromatograph fitted with either glass columns packed with OV-17 (2%) on Chromosorb W HP (80–100 mesh) or fused silica capillary column of SE 54.

Since pentoses are rare constituents in LPS, xylose is frequently used as internal standard for the quantitation of LPS sugar constituents by GLC; in other cases inositol can be used.

Recently, a one-step procedure for hydrolysis (or more accurately methanolysis) and derivatisation of the majority of LPS constituents (sugars and fatty acids) has been described (Bryn and Jantzen, 1982). For this analysis, the LPS is treated with 2 M dry HCl in methanol at 85°C for 18 h in Teflon-lined screw-capped vials. After concentrating the methanolysate, trifluoroacetyl derivatives are prepared with trifluoroacetic anhydride (50% in acetonitrile) by heating to boiling for about 2 min. After 10 min at room temperature the samples are diluted with acetonitrile and subjected to GLC analysis on a fused silica capillary column coated with SE-30 stationary phase. Although this method is simple and provides semiquantitative data, it suffers from some disadvantages. Phosphorylated LPS constituents cannot be detected because of the mild hydrolytic conditions applied; several peaks are observed for each monosaccharide (α,β-pyranoses and -furanoses), thus making the chromatograms complex and, finally, during the evaporation step considerable loss of the more volatile fatty acid methyl esters is unavoidable. In spite of this, the method is most valuable for routine analyses of a large number of LPS samples and is also used for monitoring during purification of LPS.

Amino sugars and aminouronic acids can be detected and identified preferentially by high-voltage paper electrophoresis or on the amino acid analyser. High-voltage paper electrophoresis at different pH-values (Kickhöfen and Warth, 1968; Mayer and Westphal, 1968) can also be used for preparative isolation of charged sugars (amino sugars and uronic acids) or oligosaccharides. The classical colorimetric methods such as the Morgan–Elson (Reissig et al., 1955) and Elson–Morgan (Rondle and Morgan, 1955) reactions for amino sugar quantitation are used in special cases (Wheat, 1966) but are largely replaced by qualitative and quantitative analyses on the amino acid analyser. A GLC method (using capillary WCOT column coated with POLY A 103) has recently been worked out (Kontrohr and Kocsis, 1983) to separate and identify the various amino sugars (2-amino-2-deoxyhexoses and 2-amino-2-deoxyhexuronic acids). The polysaccharide containing these amino sugars is first depolymerised by methanolysis followed by reduction of an aliquot with sodium borohydride, thus converting the aminouronic acid methylglycoside methyl ester into amino sugar methylglycoside. This allows the differentiation between amino sugars and aminouronic acids in the same hydrolysate.

Since 2-keto-2-deoxy-D-manno-octonate (KDO) is attached to lipid A by an acid labile ketosidic linkage, mild hydrolytic conditions can be applied for its liberation (Unger, 1981). Hydrolysis with 1% acetic acid (100°C, 1–3 h) is usually sufficient for its release but stronger hydrolysis conditions are in some cases necessary, e.g., with *Chromatium vinosum* and *Thiocapsa roseopersicina* LPS (Hurlbert *et al.*, 1976, 1978).

A recent systematic kinetic study for the optimal release of KDO has provided interesting information (Brade *et al.*, 1983). Using this method, differently substituted KDO residues in *Salmonella* LPS have been separately determined and quantitated. During the first mild hydrolysis (0.1 M acetate buffer, pH 4.4; 100°C, 1 h), the ketosidic linkage of a KDO-disaccharide unit is selectively broken (Fig. 2) leaving the more stable glycosidic linkage between heptose and KDO unhydrolysed. The latter linkage is then quantitatively split by 1 M HCl (100°C, 1–4 h). To account for the degradation of KDO during hydrolysis (formation of KDO lactones and anhydro derivatives, Volk *et al.*, 1972) the use of an internal standard of KDO, preferably the methyl ketoside of KDO, is recommended during hydrolysis.

KDO is quantitated by the classical periodate–thiobarbituric acid assay in one of the various, more recent modifications (Karkhanis *et al.*, 1978; Skoza and Mohos, 1976). To a suitably hydrolysed sample, 40 mM H_5IO_6 in 60 mM H_2SO_4 is added. The oxidation is performed at room temperature for 30 min and is then stopped by adding 0.2 M NaAsO$_2$ in 0.5 M HCl. After reduction of the excess periodate (disappearance of the brown colour), freshly prepared 0.6% aqueous thiobarbituric acid solution is added and the mixture incubated in a boiling water bath for 15 min. Hot DMSO is added to each tube and well mixed; after cooling, the absorbance of the pink colour is measured at 549 nm against a reagent blank.

Besides the common LPS sugar constituents encountered frequently, many reports describe the isolation and identification of rather rare sugars of various sugar classes (Table I). In the Enterobacteriaceae family lipophilic sugars are mostly deoxy- or dideoxyhexoses ("deoxy-principle versus methoxy-principle"; Westphal and Lüderitz, 1960). O-Methyl ethers of sugars occur naturally in O antigens of photosynthetic prokaryotes (Rhodospirillaceae and Chromatiaceae, as well as in cyanobacteria) (Drews *et al.*, 1978). Their occurrence in Enterobacteriaceae is restricted to a few strains (Björndal *et al.*, 1971; Lindberg *et al.*, 1972). Methoxy sugars either can be part of the repeating unit or can occur in trace amounts only and are then localised on non-reducing (and in rare cases also reducing) terminals (Tharanathan *et al.*, 1978b). The application of sensitive techniques, such as gas–liquid chromatography combined with mass spectrometry (GLC–MS) and single-ion detection (mass fragmentography), has led to their unequivocal detection and identification.

2. Fatty acids

The analytical methods for the qualitative and quantitative analyses of the fatty acids in lipid A have been summarised recently by Wollenweber and Rietschel (1984). Of all the methods described, GLC of fatty acid methylesters (often in combination with mass spectrometry) is the most widely used method.

(a) Total fatty acids.   For the liberation of ester- and amide-linked fatty acids, hydrolysis with acid (HCl or H_2SO_4) is preferred over alkaline hydrolysis (NaOH or N_2H_4). Under alkaline conditions an alkali-catalysed β-elimination of fatty acids from 3-acyloxyacyl residues may generate considerable amounts of α, β-unsaturated fatty acids, and, furthermore, the liberation of amide-linked fatty acids may not be quantitative under alkaline conditions. A one-step reaction involving hydrolysis of O- and N-linked fatty acids followed by their derivatisation into volatile methyl esters by acid-catalysed transesterification appears simple and accurate. LPS is treated with 2 N methanolic HCl and heated together with a standard of heptadecanoic acid (17:0) at 85°C for 18 h. The hydrolysate is concentrated to about half of the volume (in a N_2 stream) and an equal volume of half-saturated NaCl is added. The volatility of the methyl esters of fatty acids, e.g., C10:0 and C12:0 has to be considered. The resulting methyl esters are extracted thrice with either distilled hexane or chloroform. The combined organic phase is concentrated and subjected to GLC–MS. For GLC of fatty acid methylesters the following columns are recommended: SE-30 (3 or 10%) on Gas Chrom Q (100–120 mesh) at 170°C (isothermal or temperature programmed); EGSS-X (15%) on Gas Chrom P (100–120 mesh) at 140°C; Castrowax (2.5%) on Chromosorb W (80–100 mesh) at 175°C; or capillary columns coated with CP-Sil 5 or SE 54.

(b) Ester-linked fatty acids.   A variety of alkaline reagents are employed for the selective release and transesterification of ester-bound fatty acids, e.g., mild hydrazinolysis (Jay, 1978), controlled hydroxylaminolysis (Rietschel *et al.*, 1972; Snyder and Stephens, 1959), treatment with

methanolic NaOH (Koeltzow and Conrad, 1971), and treatment with NaOCH$_3$ (Rietschel *et al.*, 1972). The reaction with sodium methylate is preferred amongst these methods. One has to be aware, however, that unsaturated fatty acids can be artificially formed during the reaction (Wilkinson, 1974; Rietschel, 1983a). In this procedure dried material (LPS or lipid A) in Teflon-lined screw cap vials is treated with NaOCH$_3$ in absolute methanol (0.25 N) and the mixture allowed to react at 37°C for 10 h. Continuous stirring is needed. Heptadecanoic acid is often used as internal standard. After centrifugation (10 min, 2000 g) the supernatant is acidified with diluted HCl (pH 5–6) and the fatty acid methyl esters are extracted with chloroform (three times). The combined extracts are dried (Na$_2$SO$_4$), evaporated to dryness, and analysed by GLC–MS. The fatty acids released by β-elimination are further characterised by methylation with diazomethane.

An ethereal solution of diazomethane is prepared by the reaction (0°C) of diethyl ether (10 ml), 40% KOH (2 ml), and nitrosomethyl urea (1 g) (*N. B.*, a carcinogen). The yellow ether layer is decanted and dried with Na$_2$SO$_4$. To a solution of the methanolysate (supernatant from the above experiment) in dry ether (1 ml), methanol (0.1 ml) is added as a catalyst (Schlenk and Gellermann, 1960) and then the ethereal diazomethane solution (0.1 ml) is successively added until the yellow colour persists for a few minutes. After 10 min, the solvent and the excess of diazomethane are removed in a N$_2$ stream (under a hood). The resulting methyl esters are dissolved in chloroform and subjected to GLC–MS analysis.

(c) Amide-linked fatty acids.   The O-deacylated LPS or LPS-OH, e.g., the sediment of the sodium methylate treatment (see Section 2,b), contains the amide-linked fatty acids. The sediment is purified by dissolving it in water (0.5 ml) with the aid of ultrasonication, the pH is adjusted to 2–4 with dilute acetic acid, and the LPS-OH precipitated at 4°C by the addition of cold ethanol (about 8 ml). Light petroleum ether may be added (0.5 ml) to induce precipitation. After centrifugation, the precipitate is washed with cold acetone (3 × 5 ml), taken up in water (2 ml) and lyophilised (yield is about 50%). To liberate the amide-linked fatty acids, the dry sample is hydrolysed with methanolic HCl (2 N, see Section 2,a) and analysed by GLC–MS.

(d) Amide-linked 3-acyloxyacyl residues.   The hydroxyl functions of β-hydroxy fatty acids (ester- and amide-linked) are often esterified with residues of fatty acids (Wollenweber *et al.*, 1984). An elegant method has recently been worked out to determine such amide-linked acylated hydroxy fatty acids ("double esters") (Wollenweber *et al.*, 1982; Kraska *et al.*, 1976). The LPS is methylated with methyl iodide in the presence of

silver salts followed by a mild hydrolysis of the resulting methyl-acylimidate linkage to yield the acyloxyacyl residues in the form of the free methyl ester. Briefly, to a mixture of free lipid A, freshly prepared silver oxide and silver trifluoromethanesulphonate in Teflon-lined screw cap vials, absolute petroleum ether (40–60°C) is added. The vial is N_2 flushed, closed, and sonicated (20 min in the dark). Freshly distilled methyl iodide is then added and sonication is repeated (2 h, 40 min, in the dark). The contents are transferred to a centrifuge tube containing an aqueous solution of sodium thiosulphate (0.5 M), standard heptadecanoic acid methyl ester and 3-O-(tetradecanoyl)tetradecanoic acid ethyl ester [3-O-(14:0)-14:0]. The mixture is acidified with 10% phosphoric acid (to pH 2), sonicated (15 min), and centrifuged (10 min, 2000 g), followed by extraction with petroleum–ether ethyl acetate (1:1). The combined extracts are washed with water (2 × 5 ml), dried (Na_2SO_4), and purified by passage through a small column (20 × 3 mm) of silica gel 60 (35–70 mesh) eluting with petroleum–ether ethyl acetate (1:1, 10 ml). The eluate is concentrated and analysed by GLC–MS. An OV-1 column (3% on Gas Chrom Q, 100–120 mesh) operated at 220°C is used.

(e) Stereochemical analysis of D- *and* L-*hydroxy fatty acids.* Routine optical rotation measurements (see Galanos *et al.*, 1977a; Rietschel *et al.*, 1977), although being simple, require relatively large amounts of material. A GLC method has therefore been worked out, in which the diastereomeric derivative of both 3- and 2-hydroxy fatty acids are separated as L-phenylethylamides according to their chain length and optical configuration.

The material (LPS, 10 mg; or lipid A, 1 mg) is hydrolysed with alkali (4 N KOH, 5 h, 100°C) and extracted with chloroform (3 × 10 ml). The combined extracts are dried (Na_2SO_4) and concentrated. The dry residue is taken up in dry ether, cooled in a stoppered test tube (10-ml size) to −15°C, and boron trifluoride–ethyletherate is added. An ether solution of diazomethane is sequentially added (in 1-ml portions) until the yellow colour remains for about 30 s. After 10 min the content is filtered and the filtrate is repeatedly washed with saturated sodium bicarbonate solution and then with water. The ether layer is collected, dried (Na_2SO_4), and evaporated to dryness. The resulting fatty acid methyl esters are saponified by adding KOH (1 N) at 70°C for 2 h. The liberated free methoxy fatty acids are extracted with chloroform and purified by chromatography on silicic acid TLC using petroleum ether–diethyl ether–acetic acid (35:15:1, v/v/v) (staining with iodine vapour). The area corresponding to methoxy fatty acids (R_f about 0.5) is extracted with chloroform into Teflon-capped vials and dried (N_2 stream). The extract is dissolved again in chloroform and reacted with thionyl chloride at 60°C for 30 min. Excess solvent and reagent are removed (N_2 stream) and the methoxy fatty acid chlorides, dissolved in dry chloroform, are treated with L-phenylethylamine. After a reaction time of 30 min at room temperature, the chloroform phase is washed successively with dilute HCl (0.01 N) and water.

After drying (Na_2SO_4) and concentration the derived methoxy fatty acid L-phenylethylamides are dissolved in chloroform and subjected to GLC–MS analysis. OV-1 (3%) on Gas Chrom Q, 100–120 mesh at 205°C is recommended.

(*f*) *Unsaturated and keto fatty acids.* Unsaturated and keto fatty acids have occasionally been found in lipid A of some non-enteric bacteria (Strittmatter *et al.*, 1983). They can be identified directly as methyl esters by GLC–MS analysis or after reduction (to saturated and β-hydroxy fatty acids) by hydrogen gas in the presence of Raney nickel as catalyst.

Recently, an elegant method for determining the position of double bonds in unsaturated fatty acids using 3-picolinyl esters has been described (Harvey, 1982). These derivatives are easy to prepare and afford characteristic fragments during mass spectrometric analysis which allow the localisation of double bonds and methyl branches in fatty acids (Holst *et al.*, 1983a). Briefly, the fatty acid is heated with thionyl chloride at 100°C for 10 min. Excess reagent is evaporated in a N_2 stream and a solution of 20% pyridylcarbinol in acetonitrile is added. After additional heating (100°C, 1 min) the samples are ready for GLC–MS analysis on an SE-30 column (3% on Gas Chrom Q, 100–120 mesh) at 220°C.

3. Phosphate determination

Phosphate (phosphorus) is determined by the method of Lowry *et al.* (1954). The method quantitates the total inorganic phosphorus. Phosphorus-containing organic compounds are first acid hydrolysed (with a mixture containing 30 ml 90% H_2SO_4, 30 ml 60% $HClO_4$, and 40 ml H_2O, at 100°C for 1 h and then at 165°C for 2 h). The liberated inorganic phosphorus is allowed to react with ammonium molybdate reagent (1 M sodium acetate, 1 ml 2.5% ammonium molybdate, and 7 ml H_2O, mixed with a 10% solution of ascorbic acid in 9 : 1 proportion) at 37°C for 90 min. The blue-coloured complex is read at 820 nm against a reagent blank. Potassium phosphate (0.01 M) solution is used as standard.

4. Amino acids

Amino acids are rather unusual constituents of lipopolysaccharides. The presence of lysine and L-alanine has been reported from *P. mirabilis* O antigens (Kotelko, 1985), where lysine is attached via its α-amino group to the carboxyl group of galacturonic acid (Gromska and Mayer, 1976). Threonine has been reported from LPS of *R. sphaeroides* (Strittmatter *et al.*, 1983), where it forms an integral part of the O antigen (Salimath *et al.*, 1984). Amino acids are liberated by acid hydrolysis (4–6 N HCl, 100°C, 12–16 h) and separated and quantified on an amino acid analyser. For qualitative detection, PC (two-dimensional technique) and especially high-voltage paper electrophoresis are of value.

5. Other constituents

O- and N-Acetyl groups are frequently observed in lipopolysaccharides. They can be determined by spectrophotometric methods (Snyder and Stephens, 1959; Ludowieg and Dorfman, 1960), but these methods are time-consuming and rather complicated. An improved GLC method is available for determining total (amide- and ester-linked) acetyl groups in LPS (Fromme and Beilharz, 1978). After hydrolysis overnight with 0.2 N HCl at 100°C, followed by pH adjustment of the hydrolysate to pH 3–4, 2 μl of the hydrolysate is directly injected into a gas–liquid chromatograph fitted with a Porapack QS glass column at 200°C (isothermal). O-Acetyl groups can be determined selectively after hydrolysis with 0.05 N NaOH (room temperature, 3 h), followed by pH adjustment (pH 3–4). N-Acetyl is calculated as the difference between total and O-acetyl content.

Short-chain fatty acids up to *n*-valeric acid can be separated and quantitated similarly. Propionic acid (20–100 μg) is used as internal standard for quantitation.

A rapid GLC–MS method employing (trideutero-) acetylation of the alditols has been described for identification of the N-acetyl groups and for the degree of N-deacetylation (Banoub and Michon, 1982). Graded hydrolysis with either 2 M trifluoroacetic acid (100°C, 4 h) or with 90% acetic acid followed by 0.25 M H_2SO_4 does not cause N-deacetylation. After reduction with sodium borohydride the sample is trideutero-acetylated prior to GLC–MS analysis.

B. Separation of O chains, core, and lipid A

1. Characterisation of R mutants

Populations of enterobacterial cells often contain spontaneously appearing rough forms (Kauffmann, 1966). Colonies of rough strains are characterised by a different colony morphology. They appear dull and rough-edged and thus are easily distinguishable from the light smooth colonies. Lüderitz *et al.* (1966) showed that in most cases rough forms have defects in the biosynthesis of their respective O antigens resulting in a lack of O antigenicity. They acquire however, a new serological specificity, the so-called R specificity (Kauffmann, 1966).

Since serologically R mutants may sometimes look smooth and morphologically defined rough forms may still contain at least part of the O antigen, a special terminology has been adopted (Schmidt and Lüderitz, 1969): the designations smooth and rough are used for characterising the colony morphology, the designations S and R forms are for strains either containing the full O antigen or completely lacking it.

It has been shown by Kauffmann (1966), Kröger (1955), and Schmidt *et al.* (1969b) that R mutants can be characterised by serological techniques, e.g., by their lacking O immunogenicity (in the rabbit), by agglutination with a 0.3% auramine solution, and by slide agglutination with saline (0.9–3.5%). These tests are valuable for discriminating between enterobacterial S and R forms, but do not discriminate between the different R chemotypes (Ra–Re, Fig. 3). For their identification, as well as for that of the different complete enterobacterial R core types (Fig. 4), phage typing is the method of choice (Schmidt and Lüderitz, 1969; Rapin and Kalckar, 1971; Mayer and Schmidt, 1973; Lindberg, 1973, 1977). Amongst the many R phages used, phages F0, ϕX174, T4, C21, 6SR, FP1, and U3 are of particular value.

2. Cleavage of polysaccharide and lipid A by mild acid hydrolysis

Lipid A can easily be split off from the polysaccharide part of LPS by mild acetic acid hydrolysis (1%, 100°C, 1–3 h) (Lüderitz *et al.*, 1971). The liberated water-insoluble lipid A precipitates and is centrifuged off and repeatedly washed with warm water (3 × 5 ml) and acetone. After lyophilisation it is obtained in a yield of 10–25% depending on the S or R type of the respective lipopolysaccharide. The crude lipid A can be purified further by extraction with chloroform–methanol (9:1, by volume) and filtration through an ultrafine fritted glass filter. After solvent evaporation and precipitation with acetone (2 volumes), partially purified lipid A is obtained. The resulting preparation of lipid A does not appear to be homogeneous, as revealed by TLC separation in different solvent systems (Nowotny, 1971; Hase and Rietschel, 1977; Tharanathan *et al.*, 1978b; Amano *et al.*, 1983). Further purification into individual subfractions may be achieved by either preparative TLC or silicic acid column chromatography.

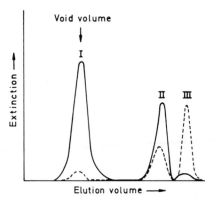

Fig. 9. Schematic representation of a typical fractionation on Sephadex G-50 of "degraded polysaccharide" from an S-form LPS. The eluate was monitored for total carbohydrate (solid line) and total phosphorus (dashed line). I, O chains with core oligosaccharide attached; II, unsubstituted core oligosaccharide; III, mainly KDO and inorganic phosphorus.

With some lipopolysaccharides higher concentrations of acetic acid (5–10%) are required for complete liberation of lipid A (Hurlbert *et al.*, 1976, 1978). It is not known whether such lipopolysaccharides contain a different linkage between KDO and lipid A.

The acetic acid supernatants contain the water-soluble (degraded) polysaccharide and can be obtained by lyophilisation. This material can be separated by subsequent chromatography on Sephadex G-50 columns (70 × 3.7 cm) into polysaccharide (O-chain repeating units) and oligosaccharide (R core) fractions (Müller-Seitz *et al.*, 1968; Schmidt *et al.*, 1969a) (see Fig. 9). The fractions can be later analysed for their composition and structure.

C. Elucidation of carbohydrate structures

1. O Chain and R core analyses

The analytical methods used for elucidation of O chain and R core structures are the same as those worked out for polysaccharides in general, except for variations in handling and preparation of the starting materials.

(a) Permethylation.   Permethylation is an indispensible method for investigation of structure, nature, and position of glycosidic linkages in oligo- and polysaccharides (Björndal *et al.*, 1970; Jansson *et al.*, 1976). Briefly, the polysaccharide is treated in the presence of alkali with a suitable methylating agent, followed by hydrolysis, transformation into volatile derivatives, and their characterisation by GLC–MS. From the position of *O*-methyl groups in the derivatives one can deduce the nature

of the substitution of each individual sugar, and, in most cases, the ring structure.

A variety of methylation methods, both in aqueous and non-aqueous media, have been published (Hirst and Percival, 1965). The method of choice, however, was developed by Hakomori (1964). An aprotic solvent, dimethyl sulphoxide (DMSO), and dimsyl sodium (sodium methyl-sulphinylmethanide, generated by reaction of anhydrous DMSO with sodium hydride at 50–60°C for 2–4 h) as the base, are used. The method is applicable to most oligo- and polysaccharides and glycoconjugates and has provided excellent results in most cases. In the case of polysaccharides insoluble or only partially soluble in DMSO, it is difficult to achieve permethylation in one step. In these cases it is advantageous to peracetylate the material before subjecting it to methylation, thus increasing its solubility in DMSO.

It is desirable to transform oligosaccharides with free reducing groups into non-reducing derivatives prior to methylation. Uronic acid-containing polysaccharides are carboxyl reduced according to Taylor *et al.* (1976) before methylation to avoid possible side reactions such as β elimination. Several modifications of the original Hakomori methylation method have been reported (Prehm, 1980; Narui *et al.*, 1982; Geyer *et al.*, 1983).

Because of possible contamination (phthalic esters) eluted from plastic caps of reagent bottles which show up in GLC and MS, it is important to distill the reagents and solvents used for methylation.

(i) Methylation step (Jansson et al., 1976). A 2 *M* dimsyl sodium solution is prepared in a Teflon-capped serum vial under a N_2 atmosphere. For this, sodium hydride (5 g of a 50% suspension in oil, washed three times with 5 ml dry petroleum ether) is reacted with dry DMSO (dried by vacuum distillation over calcium hydride, keeping the distillate over a molecular sieve, 4 Å) at 50–60°C for 2–4 h. The liberated H_2 is exhausted through an injection needle. The resulting greenish-grey solution is stable for at least 3 months, when kept frozen under anhydrous conditions.

The methylation is performed as follows: the dimsyl reagent (twice the volume of DMSO) is added to a solution of the substrate (vacuum-dried oligo- or polysaccharide, 0.5–2 mg in about 0.5–1 ml dry DMSO) and the mixture is sonicated for 15–30 min. After a second sonication and a reaction time of about 4 h at room temperature the vial is cooled in an ice bath and methyl iodide (1–2 ml) is added dropwise using a syringe. The reaction mixture is left stirring for 2 h in the dark. Excess reagent is flushed off with N_2 and the following purification steps are performed. For polymeric material, the solution is distributed between chloroform (5 ml) and water (10 ml). The organic phase is separated, dried with Na_2SO_4, concentrated, and passed over a Sephadex LH-20 column. For oligomeric material, the solution is purified by an LH-20 column (40 × 2 cm), using a 1:1 (by volume) mixture of ethanol and chloroform as eluent.

(ii) Hydrolysis. Hydrolysis of the permethylated material is performed by a two-step procedure (Jansson *et al.*, 1976): formolysis (2 ml of 90% formic acid at 100°C, 1 h), followed by acid hydrolysis (3 ml of 0.13 M H_2SO_4, 100°C, 16 h). For permethylated amino sugars, initial acetolysis (0.3 ml of 0.5 N H_2SO_4 in glacial acetic acid at 100°C, overnight) followed by hydrolysis (by adding 0.3 ml H_2O and heating at 80°C for 5 h) is recommended (Stellner *et al.*, 1973). The resulting partially methylated sugars are then transformed into alditol acetates which are identified and quantitated by GLC–MS.

(b) Periodate oxidation and Smith degradation. Fragmentation of polysaccharides according to Smith (Takahashi and Murachi, 1976) starts with periodate oxidation. Sodium metaperiodate selectively oxidises vicinal unsubstituted hydroxy and amino groups to carbonyl groups with opening of the C–C chain (Fig. 10). The oxidation is usually performed in aqueous solution with 10 mM sodium metaperiodate solution in the dark for various time intervals (hours and days). From the kinetics of the oxidation (Avigad, 1969) as well as from the analysis of resulting oxidation products one can often recognise the substitution pattern in the glycosyl moiety. By ring opening through the periodate oxidation and subsequent reduction of the carbonyl groups formed, the glycosidically linked sugar is transformed to an acetal-linked polyhydroxycarbonyl compound. These acetal linkages are more acid labile than glycosidic linkages and therefore can be cleaved by very mild conditions (0.1 N H_2SO_4, 25°C, 48 h).

After destroying the excess periodate by the addition of 0.1 M ethylene glycol (about 1 ml), reduction is usually performed by a large excess of sodium borohydride (at least 10 volumes). All reaction products obtained by Smith degradation have a polyol component (glycerol, threitol, or erythritol) to which mono- or oligosaccharides may be linked (Goldstein *et al.*, 1965; Hay *et al.*, 1965; Aspinall, 1977).

(c) Selective acid hydrolysis. Many bacterial polysaccharides (and glycoconjugates in general) contain relatively acid-labile glycosidic linkages in peripheral or central positions, e.g., the linkages of 3,6-dideoxyhexoses, deoxyhexoses, and furanoses. When located terminally such sugars can be removed selectively from the main chain (or the backbone structure) by partial acid hydrolysis. By methylation analysis of native and degraded polysaccharide, information on acid-labile residues can be obtained.

Acetolysis can be useful especially for the selective cleavage of 1,6-linked glycosyl residues. The native or per-O-acetylated material is

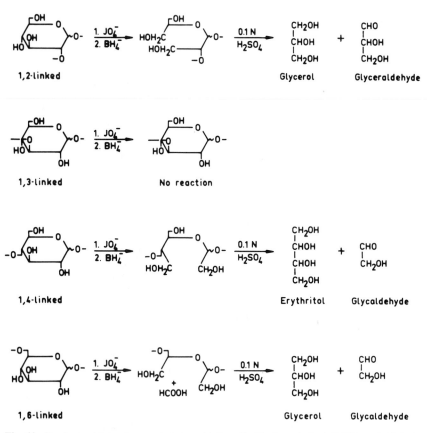

Fig. 10. Products obtained on periodate oxidation–Smith degradation of differently linked, linear glucans.

treated with a mixture of glacial acetic acid/acetic anhydride/conc. sulphuric acid (11 : 10 : 1, v/v/v, 30°C, for different time intervals) (Tai *et al.*, 1975). According to the modifications introduced by Vijay and Perdew (1982), a reaction time of 3 h is sufficient to hydrolyse peripheral 1,6-linkages while a reaction time of 7 h (extended acetolysis) results in a more effective cleavage of interior, 1,6-linkages. No significant degradation of other glycosidic linkages occurs according to these authors.

(d) Solvolysis. In many instances hydrolytic reaction with liquid or anhydrous hydrogen fluoride has been successfully applied for selective depolymerisation of oligo- or polysaccharides (Mort and Lamport, 1977; Knirel *et al.*, 1982a, b), e.g., for the structural elucidation of the 2,3-

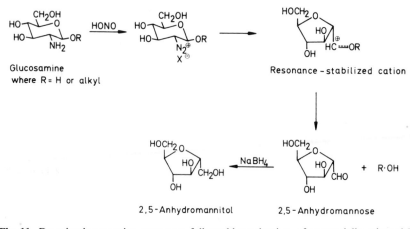

Glucosamine
where R = H or alkyl

Resonance – stabilized cation

2,5-Anhydromannitol   2,5-Anhydromannose

Fig. 11. Deamination reaction sequence, followed by reduction, of equatorially oriented 2-amino-2-deoxyglycosides; according to Williams (1975).

diacetamido-2,3-dideoxyhexuronic acids in the somatic O antigens of *P. aeruginosa* (Dmitriev *et al.*, 1982; Knirel *et al.*, 1982c, 1983). Since the reaction is carried out at 0°C there is little or no degradation of the sugars liberated. The method seems to be specific and leaves peptide bonds and glycopeptide linkages of amino sugars intact.

In addition, hydrogen fluoride (50%) is also used to dephosphorylate LPS core oligosaccharides (Prehm *et al.*, 1975; Feige and Radziejewska-Lebrecht, 1979). This reaction is carried out either at 4°C for 4 days or at −23°C for 40 days followed by lyophilisation to remove the excess hydrogen fluoride. Dephosphorylation of permethylated core polysaccharide with hydrogen fluoride and subsequent permethylation with C^2H_3I allows the positional location of phosphate(s) in core oligosaccharides (Radziejewska-Lebrecht *et al.*, 1981).

(e) Deamination. Nitrous acid deamination of amino sugars in polysaccharides may lead to fragmentation of the polymer depending on the amino sugars present. Deamination proceeds with rearrangement of the amino sugars and formation of modified products (Williams, 1975). Aminodeoxy glycosides with an axially oriented amino group undergo a series of reactions on deamination, the dominating one being conversion of D-mannosamine into D-glucose without concomitant scission of glycosidic linkage. Aminodeoxy glycosides with equatorial orientation of the amino group, like D-glucosamine and D-galactosamine are degraded to 2,5-anhydromannose and 2,5-anhydrotalose, respectively (see Fig. 11). Amino sugars with free amino groups occur only rarely in bacterial lipo-

polysaccharides, an exception being the lipid A of *R. tenue* (Tharanathan *et al.*, 1978a,b). In all other cases N-deacetylation is an essential prerequisite for the deamination reaction. The method of choice is hydrazinolysis with dry hydrazine alone or together with catalytic quantities of anhydrous hydrazine sulphate (Yosizawa *et al.*, 1966) at 105°C for 2–40 h.

An improved method of N-deacetylation has been described by Erbing *et al.* (1976): The material is treated with sodium thiophenolate as an oxygen scavenger/catalyst and 2 *M* dimsyl sodium in DMSO is added. The reaction is pei ormed overnight at 80–100°C in a sealed tube. A precipitate of diphenyldisulphide is formed during the reaction. Finally, the reaction mixture is filtered after dilution with water, dialysed, and lyophilised.

The deamination reaction is normally carried out in aqueous medium with nitrous acid generated *in situ* by the action of sodium nitrite/dilute HCl. The solvent has a marked effect on the deamination. The extent of rearrangement is greater, the more polar the solvent. The reaction is especially useful in the case of D-glucosamine because of the cleavage of the amino sugar glycosidic linkage, otherwise difficult to hydrolyse. Extremely mild conditions (pH 4 at room temperature or even below) are used and offer a valuable approach in the structural elucidation of complex glycosylaminoglycans.

(f) Determination of the anomeric configuration by chromium trioxide oxidation. The anomeric configuration of glycosidic linkages may be determined by polarimetric methods, by enzymatic hydrolysis with appropriate α- and β-glycosidases, or by analysis with distinct lectins (Bøg-Hansen, 1982). These techniques are applicable only in favourable cases. NMR spectroscopy and chromium trioxide oxidation, on the other hand, can be used successfully in most cases. The latter procedure involves oxidation with CrO_3 of fully acetylated polysaccharides. The β-pyranosidically linked sugar residues carrying an equatorial glycosidic bond at C-1 are selectively oxidised, whereas sugar residues with an axial glycosidic bond are unaffected or only slowly oxidised (Hoffmann *et al.*, 1972; Fig. 12); α- and β-linked furanosides are, however, both readily oxidised. The sample, peracetylated by treatment with acetic anhydride/pyridine (Carson and Maclay, 1946), is dissolved in glacial acetic acid and mixed with powdered chromium trioxide. The mixture is sonicated at 50°C for 1–2 h and the reaction followed by TLC in ethyl acetate–light petroleum (1:2; v/v). The reaction mixture is finally diluted with water (20 ml) and extracted with chloroform (3 × 20 ml). The pooled extracts are dried (with Na_2SO_4), concentrated to dryness, and analysed for sugars after hydrolysis.

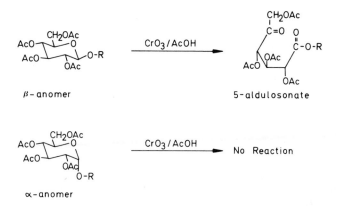

Fig. 12. Selective oxidation (with CrO₃/AcOH) of equatorially oriented acetylated glycosides; according to Hoffmann *et al.* (1972).

(g) Physicochemical methods of analysis. The application of combined GLC–MS to structural analysis of oligo- and polysaccharides and various glycoconjugates has been covered extensively in a number of publications (Björndal *et al.*, 1970; Jansson *et al.*, 1976; Lindberg, 1973; Rauvala *et al.*, 1981; Breimer *et al.*, 1983). Only the essentials of this technique will be discussed here. Permethylation of the polysaccharide and hydrolysis and derivatisation of the permethylated substance have been described above. The partially methylated alditol acetates are separated by GLC, making use of temperature-programmed capillary columns which allow separation of almost all partially methylated alditol acetates of common glycoses (Geyer *et al.*, 1982, 1983).

On mass spectrometric analysis a fragmentation pattern is obtained which is dependent on the substituents of neighbouring C atoms. Fragmentation between two adjacent methoxy groups is preferentially obtained. When there is fission between a methoxy group and a neighbouring acetoxy group the positive charge is found with the methoxy group. In most cases GLC–MS recognises only positive charged fragments; therefore, only this latter fragment is detected. Fission between neighbouring acetoxy groups is observable to a higher extent when the alditol acetate is not carrying methoxy groups. From primary fragments, formed by direct fission of the C–C bonds, secondary fragments can arise by elimination of acetic acid, ketene, methanol, or formaldehyde.

Laser desorption mass spectrometry, performed with a LAMMA 500 microprobe mass analyser, is a useful technique for structure elucidation of underivatised samples. In laser desorption mode high yields of the quasi-molecular peaks $(M+Na)^+$ or $(M+K)^+$ are often observed. Dephosphorylated preparations of free lipid A obtained from various enterobacterial Re mutants have been recently analysed by this technique (Seydel et al., 1984). The presence of several lower mass peaks in the spectrum has been attributed to microheterogeneity and/or to degradations caused by dephosphorylation, since phosphate-free synthetic lipid A analogues are shown to possess predominantly the $(M + K)^+$ peak (Seydel et al., 1984).

Recently, mass spectrometry with fast atom bombardment (FAB–MS) has been employed successfully for structural studies of glycoconjugates and lipid A's (Aubagnac et al., 1983; Kamerling et al., 1983; Dell et al., 1983; Takayama et al., 1984). This technique allows the analysis of polar and labile components which would probably be destroyed by derivatisation. The physicochemical behavior of a substance when deposited in a glycerol matrix is of decisive importance for the generation of high-quality FAB–MS spectra (Barber et al., 1982). The value of this technique for the establishment of heterogeneity of isolated carbohydrate materials has yet to be determined.

(h) NMR techniques.   Both proton (^{1}H) and carbon (^{13}C) NMR are useful techniques for complete structural analysis of glycoconjugates (Kotowycz and Lemieux, 1973; Barker et al., 1982). These techniques allow non-destructive analysis of polysaccharides and glycoconjugates. ^{13}C NMR especially gives important information, e.g., the number and nature of anomeric C atoms, presence and intensity of methyl, carbonyl, and N-acetyl groups, and the number of glycosyl residues in the repeating units. A serious drawback has been the amount of material needed (>20 mg), but with the recently available smaller probe heads the amount can be scaled down considerably. Microheterogeneity of the samples is a serious problem with LPS, R core, and lipid A analyses. Nevertheless, a complete R core analysis based on ^{13}C NMR has recently been published (Rowe and Meadow, 1983). NMR can also serve as an indicator of the purity of samples which might be of considerable value when individual polysaccharides have to be separated. Application of NMR techniques for carbohydrate structural analysis has been extensively reviewed (Gorin, 1981; Usui et al., 1973; Kotowycz and Lemieux, 1973; Sugiyama, 1978; Barker et al., 1982).

^{31}P NMR has been used successfully for recognition of phosphate, pyrophosphate, and phosphodiester groups and also for determining the purity and homogeneity of LPS samples (Strain et al., 1983; Rosner et al., 1979; Lehmann and Rupprecht, 1977; Horton and Riley, 1981).

2. Lipid A structure

(a) Lipid A backbone analysis. Hase and Rietschel (1976b) developed a selective degradation procedure for the qualitative and quantitative characterisation of the lipid A backbone from various bacterial groups. LPS is first subjected to alkaline hydrolysis (0.17 *M* NaOH, 100°C, 1 h) to yield alkali-treated LPS (LPS-OH) which is devoid of ester-linked fatty acids. Subsequent acid hydrolysis (dilute HCl, pH 1.5, 100°C, for about 40 min) of LPS-OH precipitates lipid A-OH, which, on sodium borohydride reduction, gives lipid A-OH$_{red}$. Hydrazinolysis of this material (100°C, 40 h) followed by preparative paper electrophoresis yields a fraction migrating at pH 5.3 like a standard of chitobiitol (M$_{GlcN}$1.14). This preparation (backbone disaccharide) is then subjected to N-acetylation, permethylation, and GLC analysis on 10% SE-30 at 225°C. By this technique, in combination with MS, the α- and β-anomers of different *N*-acetylglucosaminyl-*N*-acetylglucosaminitols can be distinguished clearly. The method allows, therefore, an unequivocal determination of the anomeric configuration as well as the position of the glycosidic linkage of the lipid A backbone disaccharide (Jensen *et al.*, 1979). The permethylated reduced disaccharide can be identified further by hydrolysis, derivatization of the sugar units into alditol acetates and subsequent GLC–MS analysis (Hase and Rietschel, 1976b). Under these conditions a partial O-demethylation at C-1 of the reducing glucosamine derivative is reported (Caroff and Szabo, 1980), whereas the *N*-methyl group is preserved. Instead of the *N*-acetylacetamido-derivative reported (Hase and Rietschel, 1976a), 1,6-di-*O*-acetyl-2-deoxy-3,4,5-tri-*O*-methyl-2-(*N*-methylacetamido)-hexitol is obtained.

(i) Degradation under acidic conditions. The analysis of substituents external to the lipid A backbone may be illustrated on free lipid A of *R. tenue* (Fig. 13 a and b, Tharanathan *et al.*, 1978b), which shows three different external residues. Hydrolysis with mild acid (0.01 *N* HCl, 80°C, 10 min) and identification of the substituents released by high-voltage paper electrophoresis and GLC revealed the presence of D-arabinofuranose and 4-amino-4-deoxy-L-arabinopyranose. The configuration of the isolated 4-aminoarabinose was deducible by the following degradation sequence (Volk *et al.*, 1970): N-acetylation, periodate oxidation, alkaline hypoiodite oxidation, and N-deacetylation, followed by paper electrophoresis and elution of the neutral component. Amino acid analysis of the eluted material indicated the presence of serine, which reacted with L-amino acid oxidase. Thus, serine and, in turn, the 4-aminoarabinose possess L-configuration.

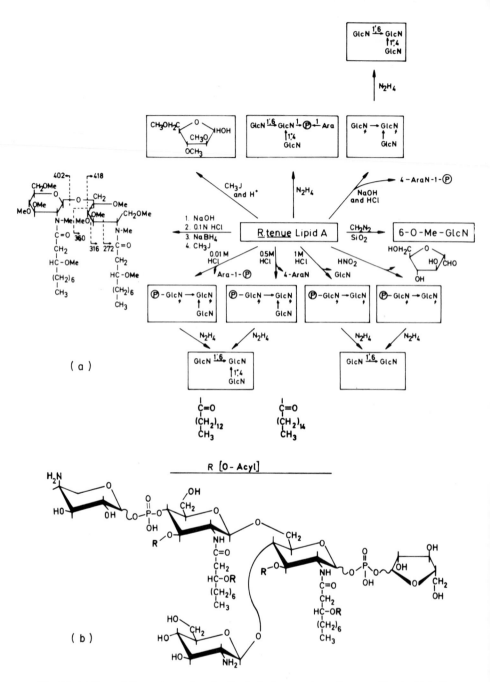

Fig. 13. (a) Degradation scheme showing the techniques and reactions used to elucidate the structure of *Rhodospirillum tenue* lipid A; according to Tharanathan *et al.* (1978b). (b) Proposed chemical structure of *R. tenue* lipid A; according to Tharanathan *et al.* (1978b).

Partial acid hydrolysis (1 N HCl, 100°C, 2.5 h) of *R. tenue* lipid A showed the presence of a third glucosamine residue (1″,4-linked to the backbone disaccharide) which could be eliminated by nitrous acid deamination as 2,5-anhydromannose, thus proving the presence of a non-acylated amino group (Fig. 13a and b).

(*ii*) *Degradation under alkaline conditions.* Treatment of *R. tenue* lipid A with mild alkali (0.17 M NaOH, 100°C, 1 h) results in the cleavage of the ester-bound fatty acids and the ester-bound substituted phosphate group which on further characterisation has been identified as 4-aminoarabinose-1-phosphate. 4-Aminoarabinose is a characteristic constituent of the lipid A's of all strains of *Salmonella* (Volk *et al.*, 1970), *R. tenue, R. purpureus, Chromobacterium violaceum* (Hase and Rietschel, 1977), and *P. mirabilis* (Kotelko, 1985), but is characteristically lacking in other species such as *E. coli, R. gelatinosa* and *R. vannielii* (Lüderitz *et al.*, 1982; Mayer and Weckesser, 1984).

Mild hydrazinolysis of *R. tenue* lipid A (anhydrous hydrazine, 100°C, 10–12 h) followed by paper electrophoresis afforded a major tetrasaccharide fraction, characterised as

$$GlcN \xrightarrow{\ 1.6\ } GlcN \xrightarrow{\ 1\ } \textcircled{P} \xleftarrow{\ 1\ } Ara f$$
$$\uparrow^{1'',4}$$
$$GlcN$$

(*b*) *Periodate oxidation studies.* Periodate oxidation destroyed both the external arabinose and glucosamine in *R. tenue* lipid A indicating, in addition to methylation data, that both substituents are non-acylated (Tharanathan *et al.*, 1978b). The same techniques (periodate oxidation and silica gel-catalysed methylation) showed the mannopyranosyl residue which is attached to the 4′ position of the lipid A backbone in *R. vannielii* to be unsubstituted (Holst *et al.*, 1983a). Caroff and Szabo (1983) have isolated and characterised by mass spectrometry 2-deoxy-2-(3-hydroxytetradecanoylamido)glycerol from the lipid A of *Bordetella pertussis,* and concluded that a 3-hydroxytetradecanoic acid is bound to the amino group of the reducing glucosamine. A similar approach has been adopted for the analysis of lipid A derived from *Rhizobium* strains (Russa *et al.*, 1984), *Rhodopseudomonas sphaeroides* (Salimath *et al.*, 1983) and *R. gelatinosa* (Tharanathan *et al.*, 1984).

(*c*) *Mild methylation with diazomethane.* Silica gel-catalysed methylation with diazomethane provides the opportunity of recognising free, unsubstituted hydroxyl groups in lipid A. The method (Ohno *et al.*, 1979) works at neutral pH and is easy to perform. Lipid A is suspended in a mixture of dichloromethane–methanol (1 : 1, v/v) together with neutral

silica gel [freshly activated by acid treatment (2 N HCl) and water washing, followed by drying at 100°C]. The reaction is started by adding a
solution of diazomethane in dichloromethane at 0–5°C. After reaction
overnight, the excess reagent is evaporated (in a N_2 stream), filtered, and
hydrolysed (4 N HCl, 100°C, 6 h). The partially methylated sugars are
then converted into alditol acetates and characterised by GLC–MS. Using this method lipid A of *R. sphaeroides* (Salimath *et al.*, 1983), *R.
gelatinosa* (Tharanathan *et al.*, 1984), *Rhizobium* (Russa *et al.*, 1984), and
Salmonella (Wollenweber, personal communication) were shown to possess unsubstituted hydroxyl groups at C-4 (GlcN I) and C-6' (GlcN II) of
the backbone disaccharide. As shown with *Salmonella* (Rietschel *et al.*,
1983b; Strain *et al.*, 1983) and *P. mirabilis* lipid A (Sidorczyk *et al.*, 1983),
the latter hydroxyl group represents the attachment site for KDO and
the polysaccharide moiety in lipopolysaccharides (Fig. 4). With *R.
sphaeroides* and *R. gelatinosa* the hydroxyl group at C-6' is methylated
only in free lipid A and not in lipopolysaccharide (Salimath *et al.*, 1984;
Tharanathan *et al.*, 1984).

IV. Concluding remarks

When confronted with the large number of techniques and analytical
methods in use for lipopolysaccharide characterisation, one realises that a
selection has to be made according to the aim of a particular study. One
might be interested in a first chemical description of a lipopolysaccharide
or in an analysis restricted to one or a few distinct constituents of taxonomic significance (KDO, heptoses, amide-linked fatty acids). For other
aims, knowledge of the partial or complete structure might be desirable.
Each aim requires a different approach. Some suggestions how such studies might be approached will be discussed in the following.

   The analytical description of an unknown lipopolysaccharide requires
its isolation and purification. The classical hot phenol–water method is
highly recommended since it is probably the method with the broadest
applicability. Glycans and RNA, which are co-extracted with LPS, can be
removed easily by ultracentrifugation or by enzymatic degradation (amy-
lase, RNase). In some cases LPS is extracted either partly or totally into
the phenol phase. Contaminating phospholipids and proteins must be removed by extraction with chloroform–methanol. PCP extraction of whole
cells or of crude LPS preparations obtained by phenol–water extraction
gives a product of high purity. In rare cases, however, ornithine-contain-
ing lipids might be co-extracted (Holst *et al.*, 1983b).

A first chemical description of LPS might include the qualitative and quantitative analyses of neutral sugars, uronic acids, or other acidic sugars and of amino sugars. Gas–liquid chromatography of suitable derivatives, high-voltage paper electrophoresis, and analysis on the amino acid analyser are the methods of choice. Heptoses can be identified by gas–liquid chromatography as their alditol acetates. The ubiquitous heptose, L-glycero-D-mannoheptose, shows in addition a faster satellite peak of its 1,6-anhydro derivative (Chaby and Szabo, 1976), which is not formed by the D-epimer. The identification of unknown sugars (deoxy-)O-, or C-methyl sugars, rare uronic acids and amino sugars) will depend on the availability of special equipment (e.g., mass spectrometry, nuclear magnetic resonance) and appropriate reference standards. KDO, present in most but not all lipopolysaccharides, can sometimes be obtained in anhydro or a substituted form and then its detection can be difficult (Caroff and Szabo, 1980). Although KDO and heptose are generally reliable markers for lipopolysaccharides, one has to realise that both sugars occur also in polymers other than lipopolysaccharides: KDO in a number of *E. coli* K antigens (e.g., K12, K14; Jann and Jann, 1983) and heptose in glycolipids (*Acholeplasma modicum* and a *Pseudomonas* species) (Wilkinson, 1977). The detection of KDO by the periodate/thiobarbituric acid method should be substantiated by co-electrophoresis with an authentic standard (Anderson, 1965). β-Hydroxy fatty acids are typical constituents of lipid A, but they are not unique to lipopolysaccharides and occur also in polar lipids, such as in ornithine–amide lipids (Wilkinson and Caudwell, 1980).

The fatty acid spectrum of a lipopolysaccharide should be identified preferentially by capillary gas chromatography of the methyl ester derivatives. Co-chromatography with an authentic standard should be performed on at least two different capillary columns, but in addition mass spectrometric analyses of the peaks is highly desirable.

Splitting of lipopolysaccharides into lipid A and "degraded polysaccharide" is a highly informative step in lipopolysaccharide characterisation. The distribution of components in the two separated fractions usually allows a safe assignment of constituents to either lipid A or polysaccharide moieties. For characterisation of lipid A it is desirable to identify the backbone sugar, its molar ratio relative to hydroxy and non-hydroxylated fatty acids, and the phosphorus content. Analysis of substituents to the backbone disaccharide (4-aminoarabinose, phosphorylethanolamine, additional neutral sugars) also provides valuable information. The detection of amide-linked fatty acids by one of the many techniques available is important.

Structural work on lipid A requires experience. The results are highly valuable not only from a taxonomic point of view (for a review, see Mayer and Weckesser, 1984; Mayer et al., 1984) but also for questions of the structural prerequisites of endotoxic activity. A recent example is the non-toxic lipid A of R. sphaeroides which is structurally very similar to the highly toxic Salmonella lipid A (Salimath et al., 1983; Strittmatter et al., 1983). Structural work on the polysaccharide moiety of endotoxins can be time consuming too and has to be restricted therefore to specified questions. The identification of the core constituents of a species and qualitative comparison with those of other taxa may be highly informative for taxonomic purposes. The availability of R mutants and their lipopolysaccharides would facilitate such study. In some cases the separation of degraded polysaccharide and core on Sephadex G-50 (Fig. 9) can be of use.

Taxonomically valuable information can sometimes be obtained without a complete analysis. The detection of constituents otherwise rarely occurring in lipopolysaccharides allows analysis to be restricted to the identification of such constituents.

Whole cells have been used in some cases for rapid detection of lipopolysaccharide-specific constituents, such as KDO, heptoses, unusual O-specific sugars, and 3-hydroxy fatty acids. Jantzen et al. (1975) analysed gas chromatographically the whole cells after acidic methanolysis. The fingerprints obtained included LPS-specific constituents and allowed, for example, recognition of well-defined groups within the Neisseriaceae family which corroborated with the classical subdivision of Neisseriaceae into the known genera.

Lipopolysaccharides, in addition to the peptidoglycan and the major outer membrane proteins, can be used also as markers for the chemical characterisation of cell wall fractions or for proving the lack of contaminants, e.g., in membrane fractions. In such cases, the KDO content or the amount of 3-hydroxy fatty acids is determined. These are highly sensitive approaches, although the quantitation depends on the ease with which KDO is released and on the amount of β-hydroxy fatty acid converted to the Δ^2-unsaturated product.

Acknowledgments

The authors express their thanks to Dr. Otto Lüderitz for valuable discussions and for critically reading the manuscript; to colleagues for making available their unpublished results, to Dr. Sumanta Basu for permission to include Fig. 8, and to Mrs. I. Kuttler for cooperation in preparing the manuscript. R.N.T. thanks the Alexander von Humboldt-Stiftung, Bonn, for the award of a fellowship.

References

Acker, G., Wartenberg, K., and Knapp, W. (1980). *Zentralbl. Bakteriol. Parasitenkd. Infektionkr. Hyg. Abt. 1: Orig. A* **247**, 229.

Acker, G., Schmidt, G., and Mayer, H. (1982). *J. Gen. Microbiol.* **128**, 1577.

Adams, G. A. (1967). *Can. J. Chem.* **45**, 422.

Ahamed, M. N., Radziejewska-Lebrecht, J., Widemann, C., and Mayer, H. (1980). *Zentralbl. Bakteriol. Parasitenkd. Infektionkr. Hyg. Abt. 1: Orig. A* **247**, 468.

Albersheim, P., Nevius, D. J., English, P. D., and Karr, A. (1967). *Carbohydr. Res.* **5**, 340.

Amano, K., Ribi, E., and Cantrell, L. (1983). *J. Biochem. (Tokyo)* **93**, 1391.

Anderson, P. J. (1966). *J. Chromatogr.* **21**, 163.

Aspinall, G. O. (1977). *Pure Appl. Chem.* **49**, 1105.

Aubagnac, J. L., Devienne, F. M., and Combabien, R. (1983). *Org. Mass Spectrom.* **18**, 173.

Avigad, G. (1969). *Carbohydr. Res.* **11**, 119.

Banoub, J. H., and Michon, F. (1982). *Carbohydr. Res.* **100**, C24.

Banoub, J. H., Shaw, D. H., and Michon, F. (1983). *Carbohydr. Res.* **123**, 117.

Barber, M., Nordoli, R. S., Elliot, G. J., Sedgwick, R. D., and Tyler, A. N. (1982). *Anal. Chem.* **54**, 645 A.

Barker, R., Nunez, H. A., Rosevear, P., and Serianni, A. S. (1982). *In* "Methods in Enzymology" (V. Ginsburg, ed.), Vol. 83, p. 58. Academic Press, New York.

Björndal, H., Hellerqvist, C. G., Lindberg, B., and Svensson, S. (1970). *Angew. Chem. Int. Ed. Engl.* **9**, 610.

Björndal, H., Lindberg, B., and Nimmich, W. (1971). *Acta Chem. Scand.* **25**, 750.

Bøg-Hansen, T. C., ed. (1982). *In* "Lectins, Biochemistry, Clinical Biochemistry," Vol. 2, de Gruyter, Berlin.

Boivin, A., and Mesrobeanu, L. (1933). *C.R. Seances Soc. Biol. Ses Fil* **112**, 76.

Brade, H., Galanos, C., and Lüderitz, O. (1983). *Eur. J. Biochem.* **131**, 195.

Breimer, M. E., Hansson, G. C., Karlsson, K. A., Larson, G., Leffler, H., Pimlott, W., Samuelsson, B. E., Strömberg, N., Teneberg, S., and Thurin, J. (1983). *Int. J. Mass Spectrom. Ion Phys.* **48**, 113.

Bryn, K., and Jantzen, E. (1982). *J. Chromatogr.* **240**, 405.

Caroff, M., and Szabo, L. (1980). *Carbohydr. Res.* **84**, 43.

Caroff, M., and Szabo, L. (1983). *Carbohydr. Res.* **114**, 95.

Carson, J. I., and Maclay, W. D. (1946). *J. Am. Chem. Soc.* **68**, 1015.

Chaby, R., and Szabo, L. (1976). *Eur. J. Biochem.* **70**, 115.

Darveau, R. P., and Hancock, R. E. W. (1983). *J. Bacteriol.* **115**, 831.

Dell, A., Morris, H. R., Egge, H., Nicolai, H. V., and Strecker, G. (1983). *Carbohydr. Res.* **115**, 41.

Dmitriev, B. A., Kocharova, N. A., Knirel, Y. A., Shashkov, A. S., Kochetkov, N. K., Stanislavsky, E. S., and Mashilova, G. M. (1982). *Eur. J. Biochem.* **125**, 229.

Drews, G., Weckesser, J., and Mayer, H. (1978). *In* "The Photosynthetic Bacteria" (R. K. Clayton and W. R. Sistrom, eds.), p. 61. Plenum, New York.

Dutton, G. G. S. (1973). *Adv. Carbohydr. Chem.* **28**, 11.

Erbing, C., Granath, K., Kenne, L., and Lindberg, B. (1976). *Carbohydr. Res.* **47**, C5.

Feige, U., and Radziejewska-Lebrecht, J. (1979). *In* "Glycoconjugates" (R. Schauer, ed.), p. 2. Thieme, Stuttgart.

Feige, U., Jann, B., Jann, K., Schmidt, G., and Stirm, S. (1977). *Biochem. Biophys. Res. Commun.* **79**, 88.

Fromme, I., and Beilharz, H. (1978). *Anal. Biochem.* **84**, 347.

Funahara, Y., and Nikaido, H. (1980). *J. Bacteriol.* **141,** 1463.

Galanos, C., and Lüderitz, O. (1975). *Eur. J. Biochem.* **54,** 603.

Galanos, C., Lüderitz, O., and Westphal, O. (1969). *Eur. J. Biochem.* **9,** 245.

Galanos, C., Lüderitz, O., Rietschel, E. T., and Westphal, O. (1977a). *Int. Rev. Biochem.* **14,** 239.

Galanos, C., Roppel, J., Weckesser, J., Rietschel, E. T., and Mayer, H. (1977b). *Infect. Immunity* **16,** 407.

Galanos, C., Lüderitz, O., and Westphal, O. (1979). *Zentralbl. Bakteriol. Parasitenkd. Infektionkr. Hyg. Abt. 1: Orig.* A **243,** 226.

Gamian, A., and Romanowska, E. (1982). *Eur. J. Biochem.* **129,** 105.

Geyer, R., Geyer, H., Kühnhardt, S., Mink, W., and Stirm, S. (1982). *Anal. Biochem.* **121,** 263.

Geyer, R., Geyer, H., Kühnhardt, S., Mink, W., and Stirm, S. (1983). *Anal. Biochem.* **133,** 197.

Gibson, J., Stackebrandt, E., Zablen, L. B., Gupta, R., and Woese, C. R. (1979). *Current Microbiology* **3,** 59.

Gmeiner, J. (1975). *Eur. J. Biochem.* **58,** 621.

Gmeiner, J., Lüderitz, O., and Westphal, O. (1969). *Eur. J. Biochem.* **7,** 370.

Gmeiner, J., Mayer, H., Kotelko, K., and Zych, K. (1977). *Eur. J. Biochem.* **72,** 35.

Goldman, R. C., and Leive, L. (1980). *Eur. J. Biochem.* **107,** 145.

Goldstein, I. J., Hay, G. W., Lewis, B. A., and Smith, F. (1965). *Meth. Carbohydr. Chem.* **5,** 361.

Gorin, P. A. J. (1981). *Adv. Carbohydr. Chem. Biochem.* **38,** 13.

Gorshkova, R. P., Komandrova, N. A., Zubkov, V. A., and Korchagina, N. I. (1983). *In* "Abstracts of Second European Symposium on Carbohydrates and Glycoconjugates," p. C-26. Budapest.

Gromska, W., and Mayer, H. (1976). *Eur. J. Biochem.* **63,** 391.

Hais, I. M., and Macek, K., eds. (1958/1960). *In* "Handbuch der Papierchromatographie." VEB Gustav Fischer Verlag, Jena.

Hakomori, S. (1964). *J. Biochem. (Tokyo)* **55,** 205.

Harvey, D. J. (1982). *Biomed. Mass Spectrom.* **9,** 33.

Hase, S., and Rietschel, E. T. (1976a). *Eur. J. Biochem.* **63,** 93.

Hase, S., and Rietschel, E. T. (1976b). *Eur. J. Biochem.* **63,** 101.

Hase, S., and Rietschel, E. T. (1977). *Eur. J. Biochem.* **75,** 23.

Hay, G. W., Lewis, B. A., and Smith, F. (1965). *Meth. Carbohydr. Chem.* **5,** 357.

Hinshaw, L., ed. (1984). Handbook of Endotoxin, Vol. 2. Elsevier North Holland.

Hirst, E. L., and Percival, E. (1965). *Meth. Carbohydr. Chem.* **5,** 287.

Hoffmann, J., Lindberg, B., and Svensson, S. (1972). *Acta Chem. Scand.* **26,** 661.

Holst, O., Borowiak, D., Weckesser, J., and Mayer, H. (1983a). *Eur. J. Biochem.* **137,** 325.

Holst, O., Weckesser, J., and Mayer, H. (1983b). *FEMS Microbiol. Lett.* **19,** 33.

Horton, D., and Riley, D. A. (1981). *Biochim. Biophys. Acta* **640,** 727.

Hurlbert, R. E., and Hurlbert, I. M. (1977). *Infect. Immun.* **16,** 983.

Hurlbert, R. E., Weckesser, J., Mayer, H., and Fromme, I. (1976). *Eur. J. Biochem.* **68,** 365.

Hurlbert, R. E., Weckesser, J., Tharanathan, R. N., and Mayer, H. (1978). *Eur. J. Biochem.* **90,** 241.

Imoto, M., Kusumoto, S., Shiba, T., Naoki, H., Iwashita, T., Rietschel, E. T., Wollenweber, H. W., Galanos, C., and Lüderitz, O. (1983). *Tetrahedron Lett.* **24,** 4017.

Jann, B., Reske, K., and Jann, K. (1975). *Eur. J. Biochem.* **60,** 239.

Jann, K., and Jann, B. (1983). *Prog. Allergy* **33**, 53.

Jann, K., and Westphal, O. (1975). *In* "The Antigens" (M. Sela, ed.), Vol. 3, p. 1. Academic Press, New York.

Jann, K., Goldemann, G., Weisgerber, C., Wolf-Ullisch, C., and Kanegasaki, S. (1982). *Eur. J. Biochem.* **127**, 157.

Jansson, P. E., Kenne, L., Liedgren, H., Lindberg, B., and Lönngren, J. (1976). *Univ. Stockholm Chem. Commun.* **8**, 1.

Jansson, P. E., Lindberg, A. A., Lindberg, B., and Wollin, R. (1981). *Eur. J. Biochem.* **115**, 571.

Jantzen, E., Bryn, K., Bergan, T., and Bøvre, K. (1975). *Acta Pathol. Microbiol. Scand., Sect. B* **83**, 569.

Jay, F. A. (1978). Ph.D. Thesis. University of Surrey.

Jensen, M., Borowiak, D., Paulsen, H., and Rietschel, E. T. (1979). *Biomed. Mass Spectrom.* **6**, 559.

Kadis, S., Weinbaum, G., and Ajl, S. J., eds. (1971). "Microbial Toxins," Vol. 5. Academic Press, New York.

Kamerling, J. P., Heerma, W., Vliegenthart, J. F. G., Green, B. N., Lewis, I. A. S., Strecker, G., and Spik, G. (1983). *Biomed. Mass Spectrom.* **10**, 420.

Karkhanis, D., Zeltner, J., Jackson, J., and Carlo, D. (1978). *Anal. Biochem.* **85**, 595.

Kauffmann, F. (1966). "The Bacteriology of *Enterobacteriaceae*," Munksgaard, Copenhagen.

Keleti, G., and Lederer, W. H. (1974). "Handbook of Micromethods for the Biological Sciences." Van Nostrand Reinhold, New York.

Kennedy, J. F., Barker, S. A., and Bradshaw, I. J. (1983). *Carbohydr. Res.* **122**, 178.

Kenne, L., and Lindberg, B. (1983). *In* "Polysaccharides" (G. O. Aspinall, ed.), Vol. 2, p. 287. Academic Press, New York.

Kickhöfen, B., and Warth, R. (1968). *J. Chromatogr.* **33**, 558.

Knirel, Y. A., Vinogradov, E. V., Shashkov, A. S., Dmitriev, B. A., Kochetkov, N. K., Stanislavsky, E. S., and Mashilova, G. M. (1982a). *Eur. J. Biochem.* **125**, 221.

Knirel, Y. A., Vinogradov, E. V., Shashkov, A. S., Dmitriev, B. A., and Kochetkov, N. K. (1982b). *Carbohydr. Res.* **104**, C4.

Knirel, Y. A., Vinogradov, E. V., Shashkov, A. S., Dmitriev, B. A., and Kochetkov, N. K. (1982c). *Carbohydr. Res.* **112**, C4.

Knirel, Y. A., Vinogradov, E. V., Shashkov, A. S., Dmitriev, B. A., Kochetkov, N. K., Stanislavsky, E. S., and Mashilova, G. M. (1983). *Eur. J. Biochem.* **128**, 81.

Koeltzow, D. E., and Conrad, H. E. (1971). *Biochemistry* **10**, 214.

Kontrohr, T., and Kocsis, B. (1983). *In* "Abstracts of Second European Symposium on Carbohydrates and Glycoconjugates," p. E-22. Budapest.

Kotelko, K. (1985). *Curr. Top. Microbiol. Immunol.* in press.

Kotelko, K., Gromska, W., Papierz, M., Sidorczyk, Z., Krajewska, D., and Szer, K. (1977). *J. Hyg. Epidemiol. Microbiol. Immunol.* **21**, 271.

Kotowycz, G., and Lemieux, R. V. (1973). *Chem. Rev.* **73**, 669.

Kraska, U., Pougny, J. R., and Sinay, P. (1976). *Carbohydr. Res.* **50**, 181.

Kröger, E. (1955). *Ergeb. Hyg. Bakteriol. Immunitätsforsch. Exp. Ther.* **29**, 475.

Lance, D. G., and Jones, J. K. N. (1967). *Can. J. Chem.* **45**, 1995.

Lederer, E., and Lederer, M., eds. (1957). "Chromatography," Elsevier, Amsterdam.

Lehmann, V., and Rupprecht, E. (1977). *Eur. J. Biochem.* **81**, 443.

Lehmann, V., Redmond, J., Egan, A., and Minner, I. (1978). *Eur. J. Biochem.* **86**, 487.

Leive, L. (1965). *Biochem. Biophys. Res. Commun.* **21**, 290.

Lindberg, A. A. (1973). *Annu. Rev. Microbiol.* **27**, 205.

Lindberg, A. A. (1977). *In* "Surface Carbohydrates of the Prokaryotic Cell" (I. Sutherland, ed.), p. 289. Academic Press, New York.

Lindberg, B. (1972). *In* "Methods in Enzymology" (V. Ginsburg, ed.), Vol. 28, p. 178. Academic Press, New York.

Lindberg, B., Lönngren, J., and Nimmich, W. (1972). *Acta Chem. Scand.* **26**, 2231.

Lönngren, J., and Svensson, S. (1974). *Adv. Carbohydr. Chem. Biochem.* **29**, 41.

Lowry, O. H., Roberts, N. R., Leiner, K. Y., Wu, M. L., and Farr, A. L. (1954). *J. Biol. Chem.* **207**, 1.

Lüderitz, O., Galanos, C., Risse, H. J., Ruschmann, E., Schlecht, S., Schmidt, G., Schulte-Holthausen, H., Wheat, R., Westphal, O., and Schlosshardt, J. (1966). *Ann. N.Y. Acad. Sci.* **133**, 349.

Lüderitz, O., Westphal, O., Staub, A. M., and Nikaido, H. (1971). *In* "Microbial Toxins" (G. Weinbaum, S. Kadis, and S. J. Ajl, eds.), Vol. 4, p. 145. Academic Press, New York.

Lüderitz, O., Galanos, C., Lehmann, V., Mayer, H., Rietschel, E. T., and Weckesser, J. (1978). *Naturwissenschaften* **65**, 578.

Lüderitz, O., Freudenberg, M. A., Galanos, C., Lehmann, V., Rietschel, E. T., and Shaw, D. H. (1982). *Curr. Top. Membr. Transp.* **17**, 79.

Lüderitz, O., Tanamoto, K. I., Galanos, C., Westphal, O., Zähringer, U., Rietschel, E. T., Kusumoto, S., and Shiba, T. (1983). *ACS Symp. Ser.* **231**, 2.

Ludowieg, J., and Dorfman, A. (1960). *Biochim. Biophys. Acta* **38**, 212.

Lugtenberg, B., and van Alphen, L. (1983). *Biochim. Biophys. Acta* **737**, 51.

Madhavan, V. N., Done, J., and Vine, J. (1981). *Chem. Phys. Lipids* **28**, 79.

McConnell, M., and Wright, A. (1979). *J. Bacteriol.* **137**, 746.

Mäkelä, P. H., and Stocker, B. A. D. (1984). *In* "Handbook of Endotoxin" (E. T. Crietschen, ed.), Vol. 1. Elsevier North Holland, Amsterdam, in press.

Männel, D., and Mayer, H. (1978). *Eur. J. Biochem.* **86**, 361.

Marx, A., Petcovici, M., Nacescu, N., Mayer, H., and Schmidt, G. (1977). *Infect. Immun.* **18**, 563.

Mayer, H. (1984). *In* "Festschrift for Herman M. Kalckar" (E. Haber, ed.), Plenum, New York, in press.

Mayer, H., and Schmidt, G. (1973). *Zentralbl. Bakteriol. Parasitenkd. Infektionkr. Hyg. Abt. 1: Orig. A.* **224**, 345.

Mayer, H., and Schmidt, G. (1979). *Curr. Top. Microbiol. Immunol.* **85**, 99.

Mayer, H., and Weckesser, J. (1984). *In* "Chemistry of Endotoxin" (E. T. Rietschel, ed.), Vol. 1. Elsevier North Holland, Amsterdam.

Mayer, H., and Westphal, O. (1968). *J. Chromatogr.* **33**, 514.

Mayer, H., Bock, E., and Weckesser, J. (1984). *FEMS Microbiol. Lett.* **17**, 93.

Mayer, H., Salimath, P. V., Holst, O., and Weckesser, J. (1984). *Rev. Infect. Diseases* **6**, 542.

Morgan, W. T. J. (1965). *Meth. Carbohydr. Chem.* **5**, 80.

Mort, A. J., and Lamport, D. A. (1977). *Anal. Biochem.* **82**, 289.

Müller-Seitz, E., Jann, B., and Jann, K. (1968). *FEBS Lett.* **1**, 311.

Naide, Y., Nikaido, H., Mäkelä, P. H., Wilkinson, R. G., and Stocker, B. A. D. (1965). *Proc. Natl. Acad. Sci. U.S.A.* **53**, 147.

Narui, T., Takahashi, K., Kobayashi, M., and Shibata, S. (1982). *Carbohydr. Res.* **103**, 293.

Ng, A. K., Butler, R. C., Chen, C. L. H., and Nowotny, A. (1976). *J. Bacteriol.* **126**, 511.

Nikaido, H. (1970). *Int. J. Syst. Bacteriol.* **20**, 383.

Nikaido, H. (1973). *In* "Bacterial Membranes and Walls" (L. Leive, ed.), Vol. 1, p. 131. Dekker, New York.

Nikaido, H., and Nakae, T. (1979). *Adv. Microbiol. Physiol.* **20**, 163.
Nowotny, A. (1971). *In* "Microbial Toxins," Vol. 4, p. 309. Academic Press, New York.
Oden, S., and Hofsten, B. (1954). *Nature (London)* **173**, 449.
Ohno, K., Nishiyama, H., and Nagase, H. (1979). *Tetrahedron Lett.* **45**, 4405.
Okuda, S., Sato, M., Uchiyano, H., and Takahashi, H. (1975). *J. Gen. Appl. Microbiol.* **21**, 169.
Omar, A. S., Flammann, H. T., Borowiak, D., and Weckesser, J. (1983). *Arch. Microbiol.* **134**, 212.
Ørskov, I., Ørskov, F., Jann, B., and Jann, K. (1977). *Bacteriol. Rev.* **41**, 667.
Osborn, M. J. (1979). *In* "Bacterial Outer membranes" (M. Inouye, ed.), p. 15. Wiley Interscience, New York.
Osborn, M. J., and Rothfield, L. I. (1971). *In* "Microbial Toxins" (G. Weinbaum, S. Kadis, and S. J. Ajl, eds.), Vol. 4, p. 331. Academic Press, New York.
Palva, E. T., and Mäkelä, P. H. (1980). *Eur. J. Biochem.* **107**, 137.
Partridge, S. M. (1948). *Biochem. J.* **42**, 238.
Partridge, S. M. (1949). *Nature (London)* **164**, 443.
Prehm, P. (1980). *Carbohydr. Res.* **78**, 372.
Prehm, P., Stirm, S., Jann, B., and Jann, K. (1975). *Eur. J. Biochem.* **56**, 41.
Radziejewska-Lebrecht, J. (1983). *In* "Abstracts of Second European Symposium on Carbohydrates and Glycoconjugates," p. C-5. Budapest.
Radziejewska-Lebrecht, J., Feige, U., Mayer, H., and Weckesser, J. (1981). *J. Bacteriol.* **145**, 138.
Randerath, K., ed. (1962). "Dünnschicht-Chromatographie," Verlag Chemie, Weinheim.
Rapin, A. M. C., and Kalckar, H. M. (1971). *In* "Microbial Toxins" (G. Weinbaum, S. Kadis, and S. J. Ajl, eds.), Vol. 4, p. 267. Academic Press, New York.
Rauvala, H., Finne, J., Krusius, T., Kärkkäinen, J., and Järnefelt, J. (1981). *Adv. Carbohydr. Chem. Biochem.* **38**, 389.
Reissig, J. L., Strominger, J. L., and Leloir, I. F., (1955). *J. Biol. Chem.* **217**, 959.
Rietschel, E. T., and Lüderitz, O. (1980). *Forum Mikrobiol.* **1**, 12.
Rietschel, E. T., Gottert, H., Lüderitz, O., and Westphal, O. (1972). *Eur. J. Biochem.* **28**, 166.
Rietschel, E. T., Hase, S., King, M., Redmond, J., and Lehmann, V. (1977). *In* "Microbiology" (D. Schlessinger, ed.), p. 262. Am. Soc. Microbiology, Washington, D. C.
Rietschel, E. T., Galanos, C., Lüderitz, O., and Westphal, O. (1982a). *In* "Immunopharmacology and the Regulation of leukocyte Function" (D. R. Webb, ed.), p. 2. Dekker, New York.
Rietschel, E. T., Schade, U., Jensen, M., Wollenweber, H., Lüderitz, O., and Greisman, S. G. (1982b). *Scand. J. Infect. Dis. Suppl.* **31**, 8.
Rietschel, E. T., Sidorczyk, Z., Zähringer, U., Wollenweber, H. W., and Lüderitz, O. (1983a). *ACS Symp. Ser.* **231**, 195.
Riëtschel, E. T., Zähringer, U., Wollenweber, H. W., Tanamoto, K., Galanos, C., Lüderitz, O., Kusumoto, S., and Shiba, T. (1983b). *In* "Handbook of Natural Toxins. Vol. II. Bacterial Toxins" (A. Tu, W. H. Habig, and M. C. Hardegree, eds.), Dekker, New York.
Rietschel, E. T., Wollenweber, H. W., Brade, H., Zähringer, U., Linder, B., Seydel, U., Bradaczek, H., Barnickel, G., Labishinski, H., and Giesbrecht, P. (1984). *In* "Chemistry of Endotoxins." Elsevier North Holland, Amsterdam, in press.
Rondle, C. J. M., and Morgan, W. T. G., (1955). *Biochem. J.* **61**, 586.
Roppel, J., Mayer, H., and Weckesser, J. (1975). *Carbohydr. Res.* **40**, 31.
Rosner, M. R., Khorana, H. G., and Satterthwait, A. C. (1979). *J. Biol. Chem.* **254**, 5918.
Rottem, S., Markowitz, O., and Razin, S. (1978). *Eur. J. Biochem.* **85**, 445.

Rowe, P. S. N., and Meadow, P. M. (1983). *Eur. J. Biochem.* **132**, 329.

Russa, R., Lüderitz, O., and Rietschel, E. T. (1985). *Eur. J. Biochem.*, in preparation.

Salimath, P. V., Weckesser, J., Strittmatter, W., and Mayer, H. (1983). *Eur. J. Biochem.* **136**, 195.

Salimath, P. V., Tharanathan, R. N., Weckesser, J., and Mayer, H. (1984). *Eur. J. Biochem.* **144**, 227.

Samuelsson, K., Lindberg, B., and Brubaker, R. R. (1974). *J. Bacteriol.* **117**, 1010.

Sawardeker, J. S., Sloneker, J. H., and Jeanes, A. (1965). *Anal. Chem.* **37**, 1602.

Schlecht, S. and Westphal, O. (1970).

Schlenk, H., and Gellerman, J. L. (1960). *Anal. Chem.* **32**, 1412.

Schmidt, G., and Lüderitz, O. (1969). *Zentralbl. Bakteriol. Parasitenkd. Infektion Kr. Hyg. Abt. 1: Orig. A.* **210**, 381.

Schmidt, G., Jann, B., and Jann, K. (1969a). *Eur. J. Biochem.* **10**, 501.

Schmidt, G., Schlecht, S., Lüderitz, O., and Westphal, O. (1969b). *Zentralbl. Bakteriol. Parasitenkd. Infektionkr. Hyg. Abt. 1: Orig. A.* **210**, 483.

Schmidt, G., Fromme, I., and Mayer, H. (1970). *Eur. J. Biochem.* **14**, 357.

Schramek, S., and Mayer, H. (1982). *Infect. Immun.* **38**, 53.

Schramek, S., Radziejewska-Lebrecht, J., and Mayer, H. (1983). *In* "Abstracts of Second European Symposium on Carbohydrates and Glycoconjugates", p. c-4. Budapest.

Seewaldt, E., Schleifer, K. H., Bock, E., and Stackebrandt, E. (1982). *Acta Microbiol.* **131**, 287.

Seydel, U., Lindner, B., Wollenweber, H. W., and Rietschel, E. T. (1984). *Biomed. Mass Spectrom.* **11**, 132.

Sidorczyk, Z., Zähringer, U., and Rietschel, E. T. (1983). *Eur. J. Biochem.* **137**, 15.

Sinensky, M. (1974). *Proc. Natl. Acad. Sci.* **71**, 522.

Skoza, L., and Mohos, S. (1976). *Biochem. J.* **159**, 457.

Snyder, F., and Stephens, N. (1959). *Biochim. Biophys. Acta.* **34**, 244.

Stackebrandt, E., and Woese, C. R. (1981). *In* "Molecular and Cellular Aspects of Microbial Evolution," p. 1. Cambridge Univ. Press, Cambridge.

Stahl, E. (ed.) (1962). "Dünnschicht-Chromatographie." Springer, Berlin.

Staub, A. M. (1965). *Meth. Carbohyd. Chem.* **5**, 92.

Stellner, K., Saito, H., and Hakomori, S. (1973). *Arch. Biochem. Biophys.* **155**, 464.

Stocker, B. A. D., and Mäkelä, P. H. (1978). *Proc. R. Soc. London Ser. B.* **202**, 5.

Strain, S. M., Fesik, S. W., and Armitage, I. M. (1983). *J. Biol. Chem.* **258**, 2906.

Strittmatter, W., Weckesser, J., Salimath, P. V., and Galanos, C. (1983). *J. Bacteriol.* **155**, 153.

Sugiyama, H. (1978). *Heterocycles* **11**, 615.

Tai, T., Yamashita, K., Ogata-Arakawa, M., Koide, N., Muramatsu, T., Iwashita, S., Inoue, Y., and Kobata, A. (1975). *J. Biol. Chem.* **250**, 8569.

Takahashi, N., and Murachi, T. (1976). *Meth. Carbohydr. Chem.* **7**, 175.

Takayama, K., Quereshi, N., Ribi, E., and Cantrell, J. L. (1984). *Rev. Infect. Dis.* **6**, 439.

Taylor, R. L., Shively, J. E., and Conrad, H. E. (1976). *Meth. Carbohydr. Chem.* **7**, 149.

Tharanathan, R. N., Mayer, H., and Weckesser, J. (1978a). *Biochem. J.* **171**, 403.

Tharanathan, R. N., Weckesser, J., and Mayer, H. (1978b). *Eur. J. Biochem.* **84**, 385.

Tharanathan, R. N., Salimath, P. V., Weckesser, J., and Mayer, H. (1985). *Arch. Microbiol.* (in press).

Tomshich, S. V., Gorshkova, R. P., and Ovodov, Yu. S. (1983). *In* "Abstracts of Second European Symposium on Carbohydrates and Glycoconjugates," p. C-34. Budapest.

Trevelyan, W. E., Procter, D. P., and Harrison, J. S. (1950). *Nature (London)* **166**, 444.

Tsai, C. M., and Frasch, C. E. (1982). *Anal. Biochem.* **119**, 115.

Tsang, Y. C., Wang, C. S., and Alaupovic, P. (1974). *J. Bacterial.* **117**, 786.

Unger, F. (1981). *Adv. Carbohydr. Chem. Biochem.* **38,** 323.

Usui, T., Yamaoka, N., Matsuda, K., Tuzimura, K., Sugiyama, H., and Seto, S. (1973). *J. Chem. Soc., Perkin Trans. 1* 2425.

Vaara, M., Vaara, T., Jensen, M., Helander, I., Nurminen, M., Rietschel, E. T., Mäkelä, P. H. (1981). *FEBS Lett.* **129,** 145.

van Alphen, L., Lugtenberg, B., Rietschel, E. T., and Mombers, Ch. (1979). *Eur. J. Biochem.* **101,** 571.

Vijay, K., and Perdew, G. H., (1982). *FEBS Lett.* **139,** 321.

Volk, W. A., Galanos, C., and Lüderitz, O. (1970). *Eur. J. Biochem.* **63,** 93.

Volk, W. A., Salomonsky, N. L., and Hunt, D. (1972). *J. Biol. Chem.* **247,** 3881.

Warren, L. (1959). *J. Biol. Chem.* **234,** 1971.

Wartenberg, K., Knapp, W., Ahamed, N. M., Widermann, C., and Mayer, H. (1983). *Zentralbl. Bakteriol. Parasitenkd. Infektionkr. Hyg. Abt. 1: Orig. A.* **253,** 523.

Weckesser, J., Mayer, H., Drews, G., and Fromme, I. (1975a). *J. Bacteriol.* **123,** 449.

Weckesser, J., Mayer, H., Drews, G., and Fromme, I. (1975b). *J. Bacteriol.* **123,** 456.

Weckesser, J., Drews, G., Indira, R., and Mayer, H. (1977). *J. Bacteriol.* **130,** 629.

Weckesser, J., Drews, G., and Mayer, H. (1979). *Annu. Rev. Microbiol.* **33,** 215.

Weckesser, J., Mayer, H., Metz, E., and Biebl, H. (1983). *Int. J. Sys. Bacteriol.* **33,** 53.

Weinbaum, G., Kadis, S., and Ajl, S. J., eds. (1971). "Microbial Toxins," Vol. 4. Academic Press, New York.

Weisgerber, C., and Jann, K. (1982). *Eur. J. Biochem.* **127,** 165.

Weisshaar, R., and Lingens, F. (1983). *Eur. J. Biochem.* **137,** 155.

Westphal, O., and Jann, K. (1965). *Methods Carbohydr. Chem.* **5,** 83.

Westphal, O., and Lüderitz, O. (1960). *Angew. Chem.* **72,** 881.

Westphal, O., Lüderitz, O., and Bister, F. (1952). *Z. Naturforsch.* **7b,** 148.

Westphal, O., Westphal, U., and Sommer, T. (1977). *In* "Microbiology-1977," pp. 221–238. Am. Soc. Microbiol, Washington, D.C.

Westphal, O., Lüderitz, O., Rietschel, E. T., and Galanos, C. (1981). *Biochem. Rev.* **9,** 191.

Wheat, R. W. (1966). *In* "Methods in Enzymology" (E. F. Neufeld and V. Ginsburg, eds.), Vol. 8, p. 60. Academic Press, New York.

Wilkinson, S. G. (1974). *J. Lipid Res.* **15,** 181.

Wilkinson, S. G. (1977). *In* "Surface Carbohydrates of the Prokaryotic Cell," p. 97. Academic Press, New York.

Wilkinson, S. G., and Caudwell, P. F. (1980). *J. Gen. Microbiol.* **118,** 329.

Wilkinson, S. G., and Taylor, D. P. (1978). *J. Gen. Microbiol.* **109,** 357.

Wilkinson, S. G., Galbraith, L., and Light-Foot, G. A. (1973). *Eur. J. Biochem.* **33,** 158.

Williams, J. M. (1975). *Adv. Carbohydr. Chem. Biochem.* **31,** 9.

Wollenweber, H. W., and Rietschel, E. T. (1984). *Methods Carbohydr. Chem.* **9,** in press.

Wollenweber, H. W., Broady, K. W., and Rietschel, E. T. (1982). *Eur. J. Biochem.* **124,** 191.

Wollenweber, H. W., Schlecht, S., Lüderitz, O., and Rietschel, E. T. (1983). *Eur. J. Biochem.* **130,** 167.

Wollenweber, H. W., Seydel, U., Lindner, B., Lüderitz, O., and Rietschel, E. T. (1984). *Eur. J. Biochem.* **145,** 265.

Wong, C. G., Sung, S. S. J., and Sweeley, C. (1980). *Methods Carbohydr. Chem.* **8,** 55.

Yosizawa, Z., Sato, T., and Schmid, K. (1966). *Biochim. Biophys. Acta* **121,** 417.

7

Lipid Analysis and the Relationship to Chemotaxonomy

THOMAS G. TORNABENE

School of Applied Biology, Georgia Institute of Technology,
Atlanta, Georgia, USA

I. Introduction

All cells have a cytoplasmic membrane that serves to form a boundary between itself and the exterior world. The composition of membranes is roughly 50% lipids and 50% non-lipids. The lipid portion contains a half-dozen or more different classes of lipids with each consisting of individual molecular species that have a half-dozen or more structural variations. The relative percentage composition of individual lipids and the lipid structuralisation are generally influenced by age of the cells and other environmental parameters. The basic identities of biomembrane lipids have been organised and a seemingly coherent picture has emerged. Although the use of lipids as an aid in classification is just beginning, it will be apparent to the reader that lipid components are useful in bacterial classification.

METHODS IN MICROBIOLOGY
VOLUME 18

In the mid 1950s, the accumulation of knowledge on lipids started at a rapid rate and it is still proceeding today. The detailed data in recent reviews will not be duplicated in this chapter; only selected examples will be used. Lipid data have been published in virtually all scientific journals; however, the journals which specialise on lipids are the *Journal of the American Oil Chemists' Society*, the *Journal of Lipid Research, Biochimica Biophysica Acta, Lipids, Chemistry and Physics of Lipids, Advances in Lipid Research,* and *Progress in Chemistry of Fats and Other Lipids*. For a source of basic background knowledge of lipids, publications by Kates (1964a), Asselineau (1966), Lennarz (1970), Shaw (1974), O'Leary (1975), Gurr and James (1980), Nes and Nes (1980), Erwin (1973), and Razin and Rottem (1982) are recommended. For specialised reviews see the following list of citations: Ikawa (1967), Op den Kamp and Van Deenan (1969), Lederer (1967), Shaw (1974, 1975), Goren and Brennan (1979), Goldfine and Hagan (1972), Reaveley and Burge (1972), Sastry (1974), Goldfine (1972, 1982), Vaczi (1973), Suto and Ryozo (1974), Lechevalier and Moss (1977), Weete (1980), Langworthy *et al.* (1982), Collins and Jones (1981). For comprehensive descriptions of lipid analysis, see Kates (1972), Bergelson (1980), and Christie (1982).

II. Analytical Methods

A. Extraction procedure for total free lipids

The majority of all lipids are membrane associated and present as a complexed integration of both polar and non-polar lipids. To disrupt the hydrogen bonding, electrostatic forces, and interfacial tensions so as to free the lipids, a mixture of solvents with comparable polarity ranges are required. A most versatile and effective extraction procedure is the method of Bligh and Dyer (1959) as modified by Kates *et al.* (1964). The procedure is to add to each 40–50 mg of dried cells or the wet cell paste containing about 40–50 mg of cells (dry wt.), 1 ml of an isotonic salt solution containing divalent ions (0.25 g $MgSO_4$, 1.8 mg $MnCl_2$, 0.22 mg $ZnSO_4$, 0.08 mg $CuSO_4$, 0.005 mg $FeCl_3$ per litre). To each 1 ml of aqueous suspension, add 2.5 ml methanol followed by 1.25 ml chloroform; the miscible solvent mixture is gently shaken and left at room temperature for several hours with intermittent shaking. Centrifuge out the extracted cell debris and decant the supernatant into a glass-stoppered centrifuge tube or a separating funnel. The extracted cell residue is resuspended in 4.75 ml of methanol–chloroform–water (2 : 1 : 0.8, v/v); the mixture is again intermittently agitated and then centrifuged. To the combined supernatant extracts, add 2.5 ml each of chloroform and water; gently roll the containers to mix solvents; centrifuge the mixture or allow the mixture to separate into the methanol–water and chloroform phases by standing at room temperature. The lower chloroform phase is withdrawn and diluted with 0.5–1 ml benzene to aid in removal of traces of water; reduce the volume in a rotary evaporator (30–35°C). The sample is transferred to a tared tube, brought to dryness under a stream of N_2, and weighed. The lipid residue is immediately dissolved in a suitable volume of chloroform, capped, and placed in a refrigerator/freezer. Because of the general lability of lipids toward peroxidation and hydrolysis, one should never store lipids for extended periods of time; best results are obtained with freshly prepared material. However, if lipids must be stored, it is best to dissolve them in freshly distilled chloroform–methanol (2 : 1), flush the storage container with N_2, and store at 0–15°C. The separation of lipids by methanol–chloroform and water partitioning is often effective in removal of most non-lipid contaminants. If excessive amounts of carbohydrates or nitrogen-containing compounds are detected in the lipid extracts in subsequent analyses, the total lipid extract should first be passed through a Sephadex G-25 column to remove possible non-lipid contaminants before further analysis. The procedure is described in detail by Wells and Dittmer (1963).

B. Fractionation of lipids

The next stage in an investigation of lipid composition is the fractionation of the lipids into their various classes: neutral and polar (Fig. 1).

Solvent partitioning. If the lipid extract contains substantially large quantities of pigments, neutral lipids, or polar lipids, it is often advantageous to partition the lipids in cold acetone. The method depends on the general insolubility of most phosphatides in cold acetone.

In general, the acetone-insoluble material will contain virtually all of the lipid phosphorous and polar pigments whereas the acetone-soluble fraction will contain the neutral lipids as glycerides, sterols, sterol esters, hydrocarbons, and carotenoids. Glycolipids such as glycosyldiacyl glycerides and sterol glycosides generally appear in the acetone-soluble fraction, while polyglycosyl diglycerides, and sulphatides may appear with the phospholipids in the acetone-insoluble fraction (Kates, 1972).

Column fractionation. Total lipid mixtures can usually be separated into various classes of polarities on silicic acid columns with organic solvents. A suitable silicic acid is Unisil (100–200 or 200–325 mesh, Clarkson Chemical Co., Williamsport, Pennsylvania). The silicic acid is heated at 120°C overnight before use. The silicic acid (weight ratio of

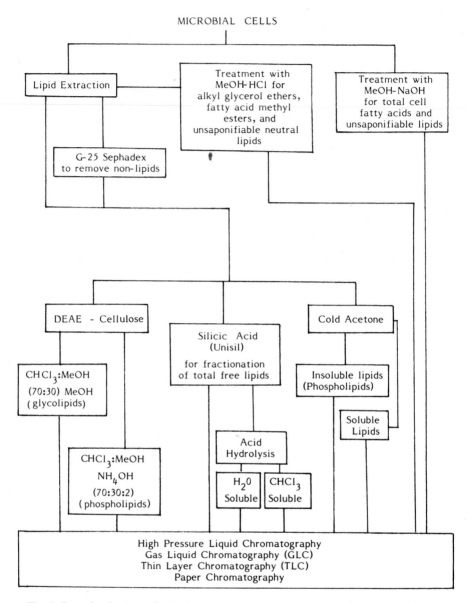

Fig. 1. Procedural scheme for the isolation, fractionation, and analysis of bacterial lipids.

silicic acid to sample, 60 : 1) is suspended in hexane and poured into a column containing hexane and the silicic acid allowed to settle into the column. The tube is tapped gently to dislodge gas bubbles. After the column is filled, it is flushed with approximately 3 bed volumes of hexane. The total lipid sample is taken to dryness under a stream of N_2 at 30–40°C and dried *in vacuo* over KOH pellets, and the desiccator is filled with N_2 before sample removal. The sample is suspended in about 1 ml of hexane and transferred to the top of the column; several small hexane washings ensure quantitative transfer of hexane-soluble materials to the column. The column is then eluted with 1 bed volume of hexane, which is collected. Since the total lipid fraction is not soluble in neutral solvents, the transfer of the total lipid sample to the column has to be repeated with each of the subsequent eluting solvents: benzene, chloroform, acetone, and methanol. Each of these solvents is used sequentially, in several small aliquots, to wash the sample into the column, followed by about 1 bed volume of elution.

The components that should be removed in the hexane eluates are the acyclic hydrocarbons; in the benzene eluate are cyclic hydrocarbons, polyunsaturated acyclic hydrocarbons, fatty acid methyl esters, sterols, fucoxanthin, and xanthophylls; in the chloroform eluate are tri-, di-, and monoglycerides, free fatty acids, and phaeophytin *a*; in the acetone eluate are glycolipids, chlorophylls *a,b,* and carotenoids; and in the methanol eluate are phospholipids, glycolipids, sulpholipids, and chlorophyll *c*.

The eluates are rotoevaporated to small volume and taken to dryness under a stream of N_2. The samples are dissolved in chloroform and transferred to respective sample tubes for analyses.

In our hands, other silicic acids such as Biosil A inadequately fractionated neutral lipids from the polar lipid while fractionation of lipids on Florisil resulted in alteration of polar lipids. When using these column materials, it is advisable to load a Biosil column with chloroform and elute with chloroform as a first solvent followed by acetone and methanol. The chloroform eluate is then taken to dryness, resolubilised in hexane, and then fractionated on a Florisil column, eluting with hexane, benzene, and chloroform to separate the neutral lipids.

Ion exchange column. Occasionally lipid extracts contain a high proportion of glycolipids, phospholipids, and glycophospholipids that cannot be distinctively separated in acetone and methanol eluates on a standard silicic acid column. Such samples should be fractionated on an ion-exchange column such as DEAE cellulose as reported by Rouser *et al.* as described by Kates (1972).

C. Thin-layer chromatography on silicic acid

Thin-layer chromatography (TLC) is a versatile and effective technique for separating lipid complexes.

TLC plates can be prepared with Silica Gel G (with binder) or Silica Gel H (without binder) on 20 × 20-cm plates. Silica Gel G is spread in distilled water slurry (40 g of gel in 90 ml H_2O), whereas Silica Gel H is spread in 0.01% Na_2CO_3 slurry. The plates are spread with a stainless-steel applicator adjusted to deliver a thickness of 0.25–0.5 mm for analytical plates and 0.75–1 mm for preparative plates. The freshly spread plates are dried and then heat activated at 120°C for 12 h before use.

Commercially prepared plates are excellent and highly desirable but expensive when needed in large quantities. Precoated TLC plates with a 0.25-mm hard layer inorganic binder can be purchased from most companies that distribute chemical supplies.

Application of lipids to the plates. Individual components should be applied in quantities of 5–20 μg. For a complex lipid sample the concentrations should range from 100 to 300 μg. The samples can be applied with disposable calibrated capillaries of 1, 5, or 10 μl capacities. The alignment of samples on the plate can be done with a commercial plastic guide or by drawing a heavy dark line on a white sheet of paper, scaling it by a mark at every 3/4 in., and placing the plate over the marked line. The plates should be developed immediately after application of the sample and exposure of the samples to air should be prevented as much as possible.

The plates are placed in an all glass tank (11 × 21 cm) with a ground glass cover. Line the tank with filter paper on three sides to aid in saturating the tank with solvent vapor. The tank must be equilibrated before use.

Lipid standards. Lipid standards are commercially available. In our laboratory, we found that lipid extracts from a fresh red potato contain an abundance of a well-balanced mixture of lipids. Following the same lipid procedure as described, the lipids are extracted and the identities verified by comparing to commercial standards. This lipid stock is then used as the reference standard.

TLC solvents. Although the number of useful developing solvents is large, the ones presented here are usually adequate for surveying most lipid mixtures.

1. Hexane–benzene (9 : 1, v/v) for separating squalene, squalane, dihydro- and tetrahydrosqualene, nonisoprenoid hydrocarbons, and vitamin K (Tornabene *et al.,* 1969).

2. Hexane–diethyl ether–acetic acid (80 : 20 : 1, v/v) or chloroform–diethyl ether (9 : 1, v/v) for separating mono- or dialkyl glycerol ethers and fatty acids (Tornabene and Langworthy, 1979).

3. Petroleum ether–diethyl ether–acetic acids (90 : 10 : 1, v/v) for separating neutral pigments, naphthoquinones, and ubiquinones (Kates, 1972).

4. Double development in one direction: Diethyl ether–benzene–ethanol–acetic acid (40 : 50 : 2 : 0.2) as first solvent (run one-half of plate, dry inside a box flushed with N_2) and hexane–diethyl ether (96 : 4) as second solvent (run to top of plate) for separating ketones, alcohols, triglycerides, diglycerides, monoglycerides, fatty acid esters, free fatty acids, and sterols (Freeman and West, 1966).

5. Chloroform–acetone–methanol–acetic acid–water (50 : 20 : 10 : 10 : 5) for separating phospho- and glycolipids (Lepage, 1967).

6. Two-dimensional polar lipid solvent: Chloroform–methanol–acetone–diethylamine–water (20 : 34 : 37.5 : 6 : 4.5) in the first dimension and chloroform–methanol–ammonium hydroxide (65 : 25 : 5) in the second dimension for separating total lipid mixtures (Evans *et al.*, 1982).

For preparative TLC, bands are scraped and eluted through sintered-glass filters with appropriate solvent mixtures such as chloroform–methanol (2 : 1, v/v).

Visualisation and characterisation of spots on TLC. A general purpose visualisation of components is accomplished by exposing the plates to iodine vapours produced by placing iodine crystals in a TLC-type tank. All compounds that are potential proton donors stain yellowish-orange. Storing the plate in a container that is flushed with N_2 will remove almost all the iodine. The samples can then be further studied. A UV lamp can be used for locating the benzoquinones and naphthoquinones.

The following is a list of destructive general stains: (1) sulphuric acid, 30–40% aqueous, or as a 5% ethanolic solution; (2) 55% H_2SO_4 containing 0.6% K_2CrO_3, (3) ammonium sulphate, 20% aqueous solution. One of these reagents is sprayed on the plate with an all-glass atomiser. The sprayed plates are charred in an oven at 120°C for 5–15 min.

Colorimetric characterisation of components consists of spraying the plate with ninhydrin for amino compounds, molybdate for phosphates, Draggendorf reagent for quarternary amines, α-naphthol for glycolipids sulphuric and acetic acid for sterols and sterol esters, or periodate-Schiff reagent for vicinal glycols according to procedures reviewed by Kates (1972).

D. Lipid analysis

Mild alkaline hydrolysis. In addition to the evaluation of the chromatographic properties of the lipids it is often helpful in the identification of the acyl glyceride and alkyl ether glycerides to deacylate the samples and to analyse the resulting chloroform-soluble and water-soluble components.

Column-fractionated and thin-layer chromatography isolated lipids are deacylated (Tornabene and Ogg, 1971) as follows: 0.2 ml chloroform, 0.3 ml methanol, and 0.5 ml of fresh 0.2 M NaOH–methanol are successively added to 0.8–16 mg of lipids in a 15-ml centrifuge tube, mixed, and left at room temperature for 15 min; 0.2 ml methanol, 0.8 ml chloroform, and 0.9 ml water are immediately added and the tubes centrifuged for 1–2 min. The upper phase of methanol–water is removed and placed in a 15-ml centrifuge tube and slightly acidified by adding 0.3 ml of Rexyn RG 50 resin. After centrifugation the supernatant is removed and made alkaline with 1 M NH$_4$OH. The sample is placed in a 40°C water bath and blown almost to dryness under a stream of N$_2$. The sample is taken up in methanol–water (10 : 9, by volume) and analysed for deacylated lipids.

The chloroform layer is washed with methanol–water until neutral. Benzene is added and the sample blown to dryness under a stream of N$_2$. The residue is analysed for fatty acids, alkyl glycerol ethers, and unsaponifiable material.

The methanol–water-soluble fraction is characterised on precoated cellulose TLC plate (Eastman chromatograms 6064) with solvents consisting of 3.8 mM EDTA and 0.7 M NH$_4$HCO$_3$ in 90 mM NH$_4$OH containing 67% by volume ethanol in the first dimension and isobutyric acid–water–concentrated NH$_4$OH (66 : 33 : 1, by volume) in the second dimension (Short et al., 1969).

The components are visualised by the ortho-tolidine method for vicinal hydroxyl groups by dipping the plate in 5 ml 0.25 M NaIO$_4$ in 95 ml acetone. After drying for 15 min the plate is dipped into a mixture containing 2.12 g o-tolidine, 6 ml acetic acid, 40 ml water, and 950 ml acetone. On drying, spots appear yellow on a blue background. Products are identified by comparing their R_f values to those of known standards. The plates are then overstained with molybdate solution for phosphoric esters by dipping the plate in a solution of 1 g ammonium molybdate in 8 ml water mixed in 3 ml HCl and 3 ml 70% perchloric acid and made up to 100 ml with acetone (Burrows et al., 1952).

Acid hydrolysis of total, neutral, and polar lipids. To analyse polar lipids that are alkyl glycerol ethers or those having sugars and/or amino compounds linked to the lipids several different procedures are followed (Fig. 1). The lipids are hydrolysed in anhydrous 2.5% methanolic HCl; water is added and lipids extracted with petroleum ether (Kates , 1964b). The water-soluble materials from a methanolic HCl digest or from an alkaline deacylation are hydrolysed in 2 N HCl for 2 h at 100°C for hydrolysis of glycosidic linkages or in 4 N HCl for 6 h at 100°C for amide linkages. The acid hydrolysates are extracted with petroleum ether to remove the lipids that were bound and released during the hydrolysis. The petroleum ether extract can be examined by TLC and by derivatisation and gas–liquid chromatography (GLC). The remaining acid hydrolysate is evaporated to dryness under N_2 followed by vacuum drying over KOH. The water-soluble material is analysed on Whatman 3MM paper chromatograms developed in pyridine–ethyl acetate–water (4 : 10 : 10, upper phase). The components with vicinal hydroxyl groups are detected with alkaline $AgNO_3$ (Trevelyan et al., 1950); the amide compounds are detected with ninhydrin (Kates, 1972). The amino compounds are further analysed on an amino acid analyser; the sugar compounds are analysed by GLC as the additol acetates (Albersheim et al., 1967; Tornabene et al., 1982). The petroleum ether extracts are analysed by silicic acid TLC for fatty acids, sterols, and alkyl glycerol ethers (Tornabene and Langworthy, 1979) in developing solvent of hexane–diethyl ether–acetic acid (80 : 20 : 1). Components suspected of being acid-stable alkyl glycerol ethers by co-migration to standards (Kates, 1978; Tornabene and Langworthy, 1979) are further digested by refluxing in 47% HI for 12 h to release alkyl iodides (Kates et al., 1965). The halides can be analysed directly by GLC or by reducing them to the alkane derivatives by refluxing in acetic acid and zinc (Panganamala et al., 1971) or by conversion to the alkyl acetate by refluxing with silver acetate in acetic acid (Kates et al., 1965) and analysed by GLC (Tornabene and Langworthy, 1979). Intact alkyl glycerol ethers obtained after acid methanolysis can be converted to trimethyl silyl ether derivatives and analysed by GLC (Tornabene and Langworthy, 1979; Kates, 1978; Langworthy et al., 1983).

Fatty acid analyses. Total cellular fatty acids can be obtained by hydrolysis of cells in 2 N NaOH in 50% aqueous methanol for 1 h at 100°C (Fig. 1). The digest is extracted with petroleum ether to remove unsaponifiables. The digest is then acidified with HCl and the saponifiable lipids are removed by extracting 3 times with chloroform. Fatty acids are also liberated from extracted free lipids by transmethylation in 2.5% methanolic HCl at 100°C for 1 h (Kates, 1964b). The fatty acids are removed by adding water and extracting with petroleum ether. The fatty acids released by alkaline or acid hydrolysis can be converted to methyl ester derivatives by hydrolysis using the 2.5% methanolic HCl method or by treatment with boron trichloride–methanol reagent (Applied Science Labs, State College, Pennsylvania) and heating the mixture for 15 min at 80–85°C, adding water, and extracting with petroleum ether.

Instrumentation. Fatty acids methyl ester derivatives, cyclic and acyclic, non-isoprenoid and isoprenoid hydrocarbons (isolated predominately in the hexane and benzene eluates from silicic acid columns), as well as lipids and sugars extracted from hydrolysates of polar lipid fractions are analysed by gas–liquid chromatography (GLC), gas–liquid chromatography/mass spectrometry (Tornabene *et al.*, 1983; Langworthy *et al.*, 1983; Holzer *et al.*, 1979), and/or high-performance liquid chromatography (Driskell *et al.*, 1982; Phillips *et al.*, 1982). Amino compounds are analysed on an automatic amino acid analyser. Isolated pigments and lipid components obtained by silicic acid column fractionation and preparative TLC are analysed spectrophotometrically with IR- and UV-visible spectrophotometers.

III. Taxonomic significance of lipids

A. Polar lipids

The most widespread and major occurring lipid class in biomembranes of eubacteria is the phospholipids consisting of the diacyl glycerol phosphate type. The composition of these diacyl glycerol derivatives in biomembranes is relatively uniform with regards to their overall stereostructures (Ikawa, 1967; Lechevalier and Moss, 1977; Goldfine, 1972, 1982) and they are generally of little value as chemotaxonomic markers. In comparative studies of specific bacterial groups, however, distribution patterns of phospholipids exhibited phyletic differences. The results from studies on phospholipids in bacteria have been compiled by Lechevalier and Moss (1977). Comparative descriptions of the occurrence and distribution of the phospholipids as well as primary references can be found in this review. There is a group of diverse bacteria that is clearly distinct from eubacteria by the fact that they do not have acyl glycerol-derived lipids. These bacteria are referred to as the archaebacteria (see Kandler, 1982) and consist of subgroups of methanogens, extreme halophiles, and thermoacidophiles. At least one other bacterium exists with no acyl glycerol lipids which is not an archaebacterium. It has been identified as a thermophilic anaerobe *Thermodesulfotobacterium commune* (Langworthy *et al.*, 1983).

The polar lipids of the archaebacteria are glycolipids and phospholipids (Langworthy *et al.*, 1974; De Rosa *et al.*, 1980a; Kushwaha *et al.*, 1981) that contain sugar and phosphoderivatives like those in most bacteria (Goldfine, 1972, 1982) but with the exception that isoprenoid chains are ether linked to glycerol. There are two types of ether-linked glycerols: diphytanyl (two C_{20} isoprenoids) diether glycerol and dibiphytanyl (two C_{40} isoprenoids) tetraether diglycerols. Both types of alkyl glycerol ethers are found in all archaebacteria with the exception of extreme halophilic bacteria (*Halobacterium, Halococcus, Sarcina*) which have only the diphytanyl diether glycerol structure (Langworthy *et al.*, 1974, 1982; Tornabene and Langworthy, 1979). *Thermoplasma* and *Sulfolobus* are two archaebacteria that are differentiated from others (methanogens and halophiles) by some of the C_{40} biphytanyl chains containing cyclopentyl rings. *Thermoplasma* has both monocyclic and bicyclic biphytanyl chains whereas *Sulfolobus* has mono-, bi-, tri-, and tetracyclopentyl C_{40} biphytanyl chains (Langworthy *et al.*, 1982; DeRosa *et al.*, 1977, 1980a,b).

A rapid and simple procedure for a presumptive test for archaebacteria had been simply subjecting lipid extracts to 2.5% methanolic HCl hydrolysis and surveying the hydrolysates by TLC for the presence of acid-stable glycerol ethers (Tornabene and Langworthy, 1979; Ross *et al.*, 1981). The recent discovery of acid-stable ether lipids in *Thermodesulfotobacterium commune*, which is not an archaebacterium (Langworthy *et al.*, 1983), demonstrated the necessity to analyse the alkyl side chains also. The glyceride lipids of *T. commune* are *sn*-1,2-dialkyl diethers and *sn*-1-alkyl glycerol ether. The alkyl chains are anteiso-C_{17}, iso-C_{16}, *n*-C_{16}, iso-C_{17}, iso-C_{18}, anteiso-C_{18}, iso-C_{19}, and *n*-C_{19} hydrocarbons ether linked to the glycerol (Langworthy *et al.*, 1983). The presence of these ether lipids suggests that this organism evolved similarly to the archaebacteria but the absence of isoprenoid side chains and the presence of alkyl hydrocarbons that are structurally common side chains in eubacterial lipids distinctly sets this organism apart from both archaebacteria and eubacteria. The ether lipids of the archaebacteria and *T. commune* are not to be confused with the acid-labile ether linkage in plasmalogens. 1-Alkenyl-2-acyl glycerolipids (plasmalogens) are major constituents that are characteristic of strict anaerobes (Goldfine and Hagen, 1972; Kim *et al.*, 1970; Kamio *et al.*, 1969; Johnston and Goldfine, 1982).

The lipids containing nitrogen that are common in bacteria are usually the phosphatidylethanolamine, phosphatidylcholine, and/or phosphatidylserine types (Goldfine, 1972). There are also the *O*-aminoacylphosphatidyl glycerols most commonly found in Gram-positive bacteria (Goldfine, 1972), *N*-acylphosphatidylserine discovered in *Rhodopseudomonas* (Donohue *et al.*, 1982), the ornithine lipids, the exact structural identities of which have not been fully elucidated, and the sphingolipids (Miyagawa *et al.*, 1978). The ornithine lipids consist of ornithine with N- and/or O-acylated fatty acids, hydroxy fatty acids, and/or alcohols. The ornithine lipids were found in large quantities in the photosynthetic bacteria of the family Rhodospirillaceae but absent in the Chromatiaceae (Imhoff *et al.*, 1982; Brooks and Benson, 1972; Park and Berger, 1967). The ornithine lipids are not restricted to Rhodospirillaceae, however, having been isolated from *Streptomyces* (Kimura and Otsuka, 1969), *Thiobacillus* (Knoche and Shively, 1972), *Desulfovibrio* (Makula and Finnerty, 1975), *Pseudomonas* (Wilkinson, 1972), *Gluconobacter* (Tahara *et al.*, 1976), *Paracoccus* (Thiele *et al.*, 1980), and *Bordetella* (Kawai and Moribayashi,

1982). The amino acid linked to the O-aminoacylphosphatidyl glycerol is lysine, alanine, or ornithine (Goldfine, 1972). A novel lipid containing a glucosaminyl moiety rather than an amino acid was identified in *Bacillus megaterium* and *Pseudomonas ovalis* (Shaw, 1975). The sphingolipids appear to be characteristic of some members of the genus *Bacteroides*. They consist of ceramide phosphorylethanolamine and ceramide phosphoryl glycerol as major components (Miyagawa *et al.*, 1978).

The existence of sulpholipids in prokaryotes is well established. The extent of its distribution, however, is not known since the presence of the sulpholipid is difficult to detect unless tagged with ^{35}S. Sulphoquinovosyldiacyl glycerides exist in varying quantities in photosynthetic bacteria and blue-green bacteria (Imhoff *et al.*, 1982; Kenyon, 1978). An analogous distribution of sulphonoquinovosyldiacyl glyceride is seen in eukaryotic photosynthetic organisms. A sulphonoquinovosyldiacyl glyceride was also detected in *Bacillus acidocaldarius* (Langworthy *et al.*, 1976). A sulphated glycodiosyl diether, a sulphated glycotriaosyl diether, and a diether analogue of phosphatidyl glycerol sulphate were identified in extremely halophilic bacteria (Kushwaha *et al.*, 1982). The anaerobic bacteria of the genus *Capnocytophaga* contain N-acylated and non-N-acylated 2-amino-3-hydroxy-15-methylhexadecane-1-sulphonic acids as major cell components (Godchaux and Leadbetter, 1980) which, at present, differentiates them from all other bacteria.

The glycolipids as a class are broadly distributed in eubacteria and archaebacteria, but not as uniformly as the phospholipids. The glycolipids have structural diversity some of which are species specific (Pask-Huges and Shaw, 1982; Oshima and Yamakawa, 1974; Langworthy *et al.*, 1982, 1974, 1976). The glycolipids have been well reviewed, however (Shaw, 1975; Brenner *et al.*, 1978; Fisher, 1978; Sastry, 1974), and additional efforts to redescribe the diverse subtleties that exist with the quantity, nature, and distribution of the hydrophilic moieties of the glycolipids will not be made here. There are a few glycolipids where the hydrophobic moieties are particularly interesting from the standpoint of being differentiating chemical markers. A novel glycolipid that consists of N-acylglucosamine linked to a pentacyclic triterpene-derived tetrahydroxypentose ($C_{35}H_{62}O_4$) was identified in *Bacillus acidocaldarius* (Langworthy *et al.*, 1976; Langworthy and Mayberry, 1976). The unique feature of this compound is the presence of the pentacyclic triterpenoid (hydroxyhopane). It is a polycyclic triterpene that differs from the tetracyclic sterol-type compound by the presence of an additional ring. The hydroxyhopane configuration has also been isolated from *Acetobacter xylinum* and seven other species of *Acetobacter* (Forster *et al.*, 1973; Carson *et al.*, 1967; Rohmer and Ourisson, 1976a,b,c) and *Acetomonas oxydans, Methylococcus cap-*

sulatus, Methylosinus trichosporium, Anabena variabilis, and *Nostoc muscorum* (Rohmer and Ourisson, 1976b; Ourisson *et al.,* 1979). The hopanoids also exist as (non-oxygenated) polycyclic triterpene hydrocarbons. These are discussed with the neutral lipids.

B. Acyclic hydrocarbons

The neutral lipids of different bacteria are frequently more diverse than the polar lipids and more commonly species specific. The occurrence of non-isoprenoid, acyclic hydrocarbons of the Micrococcaceae differentiate them from almost all other bacteria as well as differentiating genera within the Micrococcaceae (Tornabene *et al.,* 1970; Morrison *et al.,* 1971; Kloos *et al.,* 1974; Tornabene, 1981, 1982). Micrococci synthesise a distinct pattern of tetrads of methyl-branched acyclic monoolefins for each carbon fraction in the range C22 to C30 (Tornabene and Markey, 1971; Albro, 1971). Distinctive distribution patterns of hydrocarbons (e.g., C_{27}, C_{28}, C_{29} of *M. luteus;* C_{25}, C_{26}, C_{27} of *M. varians;* C_{30}, C_{31}, C_{32}, of *M. sedentarius*) clearly separate the members into subgroups (Tornabene *et al.,* 1970; Morrison *et al.,* 1971; Kloos *et al.,* 1974). The yields of hydrocarbons vary from 0.1 to 2% of dry cell weight (Tornabene, 1981; Kloos and Tornabene, unpublished results). The absence of these hydrocarbons in staphylococci provides a defined and simple test for differentiation between the staphylococci and the micrococci (Morrison *et al.,* 1971). Non-isoprenoid, acyclic hydrocarbons have also been found in *Pseudomonas maltophilia.* The hydrocarbon composition in all 22 species tested, comprising up to 2% of cell dry weight, is unlike those of micrococci, being families of branched mono-, di, and triunsaturated hydrocarbons with predominance of the C_{29} fraction (Tornabene and Peterson, 1978; Tornabene, unpublished results). These hydrocarbons have not been detected in any other pseudomonad except *P. maltophilia* (Tornabene, 1981).

The occurrence of acyclic isoprenoid hydrocarbons in bacteria is more widespread than non-isoprenoid hydrocarbons. Pristane (C-19) and phytane (C-20) have been the most publicised because of their geochemical significance, but they have limited distribution in bacteria (Han and Calvin, 1969; Tornabene *et al.,* 1979). Squalene (C-30) is widely distributed among bacteria (Amdur *et al.,* 1978; Han and Calvin, 1969; Tornabene, 1978, 1981; Tornabene *et al.,* 1969, 1979, 1982; Holzer *et al.,* 1979; Goldberg and Shechter, 1978; Week and Francesconi, 1978; Bird *et al.,* 1971b; Maudinas and Villoutriex, 1976; Suzue *et al.,* 1967). Although the quantitative data on squalene are sparse, earlier reports concerning levels in prokaryotes (0.001–0.1 mg g^{-1} cells) are incorrect. Recent reports of squalene contents are *Halobacterium* (Tornabene, 1978; Tornabene *et al.*, 1969) with 1 mg g⁻¹ of cells, *Methylococcus capsulatus* (Bird *et al.*, 1971; Bouvier *et al.*, 1976) with 5.5 mg g⁻¹ of cells, *Cellulomonas dehydrogenans* (Suzue *et al.*, 1967) with 0.5 mg g⁻¹ of cells, and in the methanogens, *Sulfolobus* and *Thermoplasma* with 10 mg g⁻¹ of cells (Tornabene *et al.*, 1979; Holzer *et al.*, 1979). These squalene concentrations exceed those in eukaryotic microorganisms (for example, *Aspergillus nidulans* which contains 0.3 mg g⁻¹ of cells) (Bird *et al.*, 1971b; Bouvier *et al.*, 1976).

In addition to squalene, the neutral lipids of nine species of methanogenic bacteria (including five methanobacilli, two methanococci, a *Methanospirillum*, one *Methanosarcina*) as well as two thermoacidophilic bacteria (*Thermoplasma* and *Sulfolobus*) contained as major components C-25 and/or C-20 acyclic isoprenoid hydrocarbons with a continuous range of hydroisoprenoid homologues. The range of acyclic isoprenoids detected were from C-14 to C-30. Apart from *Methanosarcina barkeri*, squalene and/or hydrosqualene derivatives were the predominant components in all species studied. The predominant components of *M. barkeri* were a family of C-25 homologues (Tornabene *et al.*, 1979; Holzer *et al.*, 1979). The structural differences among many of the isoprenoids found in these bacteria, collectively referred to as archaebacteria, are seen in the carbon skeletons of the individual isoprenoids. The carbon skeleton of the C-30 isoprenoid is that expected from a tail-to-tail (pyrophosphate end-to-pyrosphosphate end) condensation product of two farnesyl derivatives; however, a positional isomer of a C-30 isoprenoid that is consistent with a head-to-tail condensation route was also identified (Tornabene *et al.*, 1979; Holzer *et al.*, 1979). The C-25 isoprenoid fraction comprises constituents that result from tail-to-tail condensations of farnesyl and geranylgeranyl pyrophosphate and one isopentenyl pyrophosphate. With the exception of phytane (C-20), the remaining isoprenoids also appear to be synthesised through condensations that involve more than one biosynthetic pathway (Tornabene *et al.*, 1979; Holzer *et al.*, 1979). The distribution of the neutral lipid components and their specific variations in relative concentrations emphasised the differences between the test organisms while the generic nature of the isoprenoid hydrocarbons demonstrated similarities between this diverse collection of bacteria (Tornabene *et al.*, 1979; Holzer *et al.*, 1979).

C. Cyclic neutral lipids

The enzymatic capability of bacteria to cyclise triterpenes, such as squalene, to hopanoids or sterols is apparently limited to a small number of diverse and seemingly unrelated organisms. Whether or not this poly-cyclisation of triterpenes signals a point of relatedness or phyletic relationship cannot be determined with the paucity of data available.

Occasional reports claiming the isolation of oxygenated tetracyclic triterpenes (sterols) from bacteria are not generally substantiated. The principal exception is the bacterium *Methylococcus capsulatus*. Four major sterols were identified (Bouvier *et al.*, 1976; Bird *et al.*, 1971b) as follows: 4,4-dimethylcholesta-8(14)24-dienol; 4,4-dimethylcholest-8(24)-enol; 4-methyl-8(14)-enol; and 4-methylcholesta-8(14)24-dienol. Traces of lanosterol and lanost-8-enol (Bouvier *et al.*,1976) were also detected. A similar sterol composition was also detected in trace quantities (0.004–0.01% of cell dry weight) in the soil bacterium *Azotobacter chroococcum* (Schubert *et al.*, 1968). The sterols were identified as lanosterol, 4,4-dimethylcholesta-5,24-dienol, 4,4-dimethylcholest-8-enol, β-ergosterol, ergosta-7,22-dienol, and ergosterol. The major difference between *M. capsulatus* and other organisms that have the capacity to synthesise sterols is apparently a metabolic block preventing complete demethylation at the level of the 4-methyl position. It is interesting that *A. chroococcum* contains lanosterol, 14-dimethyllanosterol homologues, as in *M. capsulatus,* as well as ergosterol and its homologues. Minute quantities of sterol have also been reported to exist and be synthesised by *Escherichia coli* (Shubert *et al.*, 1968), *Streptomyces olivaceus* (Shubert *et al.*, 1967), and *Cellulomonas* (Week and Francesconi, 1978). The sterol contents are unlike those of *M. capsulatus* and *A. chroococcum*. *Escherichia coli* contained cholesterol, (24S)-ergost-7-enol, (24R)-stigmast-5-enol, and (24R)-stigmasta-5,22-dienol. *Streptomyces olivaceus* extracts contained only cholesterol. *Cellulomonas* contained cholesterol and β-sitosterol.

The hopanes, as polycyclic hydrocarbons, are not derivatised as the hydroxy hopanes described earlier. Two families of pentacyclic triterpene hydrocarbons were identified in *Zymomonas mobilis* (Tornabene *et al.*, 1982). One family consists of five six-carbon ring hydrocarbons and the other family consists of four six-carbon rings plus one five-carbon ring hydrocarbons. A pentacyclic triterpene hydrocarbon has also been found in *B. acidocaldarius* (DeRosa *et al.*, 1973), *Methylococcus* (Bird *et al.*, 1971a), and possibly blue-green bacteria (Gelpi *et al.*, 1970).

Cyclised isoprenoid derivatives found in bacteria, in addition to the hopanoids (pentacyclic triterpenes) and steroids (tetracyclic triterpenoids), are the benzoquinone and naphthoquinone structural types. This distribution of isoprenoid quinones in bacteria and their usefulness in chemotaxonomy have been comprehensively reviewed by Collins and Jones (1981) and Bentley and Meganathan (1982) and in chapter 11 in this volume. The majority of the aerobic Gram-negative bacteria produce only benzoquinones; most Gram-negative obligate anaerobes produce naphthoquinones; Gram-negative facultative anaerobic rods produce benzoquinones, napthoquinones, or a combination of both; blue-green bacteria produce both benzoquinones and naphthoquinones; and Gram-positive bacteria produce only naphthoquinones and never benzoquinones. Exceptions to the more common occurrence of benzoquinones in aerobic Gram-negative bacteria are the gliding bacteria and archaebacteria which produce either naphthoquinones or no isoprenoid quinones. These isoprenoid quinones function in both structure of membranes and function of election transport systems.

Carotenoids are also a class of isoprenoids. Oxygenated and non-oxygenated pigments of the C_{40} series as well as their precursors are widespread in bacteria. These types of pigments are distributed in both the neutral and polar lipid fractions eluted from a silicic acid column. The carotenoids are reviewed in Chapter 8 in this volume.

D. Fatty acids

All bacteria contain fatty acids. The majority of all fatty acids in eubacteria are acylated to glycerol, sugar, or amino compounds; relatively small amounts are free fatty acids. In the case of the archaebacteria, the fatty acids are relatively minor amounts with the largest proportion of them being free fatty acids.

There are several principal types of fatty acid patterns. There are those that have predominantly branched fatty acids, even-numbered carbon chains, odd-numbered carbon chains, or combinations of each. In additions, there are hydroxy fatty acids, cyclopropane fatty acids, and mixtures of saturated and unsaturated acids. The hydroxy fatty acids are generally bound to sugars and amino compounds and not extracted by the Bligh–Dyer methods. They can be removed by whole-cell hydrolysis methods. The major fatty acids of all bacteria are within the range of C-12 to C-20. Thus, there are a large number of varieties of fatty acid composites, and many groups of bacteria, genera, and species can be differentiated in part on the basis of their fatty acid contents. Methyl-branched fatty acids of the iso and anteiso configurations occur predominantly in Gram-positive bacteria (Kaneda, 1977; Fulco, 1983). Gram-negative exceptions are *Pseudomonas maltophilia* (suspected of being a *Xanthomonas,* Swings et al., 1983), *Legionella* (Moss, 1981), and *Bacteroides* (Kunsman, 1973). Even then, the distribution patterns are different in these bacterial groups. For example, the even-numbered carbon chains have the configuration of iso and normal while the odd-numbered carbon chains are iso and anteiso for Gram-positive micrococci, staphylococci,

TABLE I

Studies on fatty acids of various bacteria

Group	References
Alcaligenes	Dees *et al.* (1980)
Arthrobacter	Walker and Fagerson (1965)
Bacillus	DeRosa *et al.* (1974); Kaneda (1977)
Bordetella	Dees *et al.* (1980); Kawai and Moribayashi (1982)
Brucella	Dees *et al.* (1980, 1981a)
Campylobacter	Blaser *et al.* (1980); Tornabene and Ogg (1971)
Capnocytophaga	Dees *et al.* (1982)
Caulobacter	Chow and Schmidt (1974)
Clostridium	Ellender *et al.* (1970); Fugate *et al.* (1971); Moss and Lewis (1967)
Corynebacterium	Moss *et al.* (1967); Moss and Cherry (1968); Prefontaine and Jackson (1972)
Escherichia	Marr and Ingraham (1962)
Flavobacteria	Moss and Dees (1978)
Haemophilus	Moss and Dunkleburg (1969); Brice *et al.* (1979)
Lactobacillus	Uchida and Mogi (1973a,b); Uchida (1975)
Legionella	Moss (1981)
Leptospira	Kondo and Ueta (1972); Livermore and Johnson (1974)
Listeria	Raines *et al.* (1968)
Methylotrophs	Weaver *et al.* (1975); Makula (1978)
Micrococcus	Jantzen *et al.* (1974a); Tornabene *et al.* (1967); Morrison *et al.* (1971)
Mycobacterium	Hung and Walker (1970); King and Perry (1975); Thoen *et al.* (1971a,b, 1972); Guerrant *et al.* (1981)
Neisseria	Jantzen *et al.* (1974b); Lambert *et al.* (1971); Lewis *et al.* (1968); Moss *et al.* (1970); Sud and Feingold (1975); Adams *et al.* (1969); Brice *et al.* (1979)
Pasteurella	Dees *et al.* (1981b)
Propionibacteria	Moss *et al.* (1969)
Pseudomonas	Tornabene and Peterson (1978); Dees and Moss (1975); Kaltenbach *et al.* (1975); Moss *et al.* (1972, 1973); Samuels *et al.* (1973); Moss and Dees, 1975; Dees *et al.* (1979)
Sphingobacterium	Yabuuchi and Moss (1982)
Spirochaeta	Livermore and Johnson (1974)
Staphylococcus	Jantzen *et al.* (1974a)
Streptococcus	Amstein and Hartman (1973); Drucker (1972); Drucker and Owens (1973); Drucker *et al.* (1973, 1974a,b); Lambert and Moss (1976)
Treponema	Cohen *et al.* (1970); Livermore and Johnson (1974)
Vibrio	Brian and Gardner (1968)
Xanthomonas	Rietschel *et al.* (1975)
Yersinia	Tornabene (1973)
Zymomonas	Tornabene *et al.* (1982)

and *Bacillus* (Tornabene *et al.*, 1970; Kaneda, 1977). The Gram-negative bacteria do not have the repeated configuration for each even- and odd-numbered carbon chain.

Polyunsaturated fatty acids generally do not exist in bacteria except in blue-green bacteria which contain a photosynthetic apparatus like those of the eukaryotic system (Stanier and Cohen-Bazire, 1977). Cyclopropyl fatty acids are predominantly in Gram-negative bacteria but the Gram-positive exceptions include *Lactobacillus, Clostridium, Streptococcus, Aerococcus,* and *Peptococcus* (Lechevalier and Moss, 1977). Hydroxyl fatty acids (bound fatty acids) are predominantly found in Gram-negative bacteria (Lechevalier and Moss, 1977) and are those most often linked to the lipopolysaccharides of the cell wall membrane (Wilkinson, 1977; Luderitz *et al.*, 1982). Polybranched fatty acids are restricted to the acid-fast bacteria (Goren, 1972; Goren and Brennan, 1979; Barksdale and Kim, 1977).

The fatty acid compositions are routinely used to support the identification of specific bacteria and as analytical markers in diagnostic bacteriology. Because of the extensiveness of the number of studies conducted on bacterial fatty acids and their application to chemotaxonomy, the great diversity of fatty acid compositions, and the proposed differentiation of bacteria on the basis of trace to relatively minor quantities of fatty acids, it is impractical to attempt to cover all of these aspects in this chapter. Instead, the reader is directed to the various studies referenced in Table I.

The analyses of short-chain fatty acids that are secreted into the growth medium have been used routinely to identify anaerobic bacteria (Holdemann *et al.*, 1977; Sutter *et al.*, 1980). The GLC analysis of short-chain fatty acids has proven to be an effective diagnostic tool for both fermentative and nonfermentative bacteria (Moss, 1981; Moss and Nunez-Montiel, 1982).

IV. Summary

Lipids are essential to the structure and function of all biomembranes. The nature of the specific lipid compositions is diverse but uniform in the requirement for non-polar and polar lipids. Lipids are essential as enzyme cofactors, electron transport intermediates, substrate transport factors (Collins and Jones, 1981; McElhaney, 1982; Bentley and Meganathan, 1982), quenchers of singlet-state oxygen, light harvesting systems, and/or protectors of the cells from harmful light rays (Goodwin, 1980; Pfenning; 1977; Glazer, 1983; Rebeiz and Lascelles, 1982; Stanier and Cohen-Bazire, 1977). Some lipid components reflect metabolic regulation of the cell by the degree of unsaturation that occurs in isoprenoid hydrocarbons (Tornabene, 1978), and fatty acids (Fulco, 1983; Bloch, 1977), the changes in concentration of nonisoprenoid hydrocarbons (Tornabene, 1981), the predominance of phospholipid components (Raetz, 1978; Goldfine, 1982), and the relative predominance of specific fatty acids (Fulco, 1983; Goldfine, 1982). Thus, total lipid compositions or relative intensities of individual lipid components will vary with changes in media, temperature, aeration, and age of the cells. These factors have to be considered in comparative studies. The diversity of bacterial lipids, the uniqueness of particular lipid classes, and the difficulties in relating lipid composition to biomembrane functions are all evidence of the complexities of lipids in biomembranes.

The applicability of a lipid or a lipid class as a chemotaxonomic tool is currently restricted to differentiating between specific bacterial groups, genera, or species. The fatty acids are the lipid components most widely used in chemotaxonomy. Both volatile fatty acids (1) and total cell fatty acids (2) have been widely used in diagnostic bacteriology. The differentiation of eubacteria from archaebacteria is rapidly accomplished and determined by the presence or absence of acid-stable lipids (3).

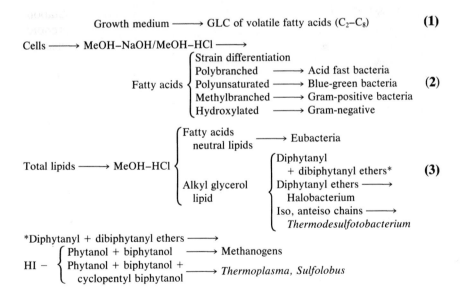

While the use of lipid classes are generally restrictive in differentiating bacterial genera and species, specific lipid types are frequently useful in chemotaxonomy. The cyclic non-isoprenoid hydrocarbons isolated in a hexane eluate from a silicic acid column can be useful as a rapid diagnostic test (4) to differentiate members of the taxonomic family Micrococcaceae. It is also a procedure that can differentiate between *Pseudomonas* spp. and *Pseudomonas maltophilia* (5). The components isolated in a benzene eluate usually consist of a greater variety of lipids than those isolated in a hexane eluate; however, the principal components are frequently cyclic and non-cyclic isoprenoid compounds. Distinct patterns of acyclic isoprenoids in the range from C_9 to C_{40} have been found in the archaebacteria. Whether the distinct isoprenoid patterns and schemes reported in differentiating *Methanogen* species, *Thermoplasma,* and *Sulfolobus* are meaningful as a taxonomic tool has not been determined (Holzer *et al.,* 1979; Tornabene *et al.,* 1979). The necessary metabolic regulation studies on the hydrogenation–dehydrogenation of isoprenoids in these organisms have not yet been conducted. Naphthoquinones and benzoquinones are also found in benzene eluates. These components are widely distributed among diverse bacteria; the distribution pattern of these components and their structural variations reveals the phylogenetic differentiation among these bacteria.

$$\text{Hexane eluate} \begin{cases} \text{None} \longrightarrow \text{Staphylococci} \\ \text{Branched monoenes} \begin{cases} C_{27}, C_{28}, C_{29} \longrightarrow M.\ luteus \\ C_{25}, C_{26}, C_{27} \longrightarrow M.\ varians \\ C_{30}, C_{31}, C_{32} \longrightarrow M.\ sedentarius \end{cases} \end{cases} \quad (4)$$

$$\text{Hexane eluate} \begin{cases} \text{None} \longrightarrow Pseudomonas\ \text{spp.} \\ \text{Branched polyenes } C_{29:1}, C_{29:2}, C_{29:3} \longrightarrow P.\ maltophilia \end{cases} \quad (5)$$

The polar lipids that elute from a silicic acid column in acetone or methanol are principally complex lipids of the glycolipid, phospholipid, and sulpholipid classes. Analysis of these lipids usually requires conversion of the lipid into the individual substructures by acid/alkaline hydrolysis. Many novel compounds have been identified as a result. Whether or not the novel compounds identified are distributed among a variety of bacteria or within specific bacterial types has not yet been determined. The studies involving glycolipids as a taxonomic tool have yielded encouraging results.

References

Adams, G. A., Tornabene, T. G., and Yaguchi, M. (1969). *Can. J. Microbiol.* **15,** 365–374.
Albersheim, P., Nevins, K. J., English, P. D., and Kerr, A. (1967). *Carbohydr. Res.* **5,** 304–345.

Albro, P. W. (1971). *J. Bacteriol.* **108**, 213–218.
Amdur, B. H., Szabo, E. I., and Socransky, S. S. (1978). *J. Bacteriol.* **135**, 161–163.
Amstein, C. F., and Hartman, P. A. (1973). *J. Bacteriol.* **113**, 38–41.
Asselineau, J. (1966). "The Bacterial Lipids." Hermann, Paris.
Barksdale, L., and Kim, K. S. (1977). *Bacteriol Rev.* **41**, 217–372.
Bentley, R., and Meganathan, R. (1982). *Microbiol. Rev.* **46**, 241–280.
Bergelson, L. D. (1980). "Lipid Biochemical Preparations." Elsevier/North Holland, Amsterdam.
Bird, C. W., Lynch, J. M., Pirt, S. J., and Reid, W. W. (1971a). *Tetrahedron Lett.* 3189–3190.
Bird, C. W., Lynch, J. M., Pirt, S. J., Reid, W. W., Brooks, C. J. W., and Middleditch, B. S. (1971b). *Nature (London)* **230**, 473–478.
Blaser, M. J., Moss, C. W., and Weaver, R. E. (1980). *J. Clin. Microbiol.* **11**, 448–451.
Bligh, E. G., and Dyer, W. J. (1959). *Can. J. Biochem. Physiol.* **37**, 911–917.
Bloch, K. (1977). *Annu. Rev. Biochem.* **46**, 263–298.
Bouvier, P., Rohmer, M., Benveniste, P., and Ourisson, G. (1976). *Biochem. J.* **159**, 267–273.
Brenner, D. J., Farmer, J. J., Fanning, G. R., Steigerwalt, A. G., and Glyken, P. (1978). *Int. J. Syst. Bacteriol.* **28**, 269–282.
Brian, B. L., and Gardner, E. W. (1968). *J. Infect. Dis.* **118**, 47–53.
Brice, J. L., Tornabene, T. G., and LaForce, F. M. (1979). *J. Infect. Dis.* **140**, 443–452.
Brooks, J. L., and Benson, A. A. (1972). *Arch. Biochem. Biophys.* **152**, 347–355.
Burrows, S., Grylls, F. S. M., and Harrison, J. J. (1952). *Nature (London)* **170**, 800–801.
Carson, J. H., Sowden, L. C., and Colvin, J. R. (1967). *Can. J. Microbiol.* **13**, 837–841.
Chow, T. C., and Schmidt, J. M. (1974). *J. Gen. Microbiol.* **83**, 369–373.
Christie, W. W. (1982). "Lipid Analysis," 2nd Ed. Pergamon, New York.
Cohen, P. G., Moss, C. W., and Farshtchi, D. (1970). *Brit. J. Ven. Dis.* **46**, 10–12.
Collins, M. D., and Jones, D. (1981). *Microbiol Rev.* **45**, 316–354.
Dees, S. B., and Moss, C. W. (1975). *J. Clin. Microbiol.* **1**, 414–419.
Dees, S. B., Moss, C. W., Weaver, R. E., and Hollis, D. (1979). *J. Clin. Microbiol.* **10**, 206–209.
Dees, S. B., Thanabalasundrum, S., Moss, C. W., Hollis, D. G., and Weaver, R. E. (1980). *J. Clin. Microbiol.* **11**, 664–668.
Dees, S. B., Hollis, D. G., Weaver, R. E., and Moss, C. W. (1981a). *J. Clin. Microbiol.* **14**, 111–112.
Dees, S. B., Powell, J., Moss, C. W., Hollis, D. G., and Weaver, R. E. (1981b). *J. Clin. Microbiol.* **14**, 612–616.
Dees, S. B., Karr, D. E., Hollis, D., and Moss, C. W. (1982). *J. Clin. Microbiol.* **16**, 779–783.
DeRosa M., Gambacorta, A., and Minale, L. (1973). *Phytochemistry* **12**, 1117–1123.
DeRosa, M., Gambacorta, A., and Bu'lock, J. D. (1974). *J. Bacteriol.* **117**, 212–214.
DeRosa, M., DeRosa, S., Gambacorta, A., Minale, L., and Bu'lock, J. D. (1977). *Phytochemistry* **16**, 1961–1965.
DeRosa, M., Esposito, E., Ganbacorta, A., Nicolaus, B., and Bu'lock, J. D. (1980a). *Phytochemistry* **19**, 821–825.
DeRosa, M., Gambacorta, A., Nicolaus, B., Sodano, S., and Bu'lock, J. D. (1980b). *Phytochemistry* **19**, 833–836.
Donohue, T. J., Cain, B. D., and Kaplan, S. (1982). *Biochemistry* **21**, 2765–2773.
Driskell, W. J., Neese, J. W., Bryant, C. C., and Bashor, M. M. (1982). *J. Chromatogr.* **231**, 439–444.

Drucker, D. B. (1972). *Microbios* **5**, 109–112.

Drucker, D. B., and Owen, I. (1973). *Can. J. Microbiol.* **19**, 247–250.

Drucker, D. B., Griffith, C. J., and Melville, T. H. (1973). *Microbios* **7**, 17–23.

Drucker, D. B., Griffith, C. J., and Melville, T. H. (1974a). *Microbios* **9**, 187–189.

Drucker, D. B., Griffith, C. J., and Melville, T. H. (1974b). *Microbios* **10**, 183–185.

Ellender, R. D., Hidalgo, R. J., and Grumbles, L. C. (1970). *Am. J. Vet. Res.* **31**, 1863–1866.

Erwin, J. A. (1973). "Lipids and Biomembranes of Eukaryotic Microorganisms." Academic Press, New York.

Evans, R. W., Kates, M., Ginzburg, M., and Ginsburg, B.-Z. (1982). *Biochim. Biophys. Acta* **712**, 186–195.

Fisher, W., Nakano, M., Laine, R. A., and Bohrer, W. (1978). *Biochim. Biophys. Acta* **528**, 288–297.

Forster, H. J., Biemann, K., Haigh, W. G., Tattrie, N. H., and Colvin, J. R. (1973). *Biochem. J.* **135**, 133–138.

Freeman, C. P., and West, D. (1966). *J. Lipid Res.* **7**, 324–327.

Fugate, K. J., Hansen, L. B., and White, O. (1971). *Appl. Microbiol.* **21**, 470–475.

Fulco, A. (1983). *Prog. Lipid Res.* **22**, 133–160.

Gelpi, E., Schneider, H., Mann, J., and Oro, J. (1970). *Phytochemistry* **9**, 603.

Glazer, A. N. (1983). *Annu. Rev. Biochem.* **52**, 125–157.

Godchaux, W., and Leadbetter, E. R. (1980). *J. Bacteriol.* **144**, 592–602.

Goldberg, I., and Shechter, I. (1978). *J. Bacteriol.* **135**, 717–720.

Goldfine, H. (1972). *Adv. Microbial Physiol.* **8**, 1–58.

Goldfine, H. (1982). *Curr. Top. Membr. Transp.* **17**, 1–43.

Goldfine, H., and Hagen, P.-O. (1972). *In* "Ether Lipids: Chemistry and Biology" (F. Snyder, ed.), pp. 329–350. Academic Press, New York.

Goodwin, T. W. (1980). "The Biochemistry of the Carotenoids," 2nd ed., Chapman and Hall, London.

Goren, M. B. (1972). *Bacteriol. Rev.* **36**, 33–64.

Goren, M. B., and Brennan, P. J. (1979). *In* "Tuberculosis" (G. P. Youmans, ed.), pp. 63–193. Saunders, Philadelphia.

Guerrant, G. O., Lambert, M. A., and Moss, C. W. (1981). *J. Clin. Microbial* **13**, 899–907.

Gurr, M. I., and James, A. T. (1980). "Lipid Biochemistry," 3rd Ed. Methuen, London.

Han, J., and Calvin, M. (1969). *Proc. Natl. Acad. Sci. U.S.A.* **64**, 436–443.

Holdemann, L. V., Cato, E. P., and Moore, W. E. C. (1977). "Anaerobic Laboratory Manual," 4th Ed. Virginia Polytechnic Institute and State University, Blacksburg, VA.

Holzer, G., Oro, J., and Tornabene, T. G. (1979). *J. Chromatogr.* **186**, 795–809.

Hung, J. G. C., and Walker, R. W. (1970). *Lipids* **5**, 720–722.

Ikawa, M. (1967). *Bacteriol. Rev.* **31**, 54–64.

Imhoff, J. F., Kushner, D. J., Kushwaha, S. C., and Kates, M. (1982). *J. Bacteriol.* **150**, 1192–1201.

Jantzen, E., Bergan, T., and Bovre, K. (1974a). *Acta Pathol. Microbiol. Scand.* **82B**, 785–798.

Jantzen, E., Bryn, K., Bergan, T., and Bovre, K. (1974b). *Acta Pathol. Microbiol. Scand.* **82B**, 767–779.

Johnston, N. C., and Goldfine, H. (1982). *J. Bacteriol.* **149**, 567–575.

Kaltenbach, C. M., Moss, C. W., and Weaver, R. E. (1975). *J. Clin. Microbiol.* **1**, 339–344.

Kamio, Y., Kanegasaki, S., and Takahashi, H. (1969). *J. Gen. Appl. Microbiol.* **15**, 439–451.

Kandler, O. (1982). "Archaebacteria," Proceedings of the 1st International Workshop on Archaebacteria, Munich, 1981. Gustav Fischer, Stuttgart/N.Y.

Kaneda, T. (1977). *Bacteriol. Rev.* **41**, 391–418.

Kates, M. (1964a). *Adv. Lipid Res.* **2**, 17–90.
Kates, M. (1964b). *J. Lipid Res.* **5**, 132–135.
Kates, M. (1972). *In* "Techniques in Lipidology" (T. S. Work and E. Work, eds.), pp. 268–618. Elsevier, Amsterdam.
Kates, M. (1978). *Prog. Chem. Fats Other Lipids* **15**, 301–342.
Kates, M., Adams, G. A., and Martin, S. M. (1964). *Can. J. Biochem.* **42**, 461–479.
Kates, M., Yengoyan, L. S., and Sastry, P. S. (1965). *Biochim. Biophys. Acta* **98**, 252–268.
Kawai, Y., and Moribayashi, A. (1982). *J. Bacteriol.* **151**, 996–1005.
Kenyon, C. N. (1978). *In* "The Photosynthetic Bacteria" (R. K. Claton and W. R. Sistrom, eds.), pp. 281–313. Plenum, New York.
Kim, K. C., Kamio, Y., and Takahashi, H. (1970). *J. Gen. Appl. Microbiol.* **16**, 321–325.
Kimura, A., and Otsuka, H. (1969). *Agric. Biol. Chem.* **33**, 781–784.
King, D. H., and Perry, J. J. (1975). JOB5. *Can. J. Microbiol.* **21**, 510–512.
Kloos, W. E., Tornabene, T. G., and Schleifer, K. H. (1974). *Int. J. Syst. Bacteriol.* **24**, 79–101.
Knoche, H., and Shively, J. M. (1972). *J. Biol. Chem.* **247**, 170–178.
Kondo, E., and Ueta, N. (1972). *J. Bacteriol.* **110**, 459–467.
Kunsman, J. E. (1973). **J. Bacteriol.** **113**, 1121–1126.
Kushwaha, S. C., Pugh, E. L., Kramer, J. K. G., and Kates, M. (1972). *Biochim Biophys. Acta* **260**, 492–506.
Kushwaha, S. C., Kates, M., and Martin, W. G. (1975). *Can. J. Biochem.* **53**, 284–292.
Kushwaha, S. C., Kates, M., Sprott, G. D., and Smith, I. C. P. (1981). *Biochim. Biophys. Acta* **664**, 156–173.
Kushwaha, S. C., Kates, M., Juez, G., Rodriquez-Valera, F., and Kushner, D. J. (1982). *Biochim. Biophys. Acta* **711**, 19–25.
Lambert, M. A., and Moss, C. W. (1976). *J. Dent. Res.* **55**, A96–A102.
Lambert, M. A., Hollis, D. G., Moss, C. W., Weaver, R. E., and Thomas M. L. (1971). *Can. J. Microbiol.* **17**, 1491–1502.
Langworthy, T. A. (1982). *Curr. Top. Membr. Transp.* **17**, 45–77.
Langworthy, T. A., and Mayberry, W. R. (1976). *Biochim. Biophys. Acta* **431**, 570–577.
Langworthy, T. A., Mayberry, W. R., and Smith, P. F. (1974). *J. Bacteriol.* **119**, 106–116.
Langworthy, T. A., Mayberry, W. R., and Smith, P. F. (1976). *Biochim. Biophys. Acta* **431**, 550–569.
Langworthy, T. A., Tornabene, T. G., and Holzer, G. (1982). *Zentralbl. Bakteriol. Parasitenkd. Infektionkr. Hyg. Abt. 1: Orig. C* **3**, 228–244.
Langworthy, T. A., Holzer, G., Zeikus, J. G., and Tornabene, T. G. (1983). *Syst. Appl. Microbiol.* **4**, 1–17.
Lechevalier, M. P., and Moss, C. W. (1977). *CRC Crit. Rev. Microbiol.* **5**, 109–210.
Lederer, E. (1967). *Chem. Phys. Lipids* **1**, 294–299.
Lennarz, W. J. (1970). *In* "Lipid Metabolism" (S. J. Wakil, ed.), pp. 155–184. Academic Press, New York.
Lepage, M. (1967). *Lipids* **2**, 244–250.
Lewis, V. J., Weaver, R. E., and Hollis, D. G. (1968). *J. Bacteriol.* **96**, 1–5.
Livermore, B. P., and Johnson, R. C. (1974). *J. Bacteriol.* **120**, 1268–1273.
Luderitz, O., Freudenberg, M. A., Galanos, C., Lehmann, V., Rietschel, E. T., and Shaw, D. H. (1982). *Curr. Top. Membr. Transp.* **17**, 79–151.
McElhaney, R. N. (1982). *Curr. Top. Membr. Transp.* **17**, 317–380.
Makula, R. A. (1978). *J. Bacteriol.* **134**, 771–777.
Makula, R. A., and Finnerty, W. R. (1975). *J. Bacteriol.* **123**, 523–529.
Marr, A. G., and Ingraham, J. L. (1962). *J. Bacteriol.* **84**, 1260–1267.

Maudinas, B., and Villoutriex, J. (1976). *C.R. Hebd. Seances Acad. Sci. Ser. D.* **278,** 2995–2997.

Miyagawa, E., Azuma, R., and Suto, T. (1978). *J. Gen. Appl. Microbiol.* **24,** 341–348.

Morrison, S. J., Tornabene, T. G., and Kloos, W. E. (1971). *J. Bacteriol.* **108,** 353–358.

Moss, C. W. (1981). *J. Chromatogr.* **203,** 337–347.

Moss, C. W., and Cherry, W. B. (1968). *J. Bacteriol.* **95,** 241–242.

Moss, C. W., and Dees, S. B. (1975). *J. Chromatog.* **112,** 595–604.

Moss, C. W., and Dees, S. B. (1978). *J. Clin. Microbiol.* **8,** 772–774.

Moss, C. W., and Dunkelberg, Jr., W. E. (1969). *J. Bacteriol.* **100,** 544–546.

Moss, C. W., and Lewis, V. J. (1967). *Appl. Microbiol.* **15,** 390–397.

Moss, C. W., and Nunez-Montiel, O. L. (1982). *J. Clin. Microbiol.* **15,** 308–311.

Moss, C. W., Dowell, V. R., Jr., Lewis, V. J., and Schekter, M. A. (1967). *J. Bacteriol.* **94,** 1300–1305.

Moss, C. W., Dowell, V. R., Farshtchi, D., Raines, L. J., and Cherry, W. B. (1969). *J. Bacteriol.* **97,** 561–570.

Moss, C. W., Kellogg, Jr., D. S., Farshy, D. C., Lambert, M. A., and Thayer, J. D. (1970). *J. Bacteriol.* **104,** 63–68.

Moss, C. W., Samuels, S. B., and Weaver, R. E. (1972). *Appl. Microbiol.* **24,** 596–598.

Moss, C. W., Samuels, S. B., Liddle, J., and McKinney, R. M. (1973). *J. Bacteriol.* **114,** 1018–1024.

Moss, C. W., Karr, D. E., and Dees, S. B. (1981). *J. Clin. Microbiol.* **14,** 692–694.

Nes, W. R., and Nes, W. D. (1980). "Lipids in Evolution." Plenum, New York.

O'Leary, W. M. (1967). "The Chemistry and Metabolism of Microbial Lipids." World Publ., Cleveland.

O'Leary, W. M. (1975). *CRC Crit. Rev. Microbiol.* **4,** 41–60.

Op den Kamp, J. A. F., and Van Deenan, L. L. M. (1969). *In* "Structural and Functional Aspects of Lipoproteins in Living Systems" (E. Tria and A. M. Scanu, eds.), pp. 227–247. Academic Press, New York.

Oshima, M., and Yamakawa, T. (1974). *Biochemistry* **13,** 1140–1146.

Osterhelt, D., and Stoeckenius, W. (1971). *Nature (London) New Biol.* **233,** 149–152.

Ourisson, G., Albrecht, P., and Rohmer, M. (1979). *Pure Appl. Chem.* **51,** 709–729.

Panganamala, R. V., Sievert, C. F., and Cornwell, D. G. (1971). *Chem. Phys. Lipids* **7,** 336–344.

Park, C.-E., and Berger, L. R. (1967). *J. Bacteriol.* **93,** 221–229.

Pask-Hughes, R. A., and Shaw, N. (1982). *J. Bacteriol.* **149,** 54–58.

Pfenning, N. (1977). *Annu. Rev. Microbiol.* **31,** 275–290.

Phillips, F. C., Erdahl, W. L., and Privett, O. S. (1982). *Lipids* **17,** 992–997.

Prefontaine, G., and Jackson, F. L. (1972). *Int. J. Syst. Bacteriol.* **22,** 210–217.

Raetz, C. R. H. (1978). *Microbiol. Rev.* **42,** 614–659.

Raines, L. J., Moss, C. W., Farshtchi, D., and Pittman, B. (1968). *J. Bacteriol.* **96,** 2175–2177.

Razin, S., and Rottem, S., eds. (1982). *Curr. Top. Membr. Transp.* **17.**

Reaveley, D. A., and Burge, R. E. (1972). *Adv. Microb. Physiol.* **7,** 1–81.

Rebeiz, C. A., and Lascelles, J. (1982). *In* "Photosynthesis: Energy Conversion by Plants and Bacteria" (Govindjee, ed.), pp. 699–780. Academic Press, New York.

Rietschel, E. Th., Luderitz, O., and Volk. W. A. (1975). *J. Bacteriol.* **122,** 1180–1188.

Rohmer, M., and Ourisson, G. (1976a). *Tetrahedron Lett.,* 3633–3636.

Rohmer, M., and Ourisson, G. (1976b). *Tetrahedron Lett.,* 3637–3640.

Rohmer, M., and Ourisson, G. (1976c). *Tetrahedron Lett.,* 3641–3644.

Ross, H. N. M., Collins, M. D., Tindall, B. J., and Grandt, W. D., (1981). *J. Gen. Microbiol.* **123**, 75–80.

Samuels, S. B., Moss, C. W., and Weaver, R. E. (1973). *J. Gen. Microbiol.* **74**, 275–279.

Sastry, P. S. (1974). *Adv. Lipid Res.* **12**, 251–310.

Schubert, K., Rose, G., and Hörhold, C. (1967). *Biochim. Biophys. Acta* **137**, 168–173.

Schubert, K., Rose, G., Wachtel, H., Hörhold, C., and Ikekawa, N. (1968). *Eur. J. Biochem.* **5**, 246–251.

Shaw, N. (1974). *Adv. Appl. Microbiol.* **17**, 63–108.

Shaw, N. (1975). *Adv. Microbiol. Physiol.* **12**, 141–167.

Short, S. A., White, D. C., and Aleem, M. I. H. (1969). *J. Bacteriol.* **99**, 142–150.

Stanier, R. Y., and Cohen-Bazire, G. (1977). *Annu. Rev. Microbiol.* **31**, 225–274.

Sud, I. J., and Feingold, D. S. (1975). *J. Bacteriol.* **124**, 713–717.

Suto, T., and Ryozo, A. (1974). *Nippon Saikingaku Zasshi* **29**, 811–849.

Sutter, V. L., Citron, D. M., and Finegold, S. M. (1980). "Wadsworth Anaerobic Bacteriology Manual," 3rd Ed. Mosby, St. Louis, MO.

Suzue, G., Tsukada, K., and Tanaka, S. (1967). *Biochim. Biophys. Acta* **144**, 186–188.

Swings, J., DeVos, P., Van den Mooter, M., and DeLey, J. (1983). *Int. J. Syst. Bacteriol.* **33**, 409–413.

Tahara, Y., Kameda, M., Yamada, Y., and Kondo, K. (1976). *Biochim. Biophys. Acta* **450**, 225–230.

Thiele, O. W., Biswas, C. J., and Hunneman, D. (1980). *Eur. J. Biochem.* **105**, 267–274.

Thoen, C. O., Karlson, A. G., and Ellefson, R. D. (1971a). *Appl. Microbiol.* **21**, 628–632.

Thoen, C. O., Karlson, A. G., and Ellefson, R. D. (1971b). *Appl. Microbiol.* **22**, 560–563.

Thoen, C. O., Karlson, A. G., and Ellefson, R. D. (1972). *Appl. Microbiol.* **24**, 1009–1010.

Tornabene, T. G. (1973). *Biochim. Biophys. Acta* **306**, 173–185.

Tornabene, T. G. (1978). *J. Mol. Evol.* **11**, 253–257.

Tornabene, T. G. (1981). *In* "Trends in the Biology of Fermentation for Fuels and Chemicals" (A. Hollaender, R. Kabson, S. Rogers, S. Pietro, R. Valentine, and R. Wolfe, eds.), pp. 421–438. Plenum, New York.

Tornabene, T. G. (1982). *Experientia* **38**, 43–46.

Tornabene, T. G., and Langworthy, T. A. (1979). *Science* **203**, 51–53.

Tornabene, T. G., and Markey, S. P. (1971). *Lipids* **6**, 190–195.

Tornabene, T. G., and Ogg, J. (1971). *Biochim. Biophys. Acta* **239**, 133–141.

Tornabene, T. G., and Peterson, S. L. (1978). *Can. J. Microbiol.* **24**, 525–532.

Tornabene, T. G., Gelpi, E., and Oro, J. (1967). *J. Bacteriol.* **94**, 333–343.

Tornabene, T. G., Kates, M., Gelpi, E., and Oro, J. (1969). *J. Lipid Res.* **10**, 294–303.

Tornabene, T. G., Morrison, S. J., and Kloos, W. E. (1970). *Lipids* **5**, 929–937.

Tornabene, T. G., Langworthy, T. A., Holzer, G., and Oro, J. (1979). *J. Mol. Evol.* **13**, 73–83.

Tornabene, T. G., Holzer, G., Bittner, A. S., and Grohmann, K. (1982). *Can. J. Microbiol.* **28**, 1107–1118.

Tornabene, T. G., Holzer, G., Lien, S., and Burris, N. (1983). *Eng. Microb. Technol.* **5**, 435–440.

Trevelyan, W. E., Procter, D. P., and Harrison, J. G. (1950). *Nature* (*London*) **166**, 444.

Uchida, K. (1975). *Agric. Biol. Chem.* **39**, 837–842.

Uchida, K., and Mogi, K. (1973a). *J. Gen. Appl. Microbiol.* **19**, 129–140.

Uchida, K., and Mogi, K. (1973b). *J. Gen. Appl. Microbiol.* **19**, 233–249.

Vaczi, L. (1973). "The Biological Role of Bacterial Lipids." Akadémiai Kiadó, Budapest.

Walker, R. W., and Fagerson, I. S. (1965). *Can. J. Microbiol.* **11**, 229–233.

Weaver, T. L., Patrick, M. A., and Dugan, P. R. (1975). *J. Bacteriol.* **124,** 602–605.

Week, O. B., and Francesconi, M. D. (1978). *J. Bacteriol.* **136,** 614–624.

Weete, J. D. (1980). "Lipid Biochemistry of Fungi and Other Organisms." Plenum, New York.

Wells, M. A., and Dittmer, J. C. (1963). *Biochemistry* **2,** 1259–1263.

Wilkinson, S. G. (1972). *Biochim. Biophys. Acta* **270,** 1–17.

Wilkinson, S. G. (1977). *In* "Surface Carbohydrates of the Procaryotic Cell" (I. W. Sutherland, ed.), pp. 97–175. Academic Press, New York.

Yabuuchi, E., and Moss, C. W. (1982). *FEMS Microbiol. Lett.* **13,** 87–91.

8

Analysis of Carotenoids and Related Polyene Pigments[1]

SYNNØVE LIAAEN-JENSEN

Organic Chemistry Laboratories, Norwegian Institute of Technology,
University of Trondheim, Trondheim, Norway

ARTHUR G. ANDREWES

Chemistry Department, Saginaw Valley State College, University Center,
Michigan, USA

I. Introduction

Approximately 500 naturally occurring carotenoids are known. A large majority of these are synthesised by microorganisms. For structures see compilations by Isler (1971) and Straub (1976) and reviews on more recent original literature (Davis, 1974; Eugster, 1979, 1982). By far the majority of carotenoids are polyene pigments of isoprenoid skeleton. Representative examples are given in Figs. 1–4. Several carotenoids have trivial names.

---

[1] Dedicated to Professor Jarl Gripenberg, Chemistry Department, Finland Institute of Technology, Helsinki, on the occasion of his seventieth birthday, in recognition of his early work on non-carotenoid polyene pigments.

METHODS IN MICROBIOLOGY
VOLUME 18

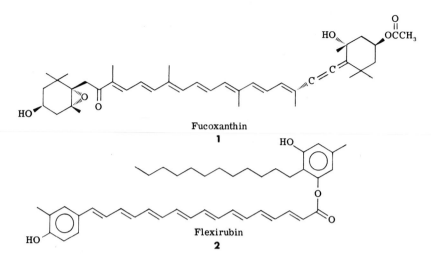

Fucoxanthin

1

Flexirubin

2

Fig. 1. A representative carotenoid (**1**) and aryl polyene (**2**). References: **1** Straub (1976); **2** Achenbach *et al.* (1976).

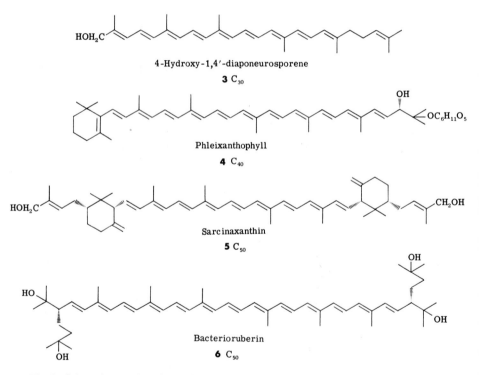

4-Hydroxy-1,4'-diaponeurosporene

3 C_{30}

Phleixanthophyll

4 C_{40}

Sarcinaxanthin

5 C_{50}

Bacterioruberin

6 C_{50}

Fig. 2. Selected examples of non-photosynthetic bacterial carotenoids. References: **3** Taylor and Davies (1974), **4** Rønneberg *et al.* (1984), **5** Hertzberg and Liaaen-Jensen (1977), **6** Johansen and Liaaen-Jensen (1979).

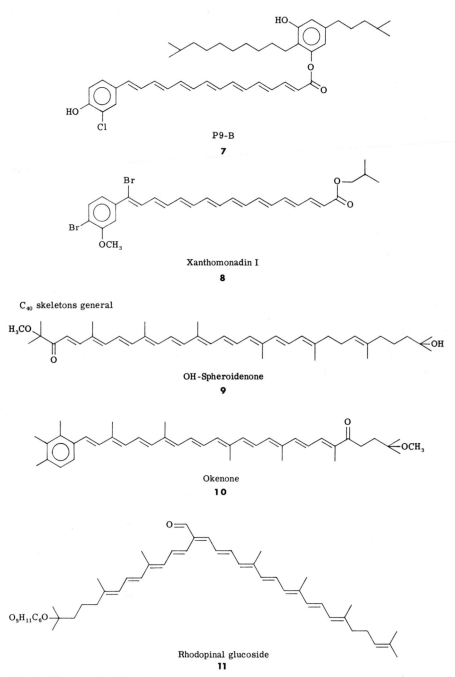

P9-B

7

Xanthomonadin I

8

C_{40} skeletons general

OH-Spheroidenone

9

Okenone

10

Rhodopinal glucoside

11

Fig. 2. (*Continued*). Selected examples of aryl polyenes and phototrophic bacterial carotenoids. References: **7** Achenbach *et al.* (1977b), **8** Andrewes *et al.* (1976), and **9, 10, 11** Straub (1976).

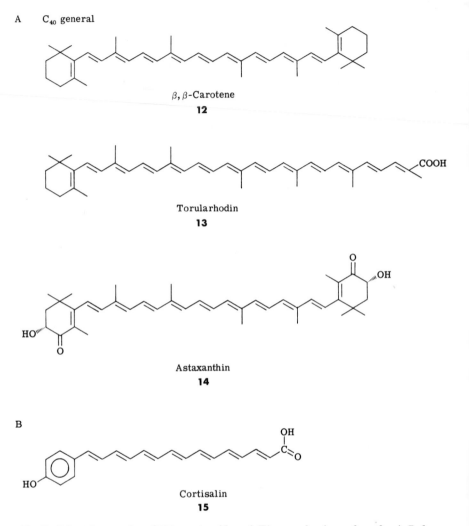

A C_{40} general

β, β-Carotene

12

Torularhodin

13

Astaxanthin

14

B

Cortisalin

15

Fig. 3. Selected examples of (A) carotenoids and (B) an aryl polyene from fungi. References: **12, 13** Straub (1976), **14** Andrewes and Starr (1976a,b), and **15** Gripenberg (1952).

A semirational nomenclature has been approved (IUPAC, 1975) and is used here for the carotenes, cf. β, β-carotene (**12,** Fig. 3).

Carotenoids are biosynthesised via the mevalonate pathway. Facts and speculations concerning their biosynthesis were recently summarised (Goodwin, 1974, 1980a). Carotenoids are produced by all photosynthetic organisms including photosynthetic bacteria, microscopic algae, and

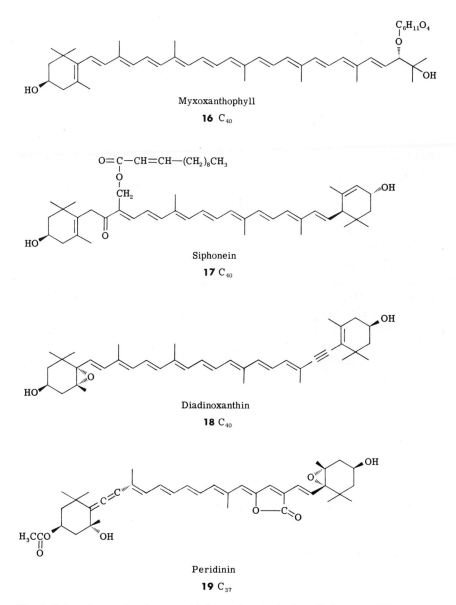

Fig. 4. Selected examples of carotenoids from microscopic algae. References: **16** Rønneberg *et al.* (1984), **17** Fiksdahl *et al.* (1984), **18** Straub (1976), **19** Johansen and Liaaen-Jensen (1980).

other microorganisms. For the occurrence of specific carotenoids in microbial organisms see Straub (1976), Baker and Murphy (1976, 1981), and Goodwin (1980a).

Related aryl polyene pigments lacking the lateral methyl groups on the polyene chain are also considered here; see examples in Figs. 1, 2B, and 3B. Such aryl polyenes are biosynthesised by routes other than the mevalonate pathway and are encountered in certain non-photosynthetic bacteria (Andrewes *et al.*, 1976; Starr *et al.*, 1977; Achenbach *et al.*, 1978; Reichenbach *et al.*, 1981). The flexirubin (2)-type pigments have the polyene chain biosynthesised mainly from acetate; the conjugated phenyl as well as the three adjacent carbon atoms along the chain are derived from tyrosine (Reichenbach *et al.*, 1981; Achenbach *et al.*, 1979c).

Microbial carotenoids were reviewed by the authors ten years ago (Liaaen-Jensen and Andrewes, 1972). Progress in the rapid analysis of carotenoids during the last decade has been particularly notable in chromatography, with the application of high-performance liquid chromatography as a routine tool, and in spectroscopy. The availability of Fourier transform NMR instruments has greatly reduced the sample size requirements. Circular dichroism can now be applied generally for stereochemical assignment of chiral carotenoids with due reference to limits set by empirical observation. Within the scope of this chapter we have chosen to refer to other relevant literature for more details when such are available, and to add supplementary references and information with emphasis on important progress.

II. Analytical methods

A. Isolation

Carotenoids and other polyenes are, as a general rule, unstable towards oxygen, peroxides, light, heat, acids, and in some cases bases. General precautions for work with carotenoids are detailed elsewhere (Liaaen-Jensen and Jensen, 1971; Davies, 1976).

Because of their long chromophores and high extinction coefficients polyene pigments can readily be isolated on the micro scale.

For carotenoids some general extraction schemes can be recommended (Liaaen-Jensen, 1971; Davies, 1976; De Ritter and Purcell, 1981). Extraction by acetone is usually employed. Ready extraction is not obtained with all types of cells. In some cases moistening of dried cells with water (Jensen, 1959) or mechanical disintegration (Andrewes and Starr, 1976a) of the cells may be advantageous prior to extraction.

Removal of chlorophyll or other saponifiable lipids by alkali treatment is advantageous for the separation of the carotenoids, provided they are alkali stable. The alkali stability can readily be checked by direct TLC or HPLC comparison of a saponified aliquot with the untreated extract.

Procedures for saponification are specified elsewhere (Liaaen-Jensen and Jensen, 1971; Davies, 1976; De Ritter and Purcell, 1981). Saponification with 5% KOH in ether–methanol at room temperature for a few hours is usually sufficient. Acetone residues from the extraction process must be absent because of the formation of aldol condensation products of the solvent which are difficult to remove later, and of aldol condensation products with aldehydic carotenoids (Schmidt et al., 1971).

Sterols and phospholipids may be precipitated from petroleum ether (Davies, 1976), or better from acetone solutions at low temperature, and removed by filtration.

Flexirubin-type polyenes are extracted with acetone directly followed by TLC (SiO₂) purification (Reichenbach et al., 1974) or column chromatography. An extraction scheme is presented by Achenbach et al. (1978).

B. Separation

1. General

Partition between immiscible solvent systems including countercurrent distribution as reviewed by Davies (1976) offers only partial separation of complex carotenoid mixtures and is generally unrewarding. For separation into components some type or combination of chromatographic techniques is necessary.

During isolation some trans–cis isomerisation of a carotenoid generally occurs. The phenomenon of cis–trans isomerisation for carotenoids and aryl polyenes has been elaborated by Zechmeister (1962). Absorption spectra in visible light and reversibility tests on iodine catalysed stereomutation in light (Zechmeister, 1962) including chromatography serve to identify cis isomers on the micro scale.

Whereas column chromatography (Strain, 1958; Strain and Sherma, 1972; Davies, 1976) was exclusively used until the mid 1950s, thin- and preparative-layer chromatography is now extensively employed (Stahl, 1969; Davies, 1976) on a scale which can provide purified samples for spectroscopy and chemical derivatization. High-performance liquid chromatography has lately proved to be an invaluable tool for separations on the nanogram scale, and further developments allowing semimicro separations are anticipated. Gas chromatography of polyenes is restricted to the perhydrogenated derivatives.

2. Thin-layer chromatography

The extensive reviews on thin layer chromatograpy by Stahl (1969), Strain and Sherma (1972), and particularly Davies (1976) cover most modern approaches. Balock et al. (1977) have reported good separation of carrot pigments on plates of Hyflo Super-cel + MgO + CaSO$_4$. Csorba et al. (1979) have described a high-speed video-densitometric determination of carotenoids and chlorophylls a and b. Very recently, Sadowski and Wójcik (1983) have separated at least 13 carotenes and xanthophylls from chloroplasts on MgO. Sherma and Latta (1978) used reverse-phase TLC to separate chloroplast pigments. Singh et al. (1973) report the separation of 17 different apocarotenoids on silica plates and a reverse-phase system on paper for separating polar apocarotenoids.

Flexirubin-type pigments are separated on kieselgel H plates. Several different developing systems have been recommended depending on the polarity of the individual pigments (Reichenbach et al., 1974; Achenbach et al., 1981). Xanthomonas pigments can be separated by TLC on silica gel developed with acetone–chloroform mixtures (Jenkins and Starr, 1982a).

3. High-performance liquid chromatography

High-performance liquid chromatography (HPLC) is the fastest developing method for carotenoid analysis. It permits rapid routine analysis on the nanogram scale and also permits the recovery of an individual component from a mixture for further characterisation. As in the other chromatographic techniques, the coincidence of a peak or a retention time is not sufficient evidence for unequivocal identification of a carotenoid. Supplementary evidence of the complete visible spectrum (readily obtained with most modern instruments using the stop-flow technique) and the mass spectrum should be obtained.

The degree of sample preparation necessary before analysis depends on the type of natural material to be analysed. Those sources, which are not inherently rich in lipids and other contaminants, may be chromatographed after simply concentrating the extracts and removing particulate matter (Mantoura and Llewellyn, 1983; Nells and Leenheer, 1983; Braumann and Grimme, 1981). Other samples rich in lipids must be saponified to obtain reproducible results (Stewart, 1977a; Zakaria et al., 1979).

A variety of different columns have been used for HPLC carotenoid analysis. Taylor and Ikawa (1980) have summarised some of the earlier carotenoid HPLC literature. The choice of column material depends on the type of carotenoid mixture to be resolved. For separation of carotenes and geometric isomers on normal adsorptive columns, MgO (Stewart and Leuenberger, 1976; Stewart, 1977b; Cabibel et al., 1981; Cadosch and Eugster, 1974); Al$_2$O$_3$ (Vecchi et al., 1981); and Ca(OH)$_2$ (Tsukida et al., 1982) have been used. Englert and Vecchi (1980) separated nine cis isomers of astaxanthin on LiChrosorb S1 60 and collected sufficient quantities of each for ^{1}H NMR analysis and assignment. Eschenmoser et al. (1983) and Märki-Fisher et al. (1983b) also separated geometric isomers of aleurixanthin and loraxanthin on Spherisorb. Xanthophylls from a number of different sources have been separated on one form or another of silica (Stewart, 1977b; Hajibrahim et al., 1978; Fiksdahl et al., 1978; Abaychi and Riley, 1979; Iriyama et al., 1978; Paanakker and Hallegraeff, 1978; Stransky, 1978). While the diol structural isomers lutein and zeaxanthin are separable on silica (Hajibrahim et al., 1978; Stransky, 1978; Fiksdahl et al., 1978), the isomeric β,ε- and β,β-carotenes are not (Iriyama et al., 1978; Fiksdahl et al., 1978). The elution solvent systems used are varied but most use a non-polar hydrocarbon (hexane, cyclohexane, or isooctane) plus a polar component which is usually acetone or a small chain alcohol.

Reverse-phase HPLC (rp-HPLC) featuring octadecylsilylated (ODS) silica has been used successfully for the analysis of chloroplast pigments (Eskins et al., 1977; Sherma and Latta, 1978; Braumann and Grimme, 1981; Ben-Amotz et al., 1982; Mantoura and Llewellyn, 1983). Carotenoids from other natural sources have also been analysed by rp-HPLC (Bandaranayake and Gentien, 1982; Baranyai et al., 1982; Bushway and Wilson, 1982). The use of rp-HPLC is a nice complement to normal adsorptive-column HPLC because the former is effective in separating the carotenes such as β,ε- and β,β-carotene, lycopene, etc. (Zakaria et al., 1979; Pfander et al., 1980), but ineffective in separating the double-bond isomeric xanthophylls lutein and zeaxanthin. The opposite situation holds for silica adsorption HPLC (Fiksdahl et al., 1978). Geometric isomers of lycopene, β,β-carotene, phytofluene, 7,8,7',8'-tetrahydro-ψ,ψ-carotene, and ε,ψ-carotene have been separated on Spherisorb ODS 5 (Märki-Fisher et al., 1983a).

One additional important application of HPLC has been toward the analysis of the optical purity of naturally occurring chiral carotenoids. Vecchi and Müller (1979) separated the three diastereomeric camphanates of astaxanthin by using a Spherisorb S-5CN stationary-phase column and hexane–isopropyl acetate–acetone as mobile phase. This basic technique has been used by Rønneberg *et al.* (1980) and Renstrøm *et al.* (1981a,b,c) for the chiroptical analysis of astaxanthin from natural sources. The same stationary phase column systems were used by Vecchi *et al.* (1982) to separate two diastereomers of 6,6'-dihydrorhodoxanthin and six diastereomers of tunaxanthin (ε,ε-carotene-3,3'-diol).

Preparative HPLC separation of dimethyl ethers of flexirubin (**2**)-type pigments on a Nucleosil 10-C_{18} column using acetone–tetrahydrofuran–water as solvent is reported by Achenbach *et al.* (1978, 1979a). Hydrogenated derivatives of the allylic polyene alcohol obtained by reductive cleavage of the methylated ester have also been subjected to HPLC (Achenbach *et al.*, 1979a).

4. Gas chromatography

Due to the low volatility of intact carotenoids and other aryl polyenes, gas chromatography (GC) can be applied only to perhydrogenated compounds and is of little value. Taylor and Davies (1975) and Taylor and Ikawa (1980) have summarised the gas chromatography of 70–80 perhydrogenated carotenoids.

The esterifying acids of natural carotenol esters may be determined by GC as methyl esters, following alkaline hydrolysis of the carotenoid and methylation (Czygan and Eichenberger, 1971; Foss *et al.*, 1978; Renstrøm and Liaaen-Jensen, 1981).

Ozonolysis products of flexirubin (**2**)-type pigments have been subjected to GC (Achenbach *et al.*, 1976). GC/MS examination of the esterifying phenols of flexirubin-type polyenes obtained after reductive cleavage of the methylated esters is reported (Achenbach *et al.*, 1979b). Hydrogenated derivatives of the allylic polyene alcohol obtained by reductive cleavage have also been studied by GC/MS (Achenbach *et al.*, 1979a).

Aryl polyenes may be oxidised efficiently with $KMnO_4$ to the substituted benzoic acids for determination of the substitution pattern of the aromatic ring (Andrewes *et al.*, 1973, 1976).

C. Spectroscopy

1. Ultraviolet-visible light absorption

Chromophore assignment of polyenes on the basis of UV-visible light spectra is of utmost importance in the analysis of such pigments. Detailed treatments of carotenoids and aryl polyenes are given by Zechmeister

(1962) and of carotenoids by Vetter *et al.* (1971) and Davies (1976). The methyl groups on the polyene chain of carotenoids are auxochromes causing marked bathochromic effects. Absorption maxima in various solvents for several natural carotenoids are tabulated by Davies (1976). For the description of spectral fine-structure the term %III/II is recommended (Ke *et al.*, 1970). Quantitative determination of carotenoids is based upon known extinction coefficients in visible light ranging from $\varepsilon \approx 100{,}000$ to 185,000. $E_{1\,cm}^{1\%}$ values for several carotenoids in different solvents have been tabulated by Davies (1976). For unknown carotenoids, ε values for carotenoids with similar chromophore, as judged from the visible spectrum, should be used. In the absence of structural information (chromophore and molecular weight) $E_{1\,cm}^{1\%} = 2500$ at λ_{max} in hexane is frequently used for carotenoids (Davies, 1976).

The electronic spectra of flexirubin-type phenolic pigments ($\varepsilon \approx 100{,}000$) are pH dependant (Reichenbach *et al.*, 1974). In alkaline solution a large bathchromic shift is observed which may be rationalised by the resonance-stabilised phenolate ion, allowing charge delocalisation to the conjugated carbonyl function.

2. Infrared spectroscopy

For the assignment of functional groups infrared spectra are usually informative (Bellamy, 1975). Representative examples are given by Vetter *et al.* (1971). Different types of carbonyl functions and allenes are readily revealed, whereas monoacetylenic carotenoids usually exhibit only very weak $C{\equiv}C$ absorption. With KBr pellets or $CHCl_3$ solutions in microcuvettes 0.2 mg of a crystalline sample is sufficient.

3. Nuclear magnetic resonance

^1H NMR spectra of carotenoids are treated by Vetter *et al.* (1971) and Englert (1981). Available high-field instruments (400 MHz) with spin-decoupling devices and Fourier transform software now also allow identification of individual olefinic protons with nanogram samples. Use of prepurified solvents and scrupulous removal of other sample impurities is imperative at this level. Lanthanide-induced shifts are successfully employed in the structural elucidation of oxygenated carotenoids (Kjøsen and Liaaen-Jensen, 1972; Fiksdahl *et al.*, 1984). Chemical shifts may be employed in the assignment of relative configuration (Liaaen-Jensen, 1980).

^1H NMR data for *Xanthomonas* polyenes (Andrewes *et al.*, 1973, 1976) and for flexirubin-type pigments are available in the literature (e.g.,

Achenbach *et al.*, 1976, 1977a,b, 1979a,b, 1981) and are discussed in a review (Achenbach *et al.*, 1978).

^{13}C NMR spectra of carotenoids are treated by Moss (1976) and Englert (1975, 1981). Generally samples of 5 mg or more are needed. Whereas ^{1}H NMR is becoming a routine tool in the analysis of known carotenoids, ^{13}C NMR is mainly used for structural elucidation. The location of *cis* double bonds in the polyene chain is best documented by ^{13}C NMR.

4. Mass spectroscopy

·Positive ion mass spectra are generally employed. Reviews on carotenoid mass spectra are given by Vetter *et al.* (1971) and Budzikiewicz (1982). A comprehensive recent treatment is given by Enzell and Wahlberg (1980).

MS data for *Xanthomonas* polyenes (Andrewes *et al.*, 1973, 1976) and for flexirubin-type are available in the literature (e.g., Achenbach *et al.*, 1976, 1977a,b, 1979a,b, 1981), and a general treatment is available (Achenbach *et al.*, 1978).

With few exceptions molecular ions of carotenoids may be obtained. Carotenoids exhibit a characteristic fragmentation pattern with elimination of toluene and xylene from the polyene chain in contrast to the aryl polyenes lacking lateral methyl groups. Specific deuterated carotenoids have been used extensively for the elucidation of fragmentation patterns.

Carotenoid glycosides are generally acetylated prior to mass spectroscopy for optimum information (Hertzberg and Liaaen-Jensen, 1967; Francis *et al.*, 1970). Trimethyl silyl ethers of tertiary carotenols are amenable to mass spectroscopy (McCormick and Liaaen-Jensen, 1966).

The bromo substituents of the *Xanthomonas* polyenes were first revealed by mass spectroscopy (Andrewes *et al.*, 1973) by the characteristic isotopic pattern of bromine. Chlorinated aryl polyenes may also be identified readily by mass spectroscopy (Achenbach *et al.*, 1977a). The presence of sulphur in a new bacterial carotenoid was recently demonstrated by high-precision mass measurements (Andrewes *et al.*, 1984). Mass spectrometry requires only nanogram quantities of material.

Analysis of carotenoids by field desorption mass spectroscopy has been reported (Watts *et al.*, 1975). Other ionisation techniques have not proved very useful so far.

5. Circular dichroism

After an introductory study of optical rotatory dispersion of chiral carotenoids (Bartlett *et al.*, 1969), circular dichroism (CD) has been preferred for the study of chiroptical properties (Liaaen-Jensen, 1980).

CD properties of carotenoids have been reviewed by Liaaen-Jensen (1980). Fundamental contributions have since appeared from Noack and Thomson (1979), Sturzenegger *et al.* (1980), Noack (1981), and Buchecker *et al.* (1982).

A chiral carotenoid is not sufficiently characterised if chiroptical properties are not quoted. Unfortunately, lack of modern CD facilities has prevented use of CD on a routine basis. Sample requirement is low, around 50–100 μg depending on the $\Delta\varepsilon$ value.

D. Chemical derivatisations

Derivatives of natural carotenoids may serve two purposes: (1) improved separation from accompanying carotenoids and (2) better characterisation. Acetylation of primary or secondary carotenols or silylation of tertiary carotenols frequently serves the former purpose.

In addition to acetylation and silylation of carotenols, the most common structural modifications comprise complex metal hydride reduction of carbonyl functions, oxidation of allylic hydroxy functions to conjugated ketones or aldehydes, methylation of allylic hydroxy groups, and epoxide–furanoid rearrangements. These and other chemical derivatisations of carotenoids have been treated by Liaaen-Jensen (1971).

Derivatisations can readily be effected on the micro scale and reactions monitored by TLC. Standard procedures have been specified (Liaaen-Jensen and Jensen, 1971). Derivatives should be characterised by chromatography and at least visible and mass spectroscopies.

The flexirubin-type pigments have also been subjected to functional group modifications. The conjugated ester function is stable towards mild saponification conditions (Reichenbach *et al.*, 1974). Alkaline hydrolysis is achieved by more vigorous conditions. Reductive cleavage of the ester with LiAlH$_4$ may be employed (Achenbach *et al.*, 1976). The phenolic hydroxy groups can be acetylated or methylated.

E. Identification

Identification of a natural carotenoid requires at least the visible spectrum, mass spectrum, and co-chromatography with an authentic sample. These tests require less than 30 μg.

An additional useful test is direct, quantitative comparison of iodine-catalysed stereomutation mixtures since identical carotenoids must produce the same equilibrium mixtures (cf. Liaaen-Jensen, 1964).

Relevant derivatisation on the micro scale and characterisation of the derivatives is commonly used for identification or structural elucidation.

Mixed melting points are rarely used as an identification criterion in this field, since more useful spectroscopic evidence may be obtained from crystalline samples.

^{1}H NMR can provide additional evidence and may add information about relative configuration, e.g., about 3,6-*cis* or 3,6-*trans* configuration in 3-hydroxylated ε rings, cf. siphonaxanthin (**17,** Fig. 4) (Fiksdahl *et al.,* 1984) or relative 2,6-stereochemistry in cyclic C_{50} carotenoids, cf. sarcinaxanthin (**5,** Fig. 2) (Hertzberg and Liaaen-Jensen, 1977).

For monochiral carotenoids the CD spectrum in comparison with a known configurational standard of the same carotenoid reveals the absolute configuration. When more chiral centres are present, the direct CD comparison is not unequivocal since not all chiral centres make a significant contribution to the Cotton effect.

Hitherto, a total X-ray analysis of a chiral carotenoid has not been successful. The ultimate structural proof or identity proof is therefore direct comparison with a synthetic sample of defined configuration. In the aryl polyene series successful X-ray analysis has been reported (Andrewes *et al.,* 1976).

For flexirubin-type pigments simplified identification tests have been recommended by Fautz and Reichenbach (1980) for screening of microorganisms. One test is based on the colour shift with alkali. The preferred test involves incorporation of labelled precursors (acetate, tyrosine, or methionine) followed by autoradiographic analysis of chromatograms, and allows distinction from phenolic and other carotenoids formed via the mevalonate pathway. Otherwise, the identification of these aryl polyenes requires isolation accompanied by spectroscopic and chemical characterisation in comparison with authentic natural or synthetic compounds.

III. Biological function and taxonomic significance

A. Biological function

The biological functions of carotenoids have been discussed in some detail by Krinsky (1971, 1978) and Goodwin (1980a). Cogdell (1978) has summarised some ideas on carotenoids in photosynthesis.

The established functions in microorganisms include (1) protection against photodynamic damage in both photosynthetic and non-photosynthetic organisms, (2) indirect participation in photosynthesis by light harvesting, and (3) involvement in phototaxis.

A biological function of the aryl polyenes has not yet been demonstrated except in the case of the xanthomonadin pigments. Early evidence suggests that these pigments do protect the bacteria from photobiological damage although the mechanism remains elusive (Jenkins and Starr, 1982b).

B. Taxonomic significance

For microorganisms with restricted morphological character, chemosystematics forms an important taxonomic tool. A recent evaluation of the use of carotenoids as chemosystematic markers besides other important secondary metabolites has been made (Liaaen-Jensen, 1979). Properties in their favour are structural diversity, widespread distribution, established major biosynthetic pathways, important functions for the organisms, and ready analysis.

Taxonomic aspects of flexirubin-type aryl polyenes have been discussed by Reichenbach *et al.* (1981), and the utility of the xanthomonadin pigments to the bacterial genus *Xanthomonas* has been reported by Starr *et al.* (1977).

In the following, a short consideration of the distribution of structurally distinct types of carotenoids and aryl polyenes within four major types of microooorganisms will be considered. Specific distribution patterns are imperative for the use of these pigments as chemosystematic markers.

1. Non-photosynthetic bacteria

Carotenoids are produced by certain non-photosynthetic bacteria with a rather unpredictable distribution pattern (Goodwin, 1980a). Characteristic structural types with limited distribution, and the occurrence of other aryl polyenes are treated below:

C_{30} *carotenoids.* *Staphylococcus aureus* synthesises unique C_{30} carotenoids (Taylor and Davies, 1974), exemplified by 4-hydroxy-4,4'-diaponeurosporene (**3**, Fig. 2). These carotenoids are not formed from C_{40} precursors but via an analogous C_{30} series (Davies and Taylor, 1982).

C_{40} *carotenoids.* C_{40} carotenoids have a scattered distribution. Several substitution patterns are encountered including aliphatic, moncyclic (**4**, Fig. 2), bicyclic, aromatic, phenolic, ketonic, hydroxylated (**4**, Fig. 2) and glycosylated (**4**, Fig. 2) derivatives.

Other aryl polyenes. The xanthomonadins are encountered in Xanthomonads (Eubacteriales) as exemplified by **8** (Fig. 2). Xanthomonadin I (**8**) is an *iso*butyl derivative of a natural ester in which the esterifying alcohol is not yet identified. Starr *et al.* (1977) and Jenkins and Starr (1982b) have presented a reasonable case for distinguishing *Xanthomonas* strains from superificially similar *Pseudomonas* strains on the basis of the presence or absence of brominated or non-brominated xanthomonadin pigments.

The flexirubin-type pigments are encountered in *Flexibacter-* and *Cytophaga*-like bacteria, in *Sporocytophaga,* and in flavobacteria with a low guanine : cytosine ratio. In contrast to carotenoids, which may also be present, the flexirubin-type pigments are located in the outer membrane of the Gram-negative cell wall. The flexirubin-type aryl polyenes [cf. flexirubin **2** (Fig. 1) and **7** (Fig. 2)] contain a polyenoic acid chromophore terminated by a *p*-hydroxyphenyl group and esterified with a dialkylated resorcinol. Variations may occur in the length of the polyene chain (6–8 double bonds), in a methyl or chloro substituent in the *meta* position of the conjugated phenyl groups, and in the alkyl substituents (chain length and branching) of the esterified resorcinol. Around 50 different flexirubin-like pigments are known (Reichenbach *et al.,* 1981).

2. Phototrophic bacteria

The chemistry of carotenoids from phototrophic bacteria was recently reviewed (Liaaen-Jensen, 1978a) and the biosynthesis and distribution thoroughly discussed (Schmidt, 1978). Other treatments are given by Goodwin (1980a,b). The carotenoids encountered reflect a close biosynthetic relationship. Characteristic features are as follows: generally aliphatic with tertiary hydroxy, methoxy, occasionally glucosyloxy groups in 1(1′) position. 1,2-Dihydro or 1,2,3- or 1,2,5- trimethylphenyl are other end groups encountered. Conjugated keto groups in the 2 or 4 position or cross-conjugated aldehyde in the 20 position are common. Selected examples are given with OH spheroidenone (**9**), okenone (**10**), and rhodopinal glucoside (**11**) in Fig. 3. It is noteworthy that these carotenoids, disregarding possible glucosyl substituents, are always achiral, provided the carotenoids of cyanobacteria are not included.

3. Fungi

The distribution of carotenoids in fungi has recently been thoroughly reviewed by Goodwin (1980a). Examples of carotenoids from microscopic fungi are given in Fig. 4, including β,β-carotene (**12**), torularhodin (**13**), and $(3R,3'R)$-astaxanthin (**14**) with interesting chirality, opposite to that of astaxanthin in other known *de novo* synthesisers (Andrewes and Starr, 1976b). In addition, several structurally distinct carotenoids have been isolated from the macroscopic fruiting bodies of particular fungi (Goodwin, 1980a).

The fungus *Cytidia salicina* (formerly *Corticium salicinum*) provides cortisalin (**15,** Fig. 3), the only known example of a fungal phenolic aryl polyene (Gripenberg, 1952). There have been no new reports on the taxonomic impact of cortisalin or cortisalin-type pigments.

4. Algae

Algal carotenoids show great structural diversity. Recent compilations are given by Liaaen-Jensen (1977, 1978b) and Goodwin (1980a). Representative examples of carotenoids from microscopic algae are given in Fig. 4, including the rhamnoside myxoxanthophyll (**16**) from Cyanophyceae (Cyanobacteria), the esterified siphonein (**17**), the acetylenic diadinoxanthin (**18**), and the norcarotenoid butenolide peridinin (**19**) with C_{37} skeleton, possessing the same end groups as the algal fucoxanthin (**1**) (Fig. 1).

Most of the 16 algal classes (Christensen, 1962; Parke and Dixon, 1976) have a rather typical carotenoid distribution pattern, and carotenoids represent, besides chlorophylls, a major criterion for the classification of microscopic algae.

Acknowledgments

A.G.A., on sabbatical leave in Trondheim from Saginaw Valley State College, gratefully acknowledges a Fellowship from the Fulbright Foundation and the hospitality of the University of Trondheim.

References

Abaychi, J. K., and Riley, J. P. (1979). *Anal. Chim. Acta* **107**, 1–11.

Achenbach, H., Kohl, W., and Reichenbach, H. (1976). *Chem. Ber.* **109**, 2490–2502.

Achenbach, H., Kohl, W., and Reichenbach, H. (1977a). *Liebigs Ann. Chem.,* 1–7.

Achenbach, H., Kohl, W., and Reichenbach, H. (1977b). *Tetrahedron Lett.,* 1061–1062.

Achenbach, H., Kohl, W., and Reichenbach, H. (1978). *Rev. Lationamer. Quim.* **9**, 111–124.

Achenbach, H., Kohl, W., Alexanian, S. and Reichenbach, H. (1979a). *Chem. Ber.* **112**, 196–208.

Achenbach, H., Kohl, W., and Reichenbach, H. (1979b). *Chem. Ber.* **112**, 1999–2011.

Achenbach, H., Böttger, A., Kohl, W., Fautz, E., and Reichenbach, H. (1979c). *Phytochemistry* **18**, 961–963.

Achenbach, H., Kohl, W., Böttger-Vetter, A., and Reichenbach, H. (1981). *Tetrahedron* **37**, 559–563.

Achenbach, H., Böttger-Vetter, A., and Hunkler, D. (1983). *Tetrahedron* **39**, 175–185.

Andrewes, A. G., and Starr, M. P. (1976a). *Phytochemistry* 15, 1003–1007.

Andrewes, A. G., and Starr, M. P. (1976b). *Phytochemistry* 15, 1009–1011.

Andrewes, A. G., Hertzberg, S., Liaaen-Jensen, S., and Starr, M. P. (1973). *Acta Chem. Scand.* 27, 2383–2395.

Andrewes, A. G., Jenkins, C. L., Starr, M. P., Shepherd, J., and Hope, H. (1976). *Tetrahedron Lett.*, 4023–4024.

Andrewes, A. G., Schmidt, K., and Liaaen-Jensen, S. (1985). To be published.

Baker, J. T., and Murphy, V. (1976). *In* "Handbook of Marine Science. Compounds from Marine Organisms," Vol. 1, pp. 15–19. CRC Press, Cleveland, Ohio.

Baker, J. T., and Murphy, V. (1981). *In* "Handbood of Marine Science, Compounds from Marine Organisms," Vol. 2, pp. 33–39. CRC Press, Cleveland, Ohio.

Balock, A. K., Buckle, K. A., and Edwards, R. A. (1977). *J. Chromatogr.* 139, 149–155.

Bandaranayake, W. M., and Gentien, P. (1982). *Comp. Biochem. Physiol.* B 72, 409–414.

Baranyai, M., Matus, Z., and Szabolcs, J. (1982). *Acta Aliment. Acad. Sci. Hung.* 10, 309–323.

Bartlett, L., Klyne, W., Mose, W. P., Scopes, P. M., Galasko, G., Mallams, A. K., Weedon, B. C. L., Szabolcs, J., and Tóth, G. (1969). *J. Chem. Soc. C*, 2527–2544.

Bellamy, L. J. (1975). "The Infra-Red Spectra of Complex Molecules," 3rd Ed. Chapman & Hall, London.

Ben-Amotz, A., Katz, A., and Avron, M. (1982). *J. Phycol.* 18, 529–534.

Braumann, T., and Grimme, L. H. (1981). *Biochim. Biophys. Acta.* 637, 8–17.

Buchecker, R., Marti, U., and Eugster, C. H. (1982). *Helv. Chim. Acta* 65, 896–912.

Budzikiewicz, H. (1982). *In* "Carotenoid Chemistry and Biochemistry" (G. Britton and T. W. Goodwin, eds.), pp. 155–166. Pergamon, Oxford.

Bushway, R. J., and Wilson, A. M. (1982). *Can. Inst. Food Sci. Technol. J.* 15, 165–169.

Cabibel, M., Lapize, F., and Ferry, P. (1981). *Sci. Aliment.* 1, 489–500.

Cadosch, H., and Eugster, C. H. (1974). *Helv. Chim. Acta* 57, 1466–1472.

Christensen, T. (1962). "Systematisk Botanik," Alger, Munksgaard, Copenhagen.

Cogdell, R. J. (1978). *Phil. Trans. R. Soc. London, Ser.* B 284, 569–579.

Csorba, I., Buzás, Z., Polyák, B., and Boross, L. (1979). *J. Chromatogr.* 172, 287–293.

Czygan, F. C., and Eichenberger, W. (1971). *Z. Naturforsch.* 26B, 264–267.

Davies, B. H. (1976). *In* "Chemistry and Biochemistry of Plant Pigments" (T. W. Goodwin, ed.), Vol. 2, pp. 38–155. Academic Press, New York.

Daraseliya, G. Y. (1983). *Mikrobiologiya* 52, 156 (in Russian); CA 98, 157551b.

Davies, B. H. (1976). *In* "Chemistry and Biochemistry of Plant Pigments" (T. W. Goodwin, ed.), Vol. 2, pp. 38–155. Academic Press, New York.

Davies, B. H., and Taylor, R. F. (1982). *Can. J. Biochem.* 60, 684–692.

Davis, J. B. (1974). *In* "Rodd's Chemistry of Carbon Compounds, Supplement to II^A/II^B Alicyclic Compounds" (M. F. Ansell, ed.), pp. 192–357. Elsevier, Amsterdam.

De Ritter, E., and Purcell, A. E. (1981). *In* "Carotenoids as Colorants and Vitamin A Precursors" (L. C. Bauernfeind, ed.), pp. 815–924. Academic Press, New York.

Englert, G. (1975). *Helv. Chim. Acta* 58, 2367–2390.

Englert, G. (1981). *In* "Carotenoid Chemistry and Biochemistry" (G. Britton and T. W. Goodwin, ed.), pp. 107–134. Pergamon, Oxford.

Englert, G., and Vecchi, M. (1980). *Helv. Chim. Acta* 63, 1711–1718.

Enzell, C. R., and Wahlberg, I. (1980). *In* "Biochemical Application of Mass Spectrometry" (G. R. Waller and O. C. Dermer, eds.), pp. 407–436. Wiley, New York.

Eschenmoser, W., Uebelhart, P., and Eugster, C. H. (1983). *Helv. Chim. Acta* 66, 82–91.

Eskins, K., Scholfield, C. R., and Dutton, H. J. (1977). *J. Chromatogr.* 135, 217–220.

Eugster, C. H. (1978). *Pure Appl. Chem.* 51, 463–506.

Eugster, C. H. (1982). In "Carotenoid Chemistry and Biochemistry" (G. Britton and T. W. Goodwin, eds.), pp. 1–26. Pergamon, Oxford.

Fautz, E., and Reichenbach, H. (1980). FEMS Microbiol. Lett. 8, 87–90.

Fiksdahl, A., Mortensen, J. T., and Liaaen-Jensen, S. (1978). J. Chromatogr. 157, 111–117.

Fiksdahl, A., Bjørnland, T., and Liaaen-Jensen, S. (1984). Phytochemistry 23, 649–655.

Foss, P. S., Green, R. W., and Richards, R. W. (1978). Aust. J. Chem. 31, 1981–1987.

Francis, G. W., Hertzberg, S., Andersen, K., and Liaaen-Jensen, S. (1970). Phytochemistry 9, 629–635.

Goodwin, T. W. (1974). In "Rodd's Chemistry of Carbon Compounds, Supplement to II^C/II^D/IIE Alicyclic Compounds" (M. F. Ansell, ed.), pp. 237–289. Elsevier, Amsterdam.

Goodwin, T. W. (1980a). "The Biochemistry of the Carotenoids, Vol. 1. Plants," 2nd Ed. Chapman & Hall, London.

Goodwin, T. W. (1980b). J. Sci. Ind. Res. 39, 682–688.

Gripenberg, J. (1952). Acta Chem. Scand. 6, 580–586.

Hajibrahim, S. K., Tibbets, P. J. C., Watts, C. D., Maxwell, J. R., Eglinton, G., Colin, H., and Guichon, G. (1978). Anal. Chem. 50, 549–553.

Hertzberg, S., and Liaaen-Jensen, S. (1967). Acta Chem. Scand. 21, 15–41.

Hertzberg, S., and Liaaen-Jensen, S. (1977). Acta Chem. Scand., Ser. B 31, 215–218.

Iriyama, K., Yoshiura, M., and Shiraki, M. (1978). J. Chromatogr. 154, 302–305.

Isler, O. (1971). "Carotenoids." Birkhäuser, Basel.

IUPAC Commission of Nomenclature of Organic Chemistry and the IUPAC-IUB Commission on Biochemical Nomenclature (1975). Pure Appl. Chem. 41, 405–431.

Jenkins, C. L., and Starr, M. P. (1982a). Curr. Microbiol. 7, 195–198.

Jenkins, C. L., and Starr, M. P. (1982b). Curr. Microbiol. 7, 323–326.

Jensen, A. (1959). Acta Chem. Scand. 13, 1259–1260.

Johansen, J. E., and Liaaen-Jensen, S. (1979). Acta Chem. Scand., Ser. B 33, 551–558.

Johansen, J. E., and Liaaen-Jensen, S. (1980). Phytochemistry 19, 441–447.

Ke, B., Imsgard, F., Kjøsen, H., and Liaaen-Jensen, S. (1970). Biochim. Biophys. Acta 210, 139–152.

Kjøsen, H., and Liaaen-Jensen, S. (1972). Acta Chem. Scand. 26, 2185–2193.

Krinsky, N. I. (1971). In "Carotenoids" (O. Isler, ed.), pp. 670–716. Birkhäuser, Basel.

Krinsky, N. I. (1978). Phil. Trans R. Soc. London, Ser. B 284, 581–590.

Liaaen-Jensen, S. (1964). Acta Chem. Scand. 18, 1562–1564.

Liaaen-Jensen, S. (1971). In "Carotenoids" (O. Isler, ed.), pp. 61–188. Birkhäuser, Basel.

Liaaen-Jensen, S. (1977). In "Marine Natural Products Chemistry" (D. J. Faulkner and W. H. Fenical, eds.), pp. 225–237. Plenum, New York.

Liaaen-Jensen, S. (1978a). In "Photosynthetic Bacteria" (R. K. Clayton and W. R. Sistrom, eds.), pp. 233–248. Plenum, New York.

Liaaen-Jensen, S. (1978b). In "Marine Natural Products. Chemical and Biological Perspectives" (P. Scheuer, ed.), Vol. 2, pp. 2–73. Academic Press, New York.

Liaaen-Jensen, S. (1979). Pure Appl. Chem. 51, 661–675.

Liaaen-Jensen, S. (1980). Fortschr. Chem. Org. Naturst. 39, 123–172.

Liaaen-Jensen, S., and Andrewes, A. G. (1972). Annu. Rev. Microbiol. 26, 225–248.

Liaaen-Jensen, S., and Jensen, A. (1971). In "Methods of Enzymology" (A. San Pietro, ed.), Vol. 23, pp. 586–602. Academic Press, New York.

McCormick, A., and Liaaen-Jensen, S. (1966). Acta Chem. Scand 20, 1989–1991.

Mantoura, R. F. C., and Llewellyn, C. A. (1983). Anal. Chim. Acta 151, 297–314.

Märki-Fisher, E., Marti, U., Buchecker, R., and Eugster, C. H. (1983a). Helv. Chim. Acta 66, 494–513.

254   S. LIAAEN-JENSEN AND A. G. ANDREWES

Märki-Fisher, E., Bütikofer, P.-A., Buchecker, R., and Eugster, C. H. (1983b). *Helv. Chim. Acta* **66**, 1175–1182.
Moss, G. P. (1976). *Pure Appl. Chem.* **47**, 97–102.
Nells, H. J. C. F., and Leenheer, A. P. (1983). *Anal. Chem.* **55**, 270–275.
Noack, K. (1981). *In* "Carotenoid Chemistry and Biochemistry" (G. Britton and T. W. Goodwin, eds.), pp. 135–154. Pergamon, Oxford.
Noack, K., and Thomson, A. J. (1979). *Helv. Chim. Acta* **62**, 1902–1921.
Pannakker, J. E., and Hallegraeff, G. M. (1978). *Comp. Biochem. Physiol.* **60B**, 51–58.
Parke, M., and Dixon, P. S. (1976). *J. Mar. Biol. Assoc. U. K.* **56**, 527–594.
Pfander, H., Schurtenberger, H., and Meyer, V. R. (1980). *Chimia* **34**, 179–180.
Reichenbach, H., Kleinig, H., and Achenbach, H. (1974). *Arch. Microbiol.* **101**, 131–144.
Reichenbach, H., Kohl, W., and Achenbach, H. (1981). *In* "The *Flavobacterium–Cytophaga* Group" (H. Reichenbach and O. B. Weeks, eds.), pp. 101–108. Verlag Chemie, Weinheim.
Renstrøm, B., and Liaaen-Jensen, S. (1981). *Comp. Biochem. Physiol.* **69B**, 625–627.
Renstrøm, B., Berger, H., and Liaaen-Jensen, S. (1981a). *Biochem. Syst. Ecol.* **9**, 249–250.
Renstrøm, B., Borch, G., and Liaaen-Jensen, S. (1981b). *Comp. Biochem. Physiol.* **69B**, 621–624.
Renstrøm, B., Borch, G., Skulberg, O., and Liaaen-Jensen, S. (1981c). *Phytochemistry* **20**, 2561–2564.
Rønneberg, H., Renstrøm, B., Aareskjold, K., Liaaen-Jensen, S., Vecchi, M., Leuenberger, F. J., Müller, R. K., and Mayer, H. (1980). *Helv. Chim. Acta* **63**, 711–715.
Rønneberg, H., Andrewes, A., Borch, G., Berger, R., and Liaaen-Jensen, S. (1985). *Phytochemistry* (in press).
Sadowski, R., and Wójcik, W. (1983). *J. Chromatogr.* **262**, 455–459.
Schmidt, K. (1978). *In* "Photosynthetic Bacteria," (R. K. Clayton and W. R. Sistrom, eds.), pp. 729–750. Plenum, New York.
Schmidt, K., Francis, G. W., and Liaaen-Jensen, S. (1971). *Acta Chem. Scand.* **25**, 2476–2486.
Séstak, Z. (1975). *J. Chromatogr. Libr.* **3**, 1039–1049.
Sherma, J., and Latta, M. (1978). *J. Chromatogr.* **154**, 73–75.
Singh, H., John, J., and Cama, H. R. (1973). *J. Chromatogr.* **75**, 146–150.
Stahl, E. (1969). *In* "Thin Layer Chromatography" (E. Stahl, ed.), 2nd Ed., pp. 259–272. Springer-Verlag, Berlin and New York.
Starr, M. P., Jenkins, C. L., Bussey, L. B., and Andrewes, A. G. (1977). *Arch. Microbiol.* **113**, 1–9.
Stewart, I. (1977a). *J. Assoc. Offic. Anal. Chem.* **60**, 132–136.
Stewart, I. (1977b). *J. Agric. Food Sci.* **25**, 1132–1137.
Stewart, I., and Leuenberger, U. (1976). *Alimenta* **15**, 33–36.
Stewart, I., and Wheaton, T. A., (1971). *J. Chromatogr.* **55**, 325–336.
Strain, H. H. (1958). "Chloroplast Pigments and Chromatographic Analysis." Pennsylvania State Univ. Press, University Park, Pennsylvania.
Strain, H. H., and Sherma, J. (1972). *J. Chromatogr.* **73**, 371–397.
Stransky, H. (1978). *Z. Naturforsch.* **3C**, 836–840.
Straub, O. (1976). "Key to Carotenoids. Lists of Natural Carotenoids." Birkhäuser, Basel.
Sturzenegger, V., Buchecker, R., and Wagniére, G. (1980). *Helv. Chim. Acta* **63**, 1074–1092.
Taylor, R. F., and Davies, B. H. (1974). *Biochem. J.* **139**, 751–760.
Taylor, R. F., and Davies, B. H. (1975). *J. Chromatogr.* **103**, 327–340.

Taylor, R. F., and Ikawa, M. (1980). *In* "Methods in Enzymology" (D. B. McCormick and L. M. Wright, eds.), Vol. 67, pp. 233–261. Academic Press, New York.

Tsukida, K., Saiki, K., Takii, T., and Koyama, Y. (1982). *J. Chromatogr.* **245**, 359–364.

Vecchi, M., and Müller, R. K. (1979). *J. High Resolut. Chromatogr. Chromatogr. Commun.* **2**, 195–196.

Vecchi, M., Englert, G., Mauer, R., and Meduna, V. (1981). *Helv. Chim. Acta* **64**, 2746–2758.

Vecchi, M., Englert, G., and Mayer H. (1982). *Helv. Chim. Acta* **65**, 1050–1058.

Vetter, W., Englert, G., Rigassi, N., and Schwieter, U. (1971). *In* "Carotenoids" (O. Isler, ed.), pp. 189–266. Birkhäuser, Basel.

Watts, C. D., Maxwell, J. R., Games, D. E., and Rossiter, M. (1975). *Org. Mass Spectrom.* **10**, 1102–1110.

Zakaria, M., Simpson, K., Brown, P. R., and Krstulovic, A. (1979). *J. Chromatogr.* **176**, 109–117.

Zechmeister, L. (1962). "*Cis–trans* Isomeric Carotenoids, Vitamins A and Arylpolyenes." Springer-Verlag, Berlin and New York.

9

Analysis of Bacteriochlorophylls*

J. OELZE

*Institut für Biologie II, Mikrobiologie, Universität Freiburg,
Freiburg, Federal Republic of Germany*

I. Introduction

The characteristic that all phototrophic bacteria have in common is the ability to perform an anoxygenic type of photosynthesis. Moreover, the phototrophic mode of energy metabolism is generally expressed in the light only under anaerobic growth conditions or under conditions of largely decreased oxygen tension (see Oelze, 1981, 1983 for reviews). This results from the fact that the formation of the central pigment of the photosynthetic apparatus, bacteriochlorophyll (Bchl), is controlled in an inverse proportion by oxygen where light enhances the inhibitory effect of oxygen (Arnheim and Oelze, 1983). The sensitivity of Bchl synthesis toward oxygen, however, may vary among different species. This applies primarily to those members of the group that are able to grow either phototrophically in the light or chemotrophically in the presence of oxygen. The role of oxygen in the control of Bchl formation is demonstrated

---

* Dedicated to Prof. Dr. N. Pfennig on the occasion of his 60th birthday in recognition of his numerous contributions on the ecology and taxonomy of phototrophic bacteria.

METHODS IN MICROBIOLOGY
VOLUME 18

by the fact that with facultative phototrophs oxygen regulates cellular Bchl levels even in darkness. Therefore, obligately phototrophic bacteria require anaerobic conditions in the light to produce Bchl while facultatively phototrophic forms produce highest Bchl levels either under anaerobic conditions in the light or under conditions of low aeration in the dark. However, it should be noted that more recently a group of marine chemotrophic bacteria has been described that require oxygen for Bchl formation (Shiba and Simidu, 1982).

Bchls absorb light quanta of defined energies. Consequently they exhibit characteristic light absorption properties *in situ,* as well as when extracted into organic solvents. On the basis of their characteristic absorption spectra in organic solvents six different Bchl chromophores designated *a* to *e* and *g* have been isolated from phototrophic bacteria (Jensen *et al.,* 1964; Gloe *et al.,* 1975; Gest and Favinger, 1983). The structure of a Bchl provisionally designated *f* was described but not isolated from natural sources (Gloe *et al.,* 1975). Detailed chemical analyses revealed that each of the alphabetically classified types of Bchl chromophores comprises a series of more than one structural form. Structural differences were identified with respect to both the substituents at the tetrapyrrole nucleus as well as the alcohol group esterifying the propionic acid side chain (Fig. 1 and Table I).

Interestingly, cultures of a single organism may produce different homologues of Bchl. Except for a predominant esterifying alcohol, however, the relative amounts of different Bchl homologues of one series have not yet been evaluated with respect to possible taxonomic implications. Considering the fact that the homologues of one series exhibit essentially identical absorption properties, this suggests that, at the

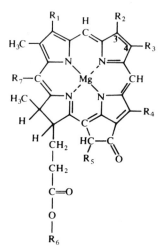

Fig. 1. Structure of bacteriochlorophylls *c, d,* and *e.* Bacteriochlorophylls *a, b,* and *g* contain no double bond between C-3 and C-4. For the various substituents (positions R^1–R^7) see Table I.

TABLE I

Different substituents in the positions R^1 through R^7 in bacteriochlorophylls and the respective near-infrared or red absorption bands

Bacteriochlorophyll	R^1	R^2	R^3	R^4	R^5	R^6 (ester)	R^7	References
a	$-CO-CH_3$	$-CH_3$	$-CH_2-CH_3$	$-CH_3$	$-CO-OCH_3$	Phytyl or geranylgeraniol	$-H$	Mittenzwei (1942); Brockmann and Kleber (1969)
b	$-CO-CH_3$	$-CH_3$	$=CH-CH_3$	$-CH_3$	$-CH-OCH_3$	Phytyl or phytadienyl	$-H$	Scheer et al. (1974); Steiner et al. (1981)
c	$-CHOH-CH_3$	$-CH_3$	$-C_2H_5$ $-C_3H_7$ $-C_4H_9$	$-C_2H_5$ $-CH_3$	$-H$	Farnesyl or stearyl	$-CH_3$	Holt et al. (1966); Gloe and Risch (1978)
d	$-CHOH-CH_3$	$-CH_3$	$-C_2H_5$ $-C_3H_7$ $-C_4H_9$ $-C_5H_{11}$	$-C_2H_5$ $-CH_3$	$-H$	Farnesyl	$-H$	Purdie and Holt (1965); Smith et al. (1983)
e	$-CHOH-CH_3$	$-CHO$	$-C_2H_5$ $-C_3H_7$ $-C_4H_9$	$-C_2H_5$	$-H$	Farnesyl	$-CH_3$	Gloe et al. (1975)
g	$-CH=CH_2$	$-CH_3$	$=CH-CH_3$	$-CH_3$	$-CO-OCH_3$	Geranylgeraniol	$-H$	Brockmann and Lipinski (1983)

present stage of knowledge, it is most useful to describe taxonomically relevant forms of Bchl in terms of their absorption properties. As mentioned already, determination of the major esterifying alcohol provides further information in categorising some bacterial species.

In situ Bchls are constituents of at least two different structurally and functionally specialised pigment–protein complexes of the photosynthetic apparatus. Some species incorporate chemically different Bchl chromophores into different functional complexes. Alternatively, quite a few species use just one kind of Bchl chromophore in constituting different complexes. These complexes are described on the basis of the distinctive Bchl absorption bands in the near infrared. Since different groups of the phototrophic bacteria produce characteristic forms of Bchl–protein complexes it seems logical that not only the Bchl chromophores but also the cellular complement of pigment–protein complexes may be of taxonomic sigificance. Consequently, this chapter describes the identification not only of Bchl chromophores but also of relevant Bchl–protein complexes.

II. Isolation and identification of bacteriochlorophyll

Bchls can be extracted from concentrated cell pastes or subcellular fractions with organic solvents that remove water before the pigments become soluble. For this purpose several water-miscible solvents have been employed such as methanol, acetone, and pyridine as well as these solvents in combinations with each other or with diethyl ether, petroleum ether, or carbon tetrachloride. Before transferring Bchls into organic solution it should be kept in mind that all chlorophylls are exceptionally labile pigments which are subject to structural alterations when exposed to factors such as light, acids, and oxidising chemicals. Therefore, to prevent destruction, all manipulations should be performed in the dark or at least under dim light. Adequate protection of Bchl *a* from photooxidation was achieved by extracting cells with methanol supplemented with a small volume (0.5%) of a solution of sodium ascorbate at a final concentration of 0.005 M (Kim, 1966). Undesired pheophytinisation of Bchl under acidic conditions can be prevented by adding ammonia or organic bases to the extraction solvent (Strain and Svec, 1966). On the other hand, complete pheophytinisation is frequently performed to transform Bchl into its more stable magnesium-free form.

The following is a selection of representative methods which allow the isolation and identification of known types of Bchl using generally available equipment and techniques. Most of the methods were developed for the isolation of one type of Bchl chromophore, but several of them may be adapted to isolate other types as well.

A. Bacteriochlorophyll *a*

1. Extraction and purification

Methanol is the most frequently used solvent for the extraction of Bchls. The cells are separated from the culture medium by centrifugation. The pelleted cell material is resuspended with methanol (plus 0.005 *M* sodium ascorbate), kept in the dark and the cold for about half an hour, and sedimented by centrifugation. The methanolic supernatant contains Bchl and carotenoids. The extraction is repeated until the sedimented cell material has become colourless. After transfer into a separating funnel diethyl ether is added to the combined methanolic extracts. Bchl is forced into the ether layer after the cautious addition of water saturated with NaCl. The ether is washed three or four times with an identical volume of water (plus 0.005 *M* sodium ascorbate), dried over Na_2SO_4, and evaporated to dryness. The residue is dissolved in a small volume of diethyl ether and spotted on thin-layer silica gel plates (Merck, No. 5721).

Thin-layer chromatographs are developed with chloroform/acetone (92 : 8, v/v). After 45–60 min a blue band containing Bchl *a* migrates about halfway between the origin and the solvent front while carotenoids and bacteriopheophytin (Bph) migrate close to the solvent front. The blue band is scraped off the thin-layer plate. Subsequently, Bchl *a* is extracted into diethyl ether or other organic solvents (see Table II). The purity of the Bchl *a* preparation is tested on the basis of its absorption spectrum and chromatography on HPTLC plates (Merck RP-8) with methanol as solvent (H. Scheer, personal communication). If necessary the separation is repeated. Alternative methods for the purification of Bchl *a* have been published by Smith and Benitez (1955), Sistrom (1966), and Strain and Svec (1966).

Another method of thin-layer chromatography was employed to separate Bchl a_p (esterified with phytol) from Bchl a_{Gg} (esterified with geranylgeraniol) (Brockmann and Knobloch, 1972). To prevent the formation of various degradation products, the pigment was transformed into Bph *a* in the presence of HCl. This can be achieved by transferring Bchl *a* into either diethyl ether, petroleum ether, or carbon tetrachloride which is subsequently washed with 20% HCl. Smith and Benitez (1955) added 20 ml of 20% HCl to Bchl *a* in 50 ml ethyl ether. The mixture was kept at 6°C in the dark for 1 h. Then the ether solution was washed with water until neutral. After pheophytinisation the organic phase was dried over Na_2SO_4, concentrated under vacuum, and spotted on silver nitrate-impregnated plates of silica gel (Merck Darmstadt, No. 5553). Gloe and Pfennig (1974) described the following method of impregnation. The precoated silica gel plates were slowly immersed in a 1% detergent solution. After complete soaking of the silica gel the plates were removed and free water was dried off with a fan. Then the moist plates were immersed in a 5% silver nitrate solution for 1 min and subsequently dried and activated at 100°C for 30 min. The impregnated plates were stored in the dark until use. After application of Bph a samples chromatography was run in the dark with carbon tetrachloride/acetone (92 : 8, v/v). Silver nitrate reacts with double bonds so that the Bph a_p containing phytol with three double bonds migrates faster than Bph a_{Gg} containing geranylgeraniol with four double bonds. After separation on silver nitrate-impregnated thin-layer plates R_f values were determined: 0.629 and 0.282 for Bph a_p and a_{Gg}, respectively (Gloe, 1977). Bchls a_p and a_{Gg} exhibit comparable differences in R_f values when separated on HPTLC thin-layer plates (Merck RP-8) with 95% aqueous methanol (H. Scheer, private communication).

2. Spectral identification and quantitative determination

Bchl a exhibits characteristic absorption properties in organic solvents. This is demonstrated in Fig. 2 for Bchl a from *Rhodopseudomonas sphaeroides* dissolved in either diethyl ether or acetone/methanol. Table

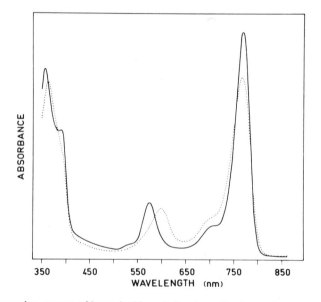

Fig. 2. Absorption spectra of bacteriochlorophyll *a* from *Rhodopseudomonas sphaeroides* in diethyl ether (solid line) and in acetone/methanol (7 : 2) (dotted line).

II compiles the wavelengths and relative peak heights of absorption bands in various solvents of Bchl *a* from representative species. Slight variations in the absorption properties are not characteristic of a given species; rather they represent the range of experimental error (compare van der Rest and Gingras, 1974).

As mentioned above, it is more convenient to analyse Bchl *a* after conversion into the more stable Bph *a*. Pheophytinisation of Bchl *a* may be performed as described above. Spectra in diethyl ether and acetone/methanol of Bph *a* from *R. sphaeroides* are shown in Fig. 3. The corresponding spectral properties in various organic solvents of Bph *a* from different species are given in Table III.

Quantitative determinations of Bchl *a* as well as Bph *a* are based on analyses performed by Weigl (1953) or Smith and Benitez (1955). Table IV compiles the long-wavelength absorption bands and molar absorption coefficients of these pigments in ether. On the basis of these data the corresponding coefficients for the pigments in other solvents were calculated (Table IV). In methanol the molar absorption coefficient for the 770-nm band of Bchl *a* is $\varepsilon = 60$ mM^{-1} cm^{-1} (Cohen-Bazire and Sistrom, 1966) and in carbon tetrachloride it is $\varepsilon = 88$ mM^{-1} cm^{-1} at 781 nm (Sauer *et al.*, 1966).

Routinely, Bchl *a* or Bph *a* is determined with crude extracts in acetone/methanol (7 : 2, v/v). Pigment concentrations are calculated with the

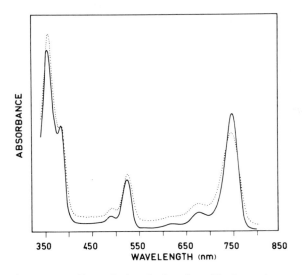

Fig. 3. Absorption spectra of bacteriopheophytin *a* from *Rhodopseudomonas sphaeroides* in diethyl ether (solid line) and in acetone/methanol (7 : 2) (dotted line).

TABLE II

Positions of absorption bands and relative absorbancies of Bchl *a*

Species	Absorption bands, λ (nm)				Solvent	Reference
Rhodospirillum	771	573	391	357	Ethyl ether	van der Rest
rubrum	(1.00)[a]	(0.299)	(0.517)	(0.78)		and Gingras
						(1974)
Rhodopseudomonas	771	574	390	356	Etliyl ether	
sphaeroides	(1.00)	(0.236)	(0.538)	(0.79)		
Chlorobium	770	580	390	360	Ethyl ether	Olson and
thiosulfatophilum						Romano
L1						(1962)
Chloroflexus	771–772	574	392	357	Ethyl ether	Pierson and
aurantiacus	(1.00)	(0.236)	(0.52)	(0.776)		Castenholz
						(1974)
Rhodospirillum	771	578	390	358	Acetone	Jensen *et al.*
rubrum	(1.00)	(0.298)	(sh)[b]	(0.962)		(1964)
Rhodopseudomonas	771	578	390	358	Acetone	Jensen *et al.*
sphaeroides	(1.00)	(0.275)	(sh)[b]	(0.98)		(1964)
Chlorobium	771.5	578	390	358	Acetone	Jensen *et al.*
thiosulfatophilum	(1.00)	(0.328)	(sh)[b]	(1.026)		(1964)
L1						
Chloroflexus	770	580	393	357	Acetone	Pierson and
aurantiacus	(1.00)	(0.303)	(0.583)	(0.960)		Castenholz
						(1974)
Rhodospirillum	772	608		365	Methanol	Smith and
rubrum	(1.00)	(0.367)		(1.284)		Benitez
						(1955)
Rhodopseudomonas	771	608		366	Methanol	Sistrom (1966)
sphaeroides Ga	(1.00)	(0.305)		(1.06)		
Chlorobium	770	606		364	Methanol	Olson and
thiosulfatophilum						Romano
L1						(1962)
Chloroflexus	771	608		366	Methanol	Pierson and
aurantiacus	(1.00)	(0.321)		(1.069)		Castenholz
						(1974)
Rhodopseudomonas	780	582	390	363	Chloroform	Sistrom (1966)
sphaeroides Ga	(1.00)	(0.28)	(0.555)	(0.81)		
Chloroflexus	780	582	390–400	363	Chloroform	Pierson and
aurantiacus	(1.00)	(0.281)	(0.531)	(0.836)		Castenholz
						(1974)

[a] Values in parentheses are relative absorbancies.
[b] Shoulder.

TABLE III

Positions of absorption bands and relative absorbancies of Bph *a*

Species	Absorption bands, λ (nm)					Solvent	Reference
Rhodospirillum	749	677	524	385	358	Ethyl ether	van der Rest
rubrum	(1.00)[a]	(0.135)	(0.396)	(0.895)	(1.627)		and Gingras (1974)
Rhodopseudomonas	748	675	523	382	354	Ethyl ether	
sphaeroides	(1.00)	(0.145)	(0.416)	(0.895)	(1.563)		
Chloroflexus	750	680	525	385	357	Ethyl ether	Pierson and
aurantiacus	(1.00)	(0.155)	(0.408)	(0.921)	(1.658)		Castenholz (1974)
Rhodopseudomonas	745	675	523	385	357	Acetone	Jensen *et al.*
sphaeroides	(1.00)	(0.229)	(0.563)	(1.083)	(2.08)		(1964)
Chlorobium	747	668	525	384	357	Acetone	Jensen *et al.*
thiosulfatophilum L	(1.00)	(0.264)	(0.626)	(1.22)	(2.198)		(1964)
Rhodospirillum	757.5	687	533	390	363	Chloroform	Smith and
rubrum	(1.00)	(0.182)	(0.413)	(0.832)	(1.567)		Benitez (1955)
Chloroflexus	757	686	531	390	362	Chloroform	Pierson and
aurantiacus	(1.00)	(0.189)	(0.426)	(0.861)	(1.648)		Castenholz (1974)

[a] Values in parentheses are relative absorbancies.

TABLE IV

Molar absorption coefficients for Bchl *a* and Bph *a* at their
long-wavelength absorption maxima

Solvent	Absorption coefficients (mM^{-1} cm^{-1})		Reference
	Bchl *a*	Bph *a*	
Ether	96	63	Weigl (1953)
	(772)[a]	(750)	
	91.15	67.6	Smith and Benitez
	(773)	(749)	(1955)
Acetone/methanol	76	—	Clayton (1966)
(7 : 2)	(767)		
	65.3	45.1	van der Rest and
	(771)	(747)	Gingras (1974)

[a] Values in parentheses are absorption maxima, λ, in nanometers.

molar absorption coefficients (Table IV) and the heights of the long-wavelength absorption band by the equation of the Beer–Lambert law. This simplification is possible because no other pigments such as carotenoids interfere with the absorption of Bchl *a* and Bph *a* in the infrared region. Crude extracts contain different amounts of water derived from the extracted biological material. Clayton (1966) reported that the presence of up to 10% water does not interfere significantly with Bchl *a* determinations. Determination of Bph *a* in crude extracts is conveniently achieved when cells are washed with dilute HCl (1–3 *N*) prior to the extraction. Subsequent addition of the usual extraction solvent results in direct removal of Bph *a*.

B. Bacteriochlorophyll *b*

Bchl *b* is especially unstable in organic solvents. This is particularly true for Bchl *b* if dissolved in acetone (Brockmann and Kleber, 1970). To diminish the destruction of Bchl *b*, light should be excluded as much as possible and sodium ascorbate should be added to solvents employed during isolation and purification of the pigment.

1. Extraction and purification

The method described above (Section A,1) for the extraction and isolation of Bchl *a* may be adopted to prepare spectrally pure samples of Bchl *b*. Alternatively, Steiner *et al.* (1981) modified a method as described by Strain and Svec (1966).

Bph *b* is obtained when Bchl *b* in methanol is treated with 1% (final concentration) of H_2SO_4 in the presence of nitrogen (Davis *et al.*, 1979).

2. Spectral identification and determination

Studies on the chemical structure of Bchl *b* were performed with pigments extraced from *Rhodopseudomonas viridis* and *Ectothiorhodospira halochloris* (Scheer *et al.*, 1974 Steiner *et al.*, 1981). Differences in the structure were reported with respect to the esterifying alcohol which is phytol (Δ-2-phytaenol) in *R. viridis* and Δ-2,10-phytadienol in *E. halochloris*. Absorption spectra in diethyl ether of Bchl *b* and Bph *b* from *R. viridis* are shown in Fig. 4. The absorption properties of Bchl *b* and Bph *b* from *R. viridis* and *E. halochloris* are compared in Table V.

Bchl *b* concentrations are estimated with crude organic extracts as well as with purified pigments on the basis of the long-wavelength absorption

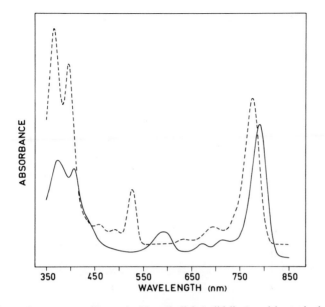

Fig. 4. Absorption spectra of bacteriochlorophyll *b* (solid line) and bacteriopheophytin *b* (dashed line) from *Rhodopseudomonas viridis* in diethyl ether.

band. According to Steiner and Scheer (private communication) the absorption coefficient of the long-wavelength absorption band in diethyl ether is 106 mM^{-1} cm^{-1}.

C. Bacteriochlorophyll *c*

Originally this pigment was introduced as "chlorobium chlorophyll 660" (Stanier and Smith, 1960). In the following years, however, the suggestion by Jensen *et al.* (1964) to rename this pigment Bchl *c* has been accepted.

1. Isolation and purification

Procedures for the extraction of Bchl *c* from cell material include the following solvents: methanol (Stanier and Smith, 1966), acetone/methanol (7:2, v/v) (Feick *et al.*, 1982), as well as acetone followed by a second extraction with acetone/carbon tetrachloride (135:65, v/v) (Gloe *et al.*, 1975). Pierson and Castenholz (1974) separated and purified Bchls *a* and *c* from *Chloroflexus aurantiacus* according to the following method. Four grams wet wt. of washed cells were extracted three times with 60-ml volumes of 85% methanol saturated with $CaCO_3$. The pigments were

TABLE V

Positions of absorption bands and relative absorbancies of Bchl *b* and Bph *b*

Species	Absorption bands, λ (nm)							Solvent	Reference
					Bacteriochlorophyll *b*				
Rhodopseudomonas viridis	791 (1.00)[a]	715 (0.127)	672 (0.100)	592 (0.187)	406 (0.673)	372 (0.729)		Ethyl ether	
Rhodopseudomonas viridis	795 (1.00)		677 (0.239)	582 (0.313)	407 (0.906)	368 (1.0417)		Acetone	Jensen *et al.* (1964)
Ectothiorhodospira halochlochloris	794 (1.00)		676 (0.13)	578 (0.25)	408 (0.73)	368 (0.81)		Ethyl ether	Steiner (1980)
					Bacteriopheophytin *b*				
Rhodopseudomonas viridis	776 (1.00)	694 (0.12)	526 (0.373)	490 (0.102)	460 (0.136)	396 (1.237)	366 (1.475)	Ethyl ether	
Rhodopseudomonas viridis	775 (1.00)		524 (0.479)	495 (0.104)		397 (1.75)	368 (2.08)	Acetone	Jensen *et al.* (1964)
Ectothiorhodospira halochloris	776 (1.00)		528 (0.5)			398 (2.37)	368 (2.26)	Ethyl ether	Steiner (1980)

[a] Values in parentheses are relative absorbancies.

transferred from methanol to ethyl ether or petroleum ether (boiling range 30–60°C) by the addition of water saturated with NaCl. The pigment solutions were washed with water, dried over Na_2SO_4, and evaporated under a stream of nitrogen. Pigments were redissolved with a mixture of ethyl ether/petroleum ether (1 : 1) and applied to powdered sugar columns. To remove carotenoids the columns were developed first with petroleum ether (boiling range 60–110°C). Following this, Bchls a and c were separated from each other with a series of mixtures of n-propanol (0.25, 0.5, 1.0, and 2.0%) in petroleum ether or a series of mixtures of ethyl ether (1 to 25%) in petroleum ether. After separation, the column material was dried under a stream of nitrogen and extruded. The bands were cut and eluted with ethyl ether. Pigments were rechromatographed two or three times until spectrally pure. The authors emphasise that only freshly opened cans of ethyl ether were used.

Gloe *et al.* (1975) purified Bchl c from *Chlorobium limicola* forma *thiosulfatophilum* by thin-layer chromatography on silica gel plates with the solvent system carbon tetrachloride/acetone (85 : 15, v/v).

Bchl c esterified with either farnesol or stearylalcohol are distinguished from one another after thin-layer chromatography on silver nitrate-impregnated silica gel plates (see Section A,1, separation of Bph a_p and a_{Gg}). Gloe *et al.* (1975) pheophytinised Bchl c (see above for the extraction procedure) by washing the crude extracts twice with 0.5 N HCl. This resulted in separation of an acetone/water phase from a carbon tetrachloride phase which contained the pigments. The latter phase was withdrawn, washed repeatedly with water until neutral, dried over Na_2SO_4, and finally spotted onto thin-layer plates. In this system Bph c_F (esterified with farnesol) exhibits an R_f value of 0.105 while Bph c_s (esterified with stearylalcohol) exhibits an R_f value comparable to that of Bph d, which is 0.202 (Gloe, 1977).

2. Spectral identification and quantitative determination

Although Bchl c comprises a mixture of several homologous compounds, no spectrally different forms have been reported (Fig. 5 and Table VI). But with the differently esterified Bphs Gloe and Risch (1978) observed the following differences (Table VII): compared to Bph c_s from *C. aurantiacus* in acetone, the long-wavelength absorption band of Bph c_F from *Chlorobium* was shifted towards shorter wavelengths by 4–5 nm and, relative to the short-wavelength band, the long-wavelength band was 25% higher. Bchls c are quantitatively determined with the specific absorption coefficients (Table VIII) published by Stainer and Smith (1960). Because

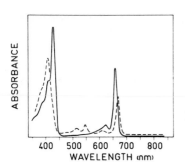

Fig. 5. Absorption spectra of bacteriochlorophyll *c* (solid line) and bacteriopheophytin *c* (dashed line) from *Chloroflexus aurantiacus* in diethyl ether. From Pierson and Castenholz (1975).

of their distinctive long-wavelength absorption maxima Bchls *a* and *c* can be determined individually with the same crude organic extract. For this purpose acetone/methanol (7 : 2, v/v) is a frequently employed solvent system. Feick *et al.* (1982) used a molar absorption coefficient of $\varepsilon_{666} = 74$ mM^{-1} cm^{-1} to determine Bchl c_s in acetone/methanol.

D. Bacteriochlorophyll *d*

Bchl *d* is the generally accepted designation of the pigment named originally "chlorobium chlorophyll 650" by Stanier and Smith (1960).

TABLE VI
Positions of absorption bands and relative absorbancies of Bchl *c*

Species	Absorption bands, λ (nm)				Solvent	Reference
Chlorobium *vibrioforme* f. *thiosulfatophilum* (NCIB 8346)	661 $(1.00)^a$	629 (0.871)	431 (1.245)	412 (0.886)	Acetone	Jensen *et al.* (1964)
Chlorobium *limicola* f. *thiosulfatophilum* (B)	663 (1.00)	627 (0.199)	434 (1.433)		Acetone	Pierson and Castenholz (1974)
Chloroflexus *aurantiacus* (strain J-10-f1)	662 (1.00)	627 (0.193)	433 (1.523)		Acetone	Pierson and Castenholz (1974)
Chlorobium *limicola* f. *thiosulfatophilum* (B)	660 (1.00)	624 (0.166)	431 (1.52)	412 (0.74)	Ethyl ether	Pierson and Castenholz (1974)
Chloroflexus *aurantiacus* (strain J-10-f1)	660 (1.00)	623 (0.171)	431 (1.609)	412 (0.821)	Ethyl ether	Pierson and Castenholz (1974)

^{a} Values in parentheses are relative absorbancies.

TABLE VII

Positions of absorption bands and relative absorbancies of Bph *c*

Species	Absorption bands, λ (nm)					Solvent	Reference
Chlorobium *vibrioforme* f. *thiosulfatophilum* (NCIB 8346)	667 (1.00)[a]	608 (0.143)	547 (0.326)	515 (0.204)	411 (2.04)	Acetone	Jensen *et al.* (1964)
Chlorobium *limicola* *thiosulfatophilum* (strain 6230, DSM-No: 249)	664 (1.00)	604 (0.231)	547 (0.338)	515 (0.262)	408 (1.538)	Acetone	Gloe *et al.* (1975)
Chloroflexus *aurantiacus* (strain J-10-f1)	669 (1.00)	608 (0.256)	548 (0.359)	514 (0.308)	408 (2.564)	Acetone	Gloe (1977)
Chlorobium *limicola* f. *thiosulfatophilum* (strain B)	667–668 (1.00)	605–615 (0.119)	547 (0.29)	516 (0.179)	410 (1.964)	Ethyl ether	Pierson and Castenholz (1974)
Chloroflexus *aurantiacus* (strain J-10-f1)	668 (1.00)	605–615 (0.112)	546 (0.272)	515 (0.178)	410 (2.036)	Ethyl ether	Pierson and Castenholz (1974)

[a] Values in parentheses are relative absorbancies.

TABLE VIII

Absorption maxima and specific absorption
coefficients of Bchl *c*[a] in different solvents[a]

Solvent	Absorption maxima, λ (nm)			
Ethyl ether	660 (112.5)[b]	622 (19.7)	432 (175.5)	412 (89.7)
Acetone	662.5 (92.6)	625 (16.9)	433 (142.8)	413 (88.8)
Methanol	670 (86.0)	620 (18.1)	435 (97.8)	419 (86.1)

[a] Data from Stainer and Smith (1960).

[b] Values in parentheses are specific absorption coeffients in liter g^{-1} cm^{-1}.

1. Isolation and purification

Methods described above in particular for Bchl c are used to extract and purify Bchl d as well. Moreover, the methods described for pheophytinisation of Bchl c are applicable to Bchl d. Separation of Bchl d from small amounts of Bchl a becomes possible due to different R_f values which are significantly higher for Bchl a (Jensen et al., 1964). The same holds true for the Bphs. Gloe et al. (1975) reported R_f values of 0.67 and 0.376 for Bphs a and d, respectively, as identified on silica gel thin-layer plates developed with carbon tetrachloride/acetone (90 : 10, v/v).

2. Spectral identification and quantitative determination

Up to eight chemically different Bchls of the d series with almost identical absorption spectra have been characterised as yet (Smith et al., 1983). Representative absorption spectra of Bchl d and Bph d are depicted in Fig. 6. The positions and relative heights of characteristic absorption bands are compiled in Table IX. Quantitative determinations of Bchl d are performed with the specific absorption coefficients (Table X) published by Stanier and Smith (1960). The long-wavelength coefficient is employed to calculate Bchl d concentrations in crude organic extracts.

E. Bacteriochlorophyll e

Isolation and purification of Bchl e and Bph e follows the methods described above for Bchl c (Gloe et al., 1975). The authors noted that after extraction from cells sometimes up to 50% of the pigment was present in the form of Bph e. Therefore, purification of Bchl e by thin-layer chromatography was suggested in order to obtain reliable spectra.

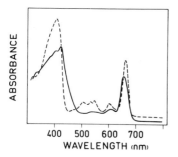

Fig. 6. Absorption spectra of bacteriochlorophyll d (solid line) and bacteriopheophytin d (dashed line) from *Chlorobium vibrioforme* in acetone. From Gloe et al. (1975).

TABLE IX

Positions of absorption bands and relative absorbancies of Bchl d and Bph d in acetone

Species	Absorption bands, λ (nm)						Reference
Bacteriochlorophyll d							
Chlorobium	651	605	577	425	406		Jensen *et al.*
thiosulfatophilum	$(1.00)^a$	(0.214)	(0.143)	(1.428)	(1.143)		(1964)
(strain L1)							
Chlorobium	654	608		424	408		Gloe *et al.*
vibrioforme	(1.00)	(0.278)		(1.639)	(1.426)		(1975)
(strain 6030,							
DSM-No 260)							
Bacteriopheophytin d							
Chlorobium	657	601		533	503	407	Jensen *et al.*
thiosulfatophilum	(1.00)	(0.227)		(0.227)	(0.25)	(2.272)	(1964)
(strain L1)							
Chlorobium	658	604	548	533	505	406	Gloe *et al.*
vibrioforme	(1.00)	(0.218)	(0.291)	(0.254)	(0.254)	(1.818)	(1975)
(strain 6030,							
DSM-No 260)							

^{a} Values in parentheses are relative absorbancies.

Bph e exhibits an R_f value of 0.181 after thin-layer chromatography on silica gel with carbon tetrachloride/acetone (90:10, v/v) as the solvent system (Gloe *et al.*, 1975). Representative absorption spectra of Bchl e and Bph e are shown in Fig. 7 and an evaluation of the absorption properties in Table XI. Quantitative estimation of the two pigments awaits the determination of absorption coefficients.

TABLE X

Absorption maxima and specific absorption coefficients of Bchl d^a

Solvent	Absorption maxima, λ (nm)					
Ethyl ether	650	612	575	530	425	406
	$(113.5)^b$	(15.8)	(8.6)	(3.4)	(146.0)	(87.2)
Acetone	654	612.5	577.5	530	427	406
	(98.0)	(16.4)	(9.2)	(3.9)	(125.9)	(87.9)
Methanol	659.0	612.5			427	411
	(82.3)	(18.0)			(83.5)	(76.5)

^{a} Data from Stanier and Smith (1960).

^{b} Values in parentheses are specific absorption coefficients in liter g^{-1} cm^{-1}.

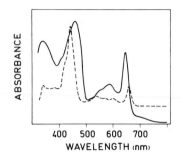

Fig. 7. Absorption spectra of bacteriochlorophyll *e* (solid line) and bacteriopheophytin *e* (dashed line) from *Chlorobium phaevibrioides* in acetone. From Gloe *et al.* (1975).

F. Bacteriochlorophyll *g*

Very recently, Gest and Favinger (1983) described a new species of the phototrophic bacteria, *Heliobacterium chlorum,* which produces the, as yet, unknown Bchl *g*. Brockmann and Lipinski (1983) isolated this pigment and characterised its structure. Geranylgeraniol was identified as the esterifying alcohol.

Because this pigment is very labile to air and light, cell pastes were extracted with acetone anaerobically in the dark. The extract was immediately treated with HCl which resulted in the formation of three major

TABLE XI

Positions of absorption bands and relative absorbancies of Bchl *e* and Bph *e* in acetone[a]

Species	Absorption bands, λ (nm)								
	Bacteriochlorophyll *e*								
Chlorobium	647	594	545	456	338				
phaeobacteroides	(1.00)[b]	(0.417)	(sh)	(4.167)	(1.417)				
(strain 2430)									
Chlorobium	646	589	550	458	338				
phaeovibrioides	(1.00)	(0.545)	(sh)	(1.515)	(1.167)				
(strain 2631)									
	Bacteriopheophytin *e*								
Chlorobium	654	596	567	534	437	418	376	358	332
phaeobacteroides	(1.00)	(0.31)	(sh)	(0.379)	(3.448)	(sh)	(0.862)	(0.758)	(0.931)
(strain 2430)									
Chlorobium	654	601	573	538	441	418	380	360	337
phaeovibrioides	(1.00)	(0.4)	(0.4)	(0.48)	(4.00)	(sh)	(0.92)	(0.88)	(1.04)
(strain 2631)									

[a] Data from Gloe *et al.* (1975).
[b] Values in parentheses are relative absorbancies; sh, shoulder.

compounds, one of which was Bph g_{Gg}. This pigment was separated from other pigments by chromatography on columns of silica gel Si 60 (0.04–0.06 mm, Merck, Darmstadt, Federal Republic of Germany) with the solvent system carbon tetrachloride/acetone (6:1, v/v). The absorption characteristics of Bchl g_{Gg} and Bph g_{Gg} are presented in Table XII.

III. Functional bacteriochlorophyll–protein complexes

Bacteriochlorophyll is the central pigment of two functionally different types of Bchl–protein complexes of the bacterial photosynthetic apparatus. Bchls contained in the first type of complex, the light-harvesting (LH) complex, become excited upon illumination. When transferred to the second type, the photochemical reaction centre (RC) complex, excitation energy can be trapped through the primary photochemical reaction leading to charge separation. This means that RC-Bchl becomes photooxidised and the primary electron acceptor reduced. Electron flow from the primary electron acceptor back to the oxidised RC-Bchl, involving several carriers, is assumed to facilitate proton movement across the membrane. This proton movement results in the formation of the electrochemical proton gradient which is utilised in the course of different energy-dependent reactions (Evans and Heathcote, 1983; Evans, 1983 for introduction).

1. Composition of bacteriochlorophyll–protein complexes

RC-Bchl complexes have been isolated from a wide variety of phototrophic bacteria (Cogdell, 1983). The most detailed studies have been

TABLE XII

Position of absorption bands and relative absorbancies in dioxane of Bchl g and Bph g from *Heliobacterium chlorum*[a]

	Absorption bands, λ (nm)				
Bacteriochlorophyll g	763	575	470[b]	418	408
	(1.00)[c]	(0.412)	(0.54)	(1.863)	(1.96)
Bacteriopheophytin g	753	518	396	368	
	(1.00)	(0.610)	(2.439)	(2.415)	

[a] Data from Brockmann and Lipinski (1983).
[b] Possibly due to an impurity.
[c] Values in parentheses are relative absorbancies.

performed with preparations from *R. sphaeroides* and *Rhodospirillum rubrum*. These centres contain per mol of complex: 4 mol of Bchl *a*, 2 mol of Bph *a*, 1–2 mol of quinone, 1 mol of carotenoid, 1 mol of iron atoms, and protein composed of three different polypeptides. Near infrared absorption spectra of these RCs are characterised by absorption bands at 760 (due to Bph *a*), 800, and 865 nm (both due to Bchl *a*). Upon photo-oxidation of RC-Bchl *a* the 865-nm band bleaches almost completely and the 800-nm band shifts a few nanometers to the blue. RCs from *R. viridis* are very similar in composition to those mentioned above, except that they contain Bchl *b* and Bph *b*. In RCs of *R. viridis* a band at 960 nm is light bleachable while a band at 830 undergoes a slight light-induced blue shift (Clayton and Clayton, 1978). In *C. aurantiacus*, on the other hand, RC complexes contain 3 mol of each, Bchl *a* and Bph *a*, per mol of minimum unit (literature compiled by Thornber *et al.*, 1983).

The amounts of Bchl present in RC make up only a minor proportion of the total cellular Bchl. By far the major part of the Bchl is associated with LH complexes. This means that the absorption spectra of whole cells or cell homogenates present largely the absorption properties of Bchls and carotenoids of LH complexes. According to their typical visible colours phototrophic bacteria are divided into two major groups, the purple and the green bacteria. In the phototrophic green bacteria the predominant long-wavelength absorption bands are due to Bchls *c*, *d*, and *e* which fulfil LH functions (Olson, 1980). Minor amounts of Bchl *a*, always present in these organisms, are assumed to participate in energy transfer as well as in the primary photochemical event. In the phototrophic purple bacteria, which produce either Bchl *a* or Bchl *b*, these pigments are constituents of both RC and LH complexes (Thornber *et al.*, 1978, for review).

Most of the phototrophic purple bacteria synthesise more than one class of LH complex. One of them, probably common to all species with Bchl *a*, exhibits an absorption peak at 870–890 nm. This complex contains 2 mol of Bchl *a* and 1–2 mol of carotenoid per mol of two different polypeptides (Thornber *et al.*, 1983). In spite of species-specific variations in the exact position of the near infrared absorption band, the complex is referred to as "B890 protein." The B890 protein appears to be associated with the RC in a fixed stoichiometry. A second class of LH complexes, named "B800–850 protein", consists of 3 mol of Bchl *a* and 1 mol of carotenoid per mol of a minimum unit of two to three different polypeptides. According to the categorisation proposed by Thornber *et al.* (1983) the class of B800–850 proteins comprises the following subtypes: type I exhibits an 850-nm absorption band which is about 1.5 times higher than the 800-nm band; type II exhibits an absorption band within the range

835–855 nm, which is of equal or lower intensity than the 800-nm band; type II occurs together with a third complex, the B800–820 protein. Complexes of the B800–850 protein class show properties typical of accessory pigment complexes, i.e., they vary in proportion to the constant composition RC-B890 protein unit. To be more specific, there is an inverse proportion of the ration of B800–850 complexes per RC-B890 complexes, on one hand, and, on the other, of light intensity as well as oxygen tension (Aagaard and Sistrom, 1972). This in turn leads to changes in the absorption spectra of cell material. As a consequence, to categorise phototrophic bacteria on the basis of their near infrared absorption spectra (i.e., the complement of Bchl complexes), spectra should be taken after growth of the organisms under different sets of growth conditions.

2. Determination of bacteriochlorophyll–protein complexes

Absorption spectra of whole cell suspensions or subcellular fractions are most frequently employed to evaluate the presence of different bulk Bchl complexes of the bacterial photosynthetic appartus. Routine spectroscopy at room temperature yields satisfactory results provided light scattering of the sample is reduced to a minimum. This is achieved by the following methods: (1) adding light-scattering material such as homogenised milk to the reference cuvette so that scattering of the sample (determined at a wavelength of no pigment absorbance) and the reference become adjusted to one another; (2) homogenising cells and measuring absorption spectra of homogenates after the sedimentation by centrifugation of the unbroken cells and large debris; (3) adjusting the density of the suspending (and reference) medium to the density of the cells in the presence 30% bovine serum albumin or 60% (w/w) sucrose (Fig. 8) (Schlegel and Pfennig, 1961; Pfennig, 1969).

While LH complexes containing Bchls c through e are easily recognised in absorption spectra on the basis of their distinctive long-wavelength absorption bands, the LH complexes containing Bchl a, with different overlapping absorption bands, require futher analyses. Crounse et al. (1963) developed the following corrections to quantify peak heights of LH-Bchl complexes in R. sphaeroides: all absorbancies are first corrected for light scattering which is assumed to increase linearly from 950 to 670 nm. This is the only correction made for the 800-nm band. Further correction of the other bands is as follows: A_{850} (corrected) $= A_{850} - A_{900}$; A_{875} (corrected) $= A_{875} - (A_{825} - A_{775})$. As the authors pointed out, this method is rather crude. So far, however, it is the best available. (Improvements of the method are in progress; Niederman, private communica-

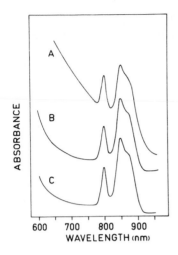

Fig. 8. Near infrared absorption spectra of whole cells of *Rhodopseudomonas sphaeroides* suspended in (A) culture medium, (B) 60% (w/w) sucrose, and (C) 30% bovine serum albumin.

tion.) From the corrected peak heights at 875 and 850 nm together with the absorption coefficients of $\varepsilon_{875} = 126 \pm 8 \text{ m}M^{-1} \text{ cm}^{-1}$ and $\varepsilon_{850} = 132 \pm 1 \text{ m}M^{-1} \text{ cm}^{-1}$, respectively, concentrations of the B890 (= B875) protein and the B800–850 protein can be calculated (Clayton and Clayton, 1981). In organisms such as *R. sphaeroides* which produce a constant composition RC-B890 protein unit the calculated concentration of B890 protein is considered representative of the concentration of the RC. Alternatively, RCs are determined after measurement of light-induced bleaching of the long-wavelength absorption band as detailed by Aagaard and Sistrom (1972).

IV. Taxonomic significance of bacteriochlorophylls

Prokaryotes of the order of the Rhodospirillales, generally referred to as the phototrophic bacteria, comprise four families: the Rhodospirillaceae, the Chromatiaceae, the Chlorobiaceae and the Chloroflexaceae. According to colours of species described originally the former two families are conveniently referred to as purple and the latter as green phototrophic bacteria. These terms are rendered legitimate by usage, although not only purple but also yellow and orange to brown, pinkish to red, and even green species belong to the families of purple bacteria, and brown species to the families of the green bacteria (Trüper and Pfennig, 1978). This overlap with respect to characteristics of visible colours is largely due to the presence of different carotenoids. Yet the formation of Bchl chromophores is applicable as a valuable parameter to distinguish phototrophic purple and green bacteria.

The majority of species of the Rhodospirillaceae and Chromatiaceae possess solely Bchl *a*. To date the presence of Bchl *b* has been reported in *R. viridis* and *R. sulfoviridis* of the Rhodospirillaceae and in *Ectothiorhodospira halochloris, E. abdelmalekii,* as well as in *Thiocapsa pfennigii* of the Chromatiaceae (Trüper and Pfennig, 1978; Imhoff and Trüper, 1981). Some further subdivison of the groups with Bchls *a* and *b* is possible after identification of the major esterifying alcohols. While most of the Bchl *a* from different species is esterified with phytol, *Rhodospirillum rubrum* esterifies all Bchl *a* with geranylgeraniol and only Bph *a* of RCs with phytol (Künzler and Pfenning, 1973; Gloe and Pfennig, 1974; Walter *et al.,* 1979). As mentioned above (Section II, B) Bchls *b* from *R. viridis* and *E. halochloris* differ in that the pigment of the former is esterified with phytol and that of the latter with Δ-2,10-phytadienol (Steiner *et al.,* 1981).

The occurrence of Bchl *a* in different functional complexes provides a further means of subdividing purple bacterial. *R. rubrum* exhibits only the constant composition RC-B890 protein unit with an LH absorption maximum at 880 nm and a minor RC absorption band at 800 nm (Fig. 9). Very recently a new isolate, *Rhodopseudomonas marina,* was described which shows *in vivo* absorption properties similar to *R. rubrum* (Imhoff, 1983). Otherwise this spectral type has been identified only with mutants. All other species of the purple bacteria which synthesise Bchl *a* are characterised by the presence of B800–850 proteins in addition to the RC-B890 protein unit. Most of these species show type I B800–850 proteins and, consequently, *in vivo* absorption spectra with characteristic bands at about 800, 850, and 800 nm. The latter is usually masked by the 850-nm band which predominates after growth under low and moderate light intensities (Fig. 9). Type II B800–850 proteins accompanied by B800–820 proteins were described in *Chromatium vinosum* and *R. acidophila* strain 7050 (Thornber *et al.,* 1983). Such Bchl *a* complexes reveal their presence by an additional absorption band at 820–830 nm after growth under low and moderate light intensities (Fig. 9). The above categorisation of purple bacteria employs known Bchl *a* complexes; however, it is likely that the system must be expanded as soon as other members of the group have been analysed. *Rhodospirillum tenue* is one example that apparently does not fit into the above classification (Fig. 9).

The members of the purple bacteria that synthesise Bchl *b* show an *in vivo* infrared absorption maximum at about 1020 nm with a minor band at about 830 nm (Fig. 10). *Ectothiorhodospira halochloris* is distinguished from those forms by additional absorption bands at about 800 nm (Imhoff and Trüper, 1977).

In Chlorobiaceae and Chloroflexaceae Bchls *c*, *d*, and *e* constitute the bulk pigments which are always accompanied by small but significant

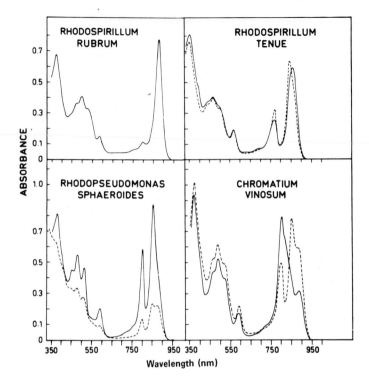

Fig. 9. Absorption spectra of membrane suspensions from representative species of the purple phototrophic bacteria producing bacteriochlorophyll *a*. Spectra taken after growth under low (——) and high (------) light intensities.

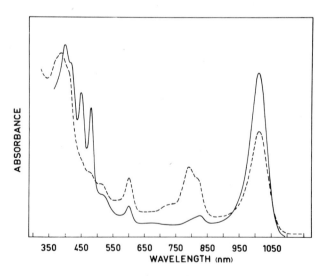

Fig. 10. Absorption spectra of cell suspensions of *Rhodopseudomonas viridis* (——) and *Ectothiorhodospira halochloris* (------).

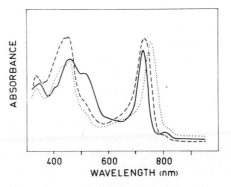

Fig. 11. Absorption spectra of cells of *Chlorobium limicola* forma *thiosulfatophilum* (Bchl *c*) (·······), *C. vibrioforme* (Bchl *d*) (------), and *C. phaeovibrioides* (Bchl *e*) (——). From Gloe *et al.* (1975).

amounts of Bchl a_p (Olson, 1980). All green-coloured species so far studied produce either Bchl *c* or *d*, while brown species produce Bchl *e* (Trüper and Pfennig, 1978). Bchls of representative species are compiled in Tables VI–XI.

As mentioned already, Bchls from Chlorobiaceae comprise a mixture of up to eight different homologues with farnesol as major esterifying alcohol. In *C. aurantiacus,* however, Bchl *c* is present as a single compound substituted with ethyl and methy groups at R^3 and R^4, respectively (see

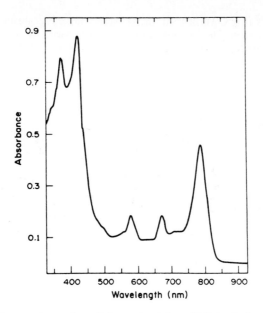

Fig. 12. Absorption spectrum of a cell-free extract from *Heliobacterium chlorum* (kindly provided by Dr. R. C. Fuller).

Table I). The major esterifying alcohol is stearylalcohol. Minor amounts of phytol and geranylgeraniol, but never farnesol, were detected (Risch *et al.*, 1979).

In situ absorption bands of Bchls *c*, *d*, and *e* cover characteristic ranges of wavelengths in the far red (Fig. 11). These are 743–760 nm for Bchl *c*, 725–745 nm for Bchl *d*, and 715–725 nm for Bchl *e*. Again, Bchl c_s of *C. aurantiacus* is an exception as it fits the range characteristic of Bchl *d* (Gloe *et al.*, 1975; Gloe and Risch, 1978).

Finally the *in situ* properties of Bchls of two more recently isolated genera should be noted. These are *Erythrobacter* (Shiba and Simidu, 1982) and *Heliobacterium* (Gest and Favinger, 1983). Species of *Erythrobacter* perform a chemo-organotrophic type of metabolism under aerobic conditions. The organisms produce Bchl a_p only in the presence of oxygen. *In situ* near infrared absorption bands are localised at 800 and 870 nm. The phototrophic *Heliobacterium chlorum*, on the other hand, contains Bchl g_{Gg} which *in situ* shows a predominant long-wavelength absorption band at 788 nm (Fig. 12).

Acknowledgments

The author thanks Dr. N. Pfennig and Dr. H. Scheer for many helpful suggestions and unpublished information. Thanks are also due to Frau M. Grest for her ever-willing help with typing the manuscript.

References

Aagaard, J., and Sistrom, W. R. (1972). *Photochem. Photobiol.* **15**, 209–225.
Arnheim, K., and Oelze, J. (1983). *Arch Microbiol.* **135**, 299–304.
Brockmann, H., Jr., and Kleber, J. (1970). *Tetrahedron Lett.* **25**, 2195–2198.
Brockmann, H., Jr., and Knobloch, G. (1972). *Arch. Mikrobiol.* **85**, 123–126.
Brockmann, H., Jr., and Lipinski, A. (1983). *Arch. Microbiol.* **136**, 17–19.
Clayton, R. K. (1966). *Photochem. Photobiol.* **5**, 669–677.
Clayton, R. K., and Clayton, B. J. (1978). *Biochim. Biophys. Acta* **501**, 478–487.
Clayton, R. K., and Clayton, B. J. (1981). *Proc. Natl. Acad. Sci. U.S.A.* **78**, 5583–5587.
Cohen-Bazire, G., and Sistrom, W. R. (1966). *In* "The Chlorophylls" (L. P. Vernon and G. R. Seely, eds.), pp. 313–341. Academic Press, New York.
Cogdell, R. J. (1983). *Annu. Rev. Plant Physiol.* **34**, 21–45.
Crounse, J., Sistrom, W. R., and Nemser, S. (1963). *Photochem. Photobiol.* **2**, 361–375.
Davis, M. S., Forman, A., Hanson, L. K., Thornber, J. P., and Fajer, J. (1979). *J. Phys. Chem.* **83**, 3325.

Evans, M. C. W. (1983). *In* "The Phototrophic Bacteria" (J. G. Ormerod, ed.), pp. 61–75. Blackwell, Oxford.

Evans, M. C. W., and Heathcote, P. (1983). *In* "The Phototrophic Bacteria" (J. G. Ormerod, ed.), pp. 35–60. Blackwell, Oxford.

Feick, R. G., Fitzpatrick, M., and Fuller, R. C. (1982). *J. Bacteriol.* **150,** 905–915.

Gest, H., and Favinger, J. L. (1983). *Arch. Microbiol.* **136,** 11–16.

Gloe, A. (1977). Doctoral Thesis, Göttingen.

Gloe, A., and Pfennig, N. (1974). *Arch. Microbiol.* **96,** 93–101.

Gloe, A., and Risch, N. (1978). *Arch. Microbiol.* **118,** 153–156.

Gloe, A., Pfennig, N., Brockmann, H., Jr., and Trowitsch, W. (1975). *Arch. Microbiol.* **102,** 103–109.

Holt, A. S., Purdie, J. W., and Wasley, J. W. F. (1966). *Can. J. Chem.* **44,** 88–93.

Imhoff, J. F. (1983). *Syst. Appl. Microbiol.* **185,** 512–521.

Imhoff, J. F., and Trüper, H. G. (1977). *Arch. Microbiol.* **144,** 115–121.

Imhoff, J. F., and Trüper, H. G. (1981). *Zentralbl. Bakteriol. Parasitenkd. Infektionkr. Hyg. Abt. 1: Orig.* C **2,** 228–234.

Jensen, A., Aasmundrud, O., and Eimhjellen, K. E. (1964). *Biochim. Biophys. Acta* **88,** 466–479.

Kim, W. S. (1966). *Biochim. Biophys. Acta* **112,** 392–402.

Künzler, A., and Pfennig, N. (1973). *Arch. Mikrobiol.* **91,** 83–86.

Mittenzwei, H. (1942). Hoppe–Seyler's *Z. Physiol. Chem.* **275,** 93–121.

Oelze, J. (1981). Subcell. Biochem., 8, 1–73.

Oelze, J. (1983). *In* "The Phototrophic Bacteria" (J. G. Ormerod, ed.), pp. 8–34. Blackwell, Oxford.

Olson, J. M. (1980). *Biochim. Biophys. Acta* **594,** 33–51.

Olson, J. M., and Romano, C. A. (1962). *Biochim. Biophys. Acta* **59,** 728–730.

Pfennig, N. (1969). *J. Bacteriol.* **99,** 597–602.

Pierson, B. K., and Castenholz, R. W. (1974). *Arch. Microbiol.* **100,** 283–305.

Purdie, J. W., and Holt, A. S. (1965). *Can. J. Chem.* **43,** 3347–3353.

Risch, N., Brockmann, H., Jr., and Gloe, A. (1979). *Liebigs Ann. Chem.* 408–418.

Sauer, K., Lindsay-Smith, J. R., and Schultz, A. J. (1966). *J. Am. Chem. Soc.* **88,** 2681–2688.

Scheer, H., Svec, W. A., Cope, B. T., Studier, M. H., Scott, R. G., and Katz, J. J. (1974). *J. Am. Chem. Soc.* **96,** 3714–3716.

Schlegel, H. G., and Pfennig, N. (1961). *Arch. Microbiol.* **38,** 1–39.

Shiba, T., and Simidu, U. (1982). *Int. J. Syst. Bacteriol.* **32,** 211–217.

Sistrom, W. R. (1966). *Photochem. Photobiol.* **5,** 843–856.

Smith, J. H. C., and Benitez, A. (1955). *In* "Modern Methods of Plant Analysis" (K. Paech and M. V. Tracy, eds.), Vol. 4, pp. 142–196. Springer-Verlag, Berlin and New York.

Smith, K. M., Goff, D. A., Fajer, J., and Barkigia, K. M. (1983). *J. Am. Chem. Soc.* **105,** 1674–1676.

Stanier, R. Y., and Smith, J. H. C. (1960). *Biochim. Biophys. Acta* **41,** 478–484.

Steiner, R. (1980). Thesis, München.

Steiner, R., Schäfer, W., Blos, I., Wieschoff, H., and Scheer, H. (1981). *Z. Naturforsch.* **36c,** 417–420.

Strain, H. H., and Svec, W. A. (1966). *In* "The Chlorophylls" (L. P. Vernon and G. R. Seely, eds.), pp. 21–66. Academic Press, New York.

Thornber, J. P., Trosper, T. L., and Strouse, C. E. (1978). *In* "The Photosynthetic Bacteria" (R. K. Clayton and W. R. Sistrom, eds.), pp. 133–160. Plenum, New York.

Thornber, J. P., Cogdell, R. K., Seftor, R. E. B., Pierson, B. K., and Tobin, E. M. (1983).
    Proceedings of the 6th Internatl. Congress on Photosynthesis, Brussels.
Trüper, H. G., and Pfennig, N. (1978). *In* "The Photosynthetic Bacteria" (R. K. Clayton
    and W. R. Sistrom, eds.), pp. 19–27. Plenum, New York.
van der Rest, M., and Gingras, G. (1974). *J. Biol. Chem.* **249,** 6446–6453.
Walter, E., Schreiber, J., Zass, E., and Eschenmoser, A. (1979). *Helv. Chim. Acta* **62,**
    899–920.
Weigl, J. W. (1953). *J. Am. Chem. Soc.* **75,** 999–1000.

10

The Analysis of Cytochromes

C. W. JONES

Department of Biochemistry, University of Leicester, Leicester, England

R. K. POOLE

Department of Microbiology, Queen Elizabeth College, University of London, London, England

I. Historical aspects

The cytochromes were discovered almost 100 years ago in mammalian tissues by an English general practitioner, Charles MacMunn (MacMunn, 1886). Unfortunately, this work was subjected to unwarranted criticism and was largely forgotten. The cytochromes were subsequently rediscovered almost 40 years later by Keilin (Keilin, 1925), who convincingly demonstrated their presence and heterogeneous nature in various animal, yeast, and bacterial cells. Interestingly, cytochromes were shown to be present in whole cells of obligately aerobic and facultatively anaerobic bacteria such as *Bacillus subtilis* and *Proteus vulgaris,* respectively, but not in cells of obligately anaerobic, fermentative bacteria such as *Clostrid-*

METHODS IN MICROBIOLOGY
VOLUME 18

ium sporogenes and *Clostridium butyricum* (Keilin, 1927), thus indicating a role in aerobic metabolism. This work was followed rapidly by extensive spectroscopic investigations of a wide range of bacterial species (see, for example, Yaoi and Tamiya, 1928; Warburg *et al.*, 1933; Yamagutchi, 1934); within a few years of their rediscovery, the cytochrome spectra of over 90 species had been reported.

The observation that bacterial cytochromes, like those in the cells of higher organisms, were able to undergo reversible spectral changes concomitant with oxidation and reduction (Keilin, 1927, 1929) pointed strongly to a role in aerobic respiration. This view was initially opposed by Warburg and his colleagues, but these objections were later withdrawn following detailed spectroscopic analyses of whole cells of the obligate aerobes *Acetobacter* and *Azotobacter*. Indeed, by measuring the relative ability of different wavelengths of light to relieve the inhibition of respiration by carbon monoxide, the resultant photochemical action spectra confirmed that certain autoxidisable cytochromes, now known as cytochrome oxidases, reacted directly with molecular oxygen (Warburg and Negelein, 1925; Kubowitz and Haas, 1932). It was later shown that cytochromes also function during anaerobic respiration (i.e., to terminal electron acceptors other than oxygen) and photosynthetic electron transfer.

This early work was carried out using two types of spectroscope, the microspectroscope ocular and the Hartridge reversion spectroscope, both of which were attached to a microscope in place of the eyepiece (Keilin, 1925; Hartridge, 1912). The sensitivity of these methods was later shown to be considerably enhanced at low temperatures, such as those of liquid air (Keilin and Hartree, 1949, 1950) or liquid nitrogen, which caused the intensification and sharpening of the absorption bands. These methods were later superceded by methods involving the use of high-resolution manual and recording spectrophotometers with split-beam facilities that allowed the accurate measurement of various types of difference spectra (e.g., reduced *minus* oxidised or reduced *plus* CO *minus* reduced) and, more recently, by extremely sophisticated dual-wavelength instruments (see, for example, Chance *et al.*, 1953; Chance, 1957; also Section III).

This chapter will concentrate on describing the analysis and functions of cytochromes during bacterial respiration.

II. Structural considerations and nomenclature

Cytochromes are specialised haemoproteins which consist of a haem prosthetic group attached to a protein (see Lemberg and Barrett, 1973). Haem is composed of four pyrrole rings joined by —CH= bridges to form

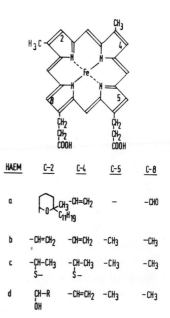

HAEM	C-2	C-4	C-5	C-8
a	$-O-\overset{-CH=CH_2}{\underset{C_{11}H_{19}}{CH_3}}$		–	–CHO
b	–CH=CH$_2$	–CH=CH$_2$	–CH$_3$	–CH$_3$
c	$-\underset{S-}{CH}-CH_3$	$-\underset{S-}{CH}-CH_3$	–CH$_3$	–CH$_3$
d	$\underset{OH}{CH}-R$	–CH=CH$_2$	–CH$_3$	–CH$_3$

Fig. 1. The haem structures of the cytochromes. Note that haem d is a dihydroporphyrin in which the 8,9 double bond is saturated.

a complex planar molecule, porphyrin, at the centre of which is a single iron atom (Fig. 1). The latter carries one electron and oscillates between the reduced (Fe^{2+}) and oxidised (Fe^{3+}) states. It is usually, but not always, in the low-spin configuration and thus forms an octahedral coordination complex which has the capacity to bind six ligands; the four equatorial coordination places are occupied by the nitrogens of the pyrrole rings, whereas the fifth and sixth coordination places are often filled with nitrogen and/or sulphur atoms from the side chains of histidyl or methionyl residues in the protein. In the cytochrome oxidases, which catalyse the terminal transfer of electrons to molecular oxygen, one of the axial coordination positions is occupied by water or oxygen. It can also be filled by stronger field ligands such as carbon monoxide or nitric oxide, which are isosteres of oxygen and hence readily inhibit cytochrome oxidase activity by competing with oxygen for the ferrous form. The C-2, C-4, C-5, and C-8 positions around the periphery of the haem molecule are available for substitution, and it is these substituents that are primarily responsible for the different absorbance properties and standard oxidation-reduction potentials (E'_θ or E_m) of the various cytochromes.

Four classes of cytochromes are generally recognised:

1. Cytochromes a. These include cytochrome oxidases aa_3 and a_1 in which the haem prosthetic group (haem a) contains a formyl side chain.

2. Cytochromes *b*. These include various non-autoxidisable cytochromes *b*, cytochrome oxidase *o*, and cytochromes P-450 and P-460 in which the haem prosthetic group (haem *b* or protohaem) is the same as that present in haemoglobin and myoglobin.

3. Cytochromes *c*. These include cytochromes *c*, c_2, c_3, c_4, c', c_{550}, c_{551}, and c_{555} in which the haem prosthetic group (haem *c* or mesohaem) is additionally covalently attached to the protein via thioether bridges between the C-2 and C-4 positions on the haem and the side chains of cysteinyl residues in the protein. The *c*-type cytochromes can be divided into three subgroups: Type I, which has a single haem attached to the protein close to the N-terminal end, the sixth coordination position being filled by methionyl-S, e.g., cytochromes *c*, c_2, c_{551}, c_{555}, and c_4; Type II, which has a single haem attached to the protein close to its C terminus and are high-spin compounds, e.g., cytochrome c'; and Type III, which has multiple haems; e.g., cytochrome c_3. Several flavocytochromes *c* are also known, particularly in photosynthetic bacteria and in pseudomonads capable of *p*-cresol oxidation; the cytochrome *c* forms an oligomeric complex with a flavoprotein.

4. Cytochromes *d*. These include cytochrome oxidase *d* (originally called cytochrome a_2) and a form of nitrite reductase, cytochrome cd_1, in which the haem prosthetic group (haem *d* or d_1) is a dihydroporphyrin or chlorin.

In general, the reduced minus oxidised difference spectrum (see Section III,B,1) of a cytochrome exhibits three major absorption bands in the visible region of the electromagnetic spectrum, viz., the α band (in the approximate range 545–650 nm), the β band (approximately 520–530 nm), and the γ or Soret band (approximately 410–450 nm). The major exceptions to this are the *a*-type cytochromes, which do not exhibit significant β bands, and the *d*-type cytochromes, which show only very weak Soret bands. The E_m (pH 7.0) values of the cytochromes lie mostly within the range −100 to +500 mV but a few specialised *c*-type cytochromes, like cytochrome c_3 of *Desulphovibrio* (Le Gall *et al.*, 1979) and cytochrome c_{550} of *Escherichia coli* and other enterobacteria (Fujita, 1966), exhibit more negative values, and the E_m (pH 3.2) of cytochrome oxidase a_1 from the acidophilic chemolithotroph *Thiobacillus ferrooxydans* is as high as 725 mV. The E_m (pH 7.0) values of the various cytochrome types are generally in the order

$$\text{low } E_m c < b < d \qquad \text{or} \qquad \text{high } E_m c < aa_3$$

Cytochrome oxidases *o* and a_1 exhibit rather variable E_m values towards the upper end of this range (see Jurtshuk *et al.*, 1975; Poole, 1983).

III. Analytical methods

Spectrophotometry, the measurement of absorption of light in the visible (and possibly near-infrared and near-ultraviolet) regions of the spectrum of electromagnetic radiation, has been the single most useful technique for the analysis of cytochromes in microbes and all cell types, since it can provide information on the identity, quantification, and functional properties of the cytochromes (see Sections III,A,B and IV). Spectrophotometry can also reveal some aspects of haemoprotein structure, although an arsenal of advanced biophysical techniques is playing a role of ever-increasing importance in this area. These techniques, notably magnetic resonance (EPR and NMR), vibrational and Mossbauer spectroscopy, circular dichroism, and magnetic susceptometry, are reviewed with special reference to haemoproteins in Fleischer and Packer (1978). A number of non-spectrophotometric techniques have also proved useful (see Sections III,C).

A. Spectrophotometer design considerations

With the exception of some c-type cytochromes (including cytochrome cd_1), bacterial cytochromes are tightly membrane-bound and their study necessitates, or is greatly facilitated by, analysis *in situ,* i.e., in membrane preparations or intact cells. Such samples are optically unfavourable, in the sense that measurement of cytochrome absorption is frustrated by the high turbidity of the cuvette contents. It is the requirement for making spectroscopic recordings of relatively small absorbance changes in highly (non-Rayleigh) light-scattering samples that necessitates the use of spectrophotometers other than the simplest single-beam instruments.

Figure 2 illustrates an attempt to record the absorbance spectrum of microbial cells in a suspension of modest turbidity. The apparent absorbance of the sample is strongly dependent on the position of the cuvette relative to the detector, and the optimal optical configurations for recording cytochrome absorption and cell turbidity are mutually exclusive. Thus, an ideal turbidimeter would receive none of the scattered light, most of which is scattered only a few degrees from the primary beam (Koch, 1970), and would allow detection of the presence of only a few scattering particles. To this end, it would be constructed with the illuminated sample far removed from the detector. In contrast, cytochrome analysis of turbid samples requires that as much of the scattered light as possible be received by the detector which should be immediately adjacent to the sample (Chance, 1957). This feature is common to most spectrophotometers suitable for cytochrome analysis, but the optical and me-

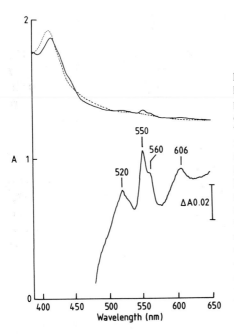

Fig. 2. Attempted "single-beam" or absolute analysis of cytochromes in turbid suspension. A suspension of *Saccharomyces cerevisiae* in 50 mM phosphate buffer (pH 7.0) was adjusted to give an A_{390} of about 1.8. The dashed line is the spectrum of the native suspension recorded with an SP1700 spectrophotometer and using buffer in the reference cuvette. The upper, solid line is the spectrum of a similar yeast sample reduced with $Na_2S_2O_4$. Note that the absorbance due to cytochrome $c(+c_1)$ at 550 nm is only about 2% of the "apparent absorbance" (largely light scattering) of the turbid suspension. Below is shown, at higher gain, the reduced *minus* oxidised difference spectrum of the same samples. The scan rate was 4 nm s^{-1} and the spectral bandwidth 2 nm.

chanical configurations are as various as the types. of spectroscopic information sought.

1. Split-beam spectrophotometry

The terminology in common usage for describing the opticomechanical configurations of spectrophotometers is confused. In this chapter, "split beam" is taken to describe the time sharing of a monochromatic beam of light (but which may be varied with respect to time) between two samples. This is perhaps the most common and accessible principle of operation and is widely used in wavelength-scanning instruments for recording the difference spectrum between two chemically differentiated states of a sample. Thus in a "reduced *minus* oxidised difference spectrum" (Fig. 2), the scanning beam is time shared (for example, using an oscillating mirror) between two samples, one appropriately reduced (traditionally in the "front" or "sample" position), whilst the oxidised sample is in the reference position. Correction of the measurements for turbidity and light scattering ("bucking out"; Chance, 1957) is assured by the comparison of two samples which are identical in all respects except for the (small) absorbance changes under study. Elaborations of this basic method include maintaining the samples at low temperature, mathematical analysis of the resultant difference spectrum, and provision for irradiating photosensitive samples during the course of the experiment; these will be described in subsequent sections.

2. Dual-wavelength (double-beam) spectrophotometry

In principle, kinetic data could be achieved by repetitive wavelength scanning in a split-beam apparatus, followed by replotting, as a function of time, the absorbance changes at the desired wavelength(s). The major problems encountered would be (1) uncertainty as to whether the observed changes were occurring in reference or sample cuvettes, (2) tedium and slowness, and (3) the possibility that light-scattering changes during the course of the reaction under study would shift successive spectra; i.e., true isosbestic points for all scans would be unlikely.

In practice, therefore, kinetic measurements are generally performed with a dual-wavelength apparatus in which a single sample is alternately illuminated with two wavelengths (λ_1, λ_2) and the difference in absorbance recorded ($\Delta A \ \lambda_1 - \lambda_2$). The criteria for the choice of the wavelength pairs are (1) sensitivity, i.e., high signal-to-noise ratio, achieved using an absorption peak of high extinction coefficient, (2) specificity, i.e., freedom from interference by other components, and (3) elimination of light-scattering changes, which are largely wavelength independent, achieved by selecting a pair of wavelengths as close as possible, e.g., 10–40 nm apart. In most instances, the α absorption bands of the cytochromes are the most suitable, being in general well-separated from the α bands of other cytochrome types and having useful extinction coefficients. The γ (Soret) bands offer greater sensitivity, but are less well separated and confused by the absorbance of flavins and the intense near-ultraviolet bands of nicotinamide nucleotides. The β bands are normally compressed into a narrow wavelength region, are generally complex due to band splitting, and are of low magnitude. For further discussion of these problems, see White and Sinclair (1970).

Two basic methods are used for delivering light of two wavelengths to the sample. In the first, and almost universally used method (time sharing), two monochromatic beams of light alternately illuminate the sample with a ''chopping'' frequency of 50–400, or even 1000, Hz. The detector response is ''gated'' to the chopping to allow temporal resolution of the absorbance at the two wavelengths. Time sharing is frequently achieved by a vibrating mirror which alternately (1) reflects light from one monochromator to the sample or (2) allows uninterrupted transmission to the sample of light from the second monochromator. An alternative method, space sharing, uses a beam of light containing both wavelengths to illuminate the sample continuously and a beam splitter behind the sample to deliver light simultaneously to a monochromator and detector for each chosen wavelength.

The dual-wavelength apparatus of classical design is well suited to monitoring the absorption changes, with time, of any component isolated by the choice of wavelength pairs. Its use can be extended readily from studies of relatively ''slow'' cytochrome reactions (limited by the speed of uniting reactants; Wilson, 1978) at room temperature by using stopped-flow or other rapid-mixing devices or by monitoring reactions at low temperatures (see later). Description of the first class is outside the scope of this chapter; for an example and references see Garland et al. (1976). An illustration of the use of this technique in measuring halftimes of cytochrome reactions in bacteria is given by Haddock et al. (1976).

An increasing number of commercially available and custom-built instruments use the dual-wavelength principle with the facility for wavelength scanning of the ''measuring'' beam (e.g., Chance and Graham, 1971; Bashford et al., 1982). Thus only one sample cuvette is used, and difference spectra can be obtained only by recording successively spectra of two samples and plotting the difference. This is currently achieved by ''memorising'' the first spectrum and calculation of differences using a microprocessor. A somewhat similar principle has been used by Wikstrom (1971) allowing manual, point-by-point plotting of a difference spectrum using a conventional dual-wavelength spectrophotometer.

3. Multiwavelength spectrophotometry

Using the time-sharing principle, multichannel instruments have been constructed that allow four or more wavelengths to illuminate the sample successively. In the most recent Johnson Foundation version (Chance et al., 1975a), high-speed time sharing (up to 500 Hz) is achieved by rotating an air-driven turbine disc, containing four appropriate filters, in front of the sample. Collection of the coded signals is effected by a single detector as close to the sample as possible.

4. Single-beam spectrophotometry

With the decline in the cost of microprocessors and microcomputers, there has been a resurgence of interest in single-beam instruments (i.e., using one light beam and one cuvette) for wavelength scanning, the spectrum of the reference sample being stored in instrument memory and subtracted from subsequent scans. Many recently introduced commercial instruments incorporate this principle, whilst a custom-built device and its application to *E. coli* cytochromes are described by Reid and Ingledew (1979).

5. Other techniques

Other principles and configurations have been developed for special applications. These include rapid wavelength scanning (e.g., 200 nm ms^{-1}; Holloway and White, 1975), microspectrophotometry of respiratory pigments in single cells (Chance, 1966), and surface reflectance spectrophotometry (Bashford *et al.*, 1982).

B. Spectrophotometric techniques and their applications

1. Difference spectra of redox states

By far the most common method of cytochrome analysis is the recording, with a split-beam spectrophotometer, of a spectrum that represents the difference between an oxidised and a reduced sample. Such spectra may provide data on the identity of the cytochrome types present and the amount of each type present, and may be obtained with suspensions of intact cells, subcellular fractions derived therefrom, or purified, or partially purified, preparations.

(*a*) *Preparation of samples.* Full reduction of a sample can often be achieved by adding a few grains of $Na_2S_2O_4$ to a cell or membrane suspension. Reduction is assumed to result from the anoxia that ensues (Jurtshuk *et al.*, 1975) and may take several minutes for completion. The dithionite anion does not cross certain biological membranes, including the cytoplasmic membrane of *Escherichia coli* (Jones and Garland, 1977). Other precautions in the use of dithionite are given by Lemberg *et al.* (1966). Alternatively, the preparation can be allowed to become anoxic by respiration of a suitable oxidisable substrate. Identification of those components reducible by various added substrates can be a valuable way of characterising and distinguishing between the cytochromes present. Sodium borohydride is also a useful reductant.

Oxidation of the reference sample may be achieved by the use of an exogenous oxidant or by vigorous aeration of the sample just prior to scanning the spectrum. Difficulty can be experienced if significant levels of endogenous reductants are present or if a respiratory chain is terminated by an oxidant with a particularly high O_2 affinity or is capable of particularly rapid rates of electron flux. Problems with cytochrome d-containing preparations of $E.$ $coli$ have been experienced in this respect (Poole and Chance, 1981). For such reasons, O_2 may be generated in the sample by adding H_2O_2 and catalase, although microbial preparations, especially of intact cells, often exhibit appreciable endogenous catalase activity. Alternatively, potassium ferricyanide (ferri/ferrocyanide, +0.43 V), hexachloroiridate ($IrCl_6^{2-}/IrCl_6^{3-}$, +1.09 V; Ingledew and Cobley, 1980), or ammonium persulphate ($S_2O_8^{2-}/2SO_4^{2-}$, +2.0 V; e.g., Liu and Webster, 1974) may be added as oxidants (E_θ' values in parentheses). The first two are highly coloured, however. Aqueous solutions of ferricyanide absorb strongly between 370 and 460 nm (λ_{max} = 420 nm), whilst hexachloroiridate solutions have complex spectra in this region with λ_{max} at 488 nm. Ammonium persulphate is colourless. Addition of these oxidants to final concentrations of about 2 mM is usually satisfactory, although it is common practice to add "a few grains" to 1- to 4-ml samples.

One should be wary of the problems of incomplete cytochrome reduction or oxidation by these reagents in quantitative work and of the danger of inadequate buffering against the pH changes that they may cause. The latter problem is especially acute in the case of dithionite, whose oxidation products are highly acidic. In addition, certain oxidases form spectrally distinct complexes with oxygen and peroxides, so that addition of these reagents does not yield the fully oxidised species. For example, cytochrome oxidase d in $E.$ $coli$ (and other bacteria) readily reacts with O_2 to give a stable, oxygenated form with a distinctive absorption band at about 650 nm (for a review, see Poole, 1983). Thus, reduced $minus$ oxidised (by aeration) difference spectra show a trough at this wavelength, whereas ferricyanide-oxidised samples are almost featureless in this region (Poole et $al.,$ 1983a). The formation of such a complex may underlie the curious (Shipp et $al.,$ 1972), apparent dependence of cytochrome d concentration upon the amount of membrane used for spectroscopy; for a discussion, see Ingledew and Poole (1984).

A baseline should be recorded prior to scanning the difference spectrum. In the usual case of a split-beam spectrophotometer, this involves scanning test and reference cuvettes when their contents have been brought to identical states. When such states have been reached (and this can be no mean task) the baseline will indicate differences of instrumental origin which will be included in subsequent difference spectra and which must be allowed for. This may be done manually (and tediously) by subtracting the baseline from the difference spectrum at, say, 2-nm intervals and replotting. In the case of a dual-wavelength scanning or single-beam apparatus, a baseline is generally obtained by computing the difference between one scan of the sample and a subsequent scan.

(b) *Instrument settings.* Except in the case of instruments specially constructed for rapid scanning, rates of wavelength scanning should be modest so as to be commensurate with the response-time capabilities of the instrument, thus avoiding spectral distortion. For many instruments $1-4$ nm s^{-1} is suitable; much slower rates over a long wavelength range can allow settling of the cells or particles from suspension, although this can be diminished by including glycerol in the samples.

Slit width (the physical width of the monochromator exit or entrance slit) in conjunction with the dispersion of the monochromator will affect both spectral resolution and the accurate measurement of peak heights. Typical reciprocal dispersion of a monochromator is 4 nm mm^{-1}; i.e., an exit slit 1-mm wide would pass light of 4-nm spectral bandwidth (that is the band of wavelengths containing the central half of the entire band of wavelength passed by the exit slit). The natural bandwidth is the width in nanometres at half the height of the sample absorptive peak; it is the ratio of spectral to natural bandwidths that dictates the precision with which peak heights can be determined. Fortunately, the natural bandwidths of most cytochromes at room temperature are about 10 nm and so, for example, spectral bandwidths of 1, 3, and 5 nm will result in underestimates of peak height of about 1, 4, and 10%, respectively. However, at low temperatures (see below), natural bandwidths are significantly decreased and narrower slits must be chosen, but with the likely consequence of increased noise.

(c) *Spectrophotometry at 77 K.* Following the pioneering work on low temperature spectrophotometry by Keilin and Hartree, this approach has since been widely used to detect differences in the absorption spectra of closely related haemoproteins, to trap unstable intermediates and steady states of oxidation and reduction, to slow down rapid rates of reaction, and to detect and measure very low concentrations of haemoproteins (for references, see Wilson, 1967). The absorbance changes are due to (1) a true temperature dependence of the absorbance of the cytochrome, resulting in narrowing, sharpening, and shifting to shorter wavelengths of the bands (Estabrook, 1956), and (2) light-scattering changes in the medium, which result in an effective increase in path length by multiple-interval reflections from ice crystals. It is worth noting that enhancement effects of related origin may be observed in highly scattering suspensions (e.g., of intact cells) at room temperature (Vincent *et al.,* 1982). The enhancement effects are strongly dependent on the suspending medium and the method used to freeze it. For example, a devitrified (i.e., poly-crystalline) 1.4 M sucrose solution can give a 25-fold intensification at $-190°C$, compared to 20°C. The presence of an organic solvent also makes the enhancement factors more reproducible than in dilute buffers (Wilson, 1967) but may inhibit electron transfer. Glycerol, however, has the added advantage of suppressing pH changes (Orii and Morita, 1977) that occur on freezing, but which can be partly circumvented by the prudent choice of buffer (Douzou, 1977). For further details of these effects and the precautions that must be taken in their use, including for example the differential enhancement of α and Soret bands, the reader is referred to the excellent accounts by Chance (1957), Estabrook (1961), Wilson (1967), and Vincent et al. (1982). A procedure for the quantifica-tion of cytochrome amounts in complex 77 K spectra is given by von Jagow et al. (1973).

Relatively few spectrophotometers have available, as standard options, devices for recording spectra at liquid nitrogen temperatures. As a result of this, and of the high costs of such attachments available commercially, many workers have designed and constructed simple, effective devices for use in a variety of instruments. Examples are given by Claisse et al. (1970) and Nuner and Payne (1973) for Cary spectrophotometers and by Edwards and Lloyd (1973) and Salmon and Poole (1980) for Pye–Unicam instruments. A description of the last device is included here. It has not been published previously and illustrates many of the features common to such accessories; the design owes much to an earlier version constructed by D. Lloyd (see above).

For Pye–Unicam spectrophotometers SP800, SP1700, and SP1800, the cuvette holder for turbid samples is removed, together with the light metal shield designed to protect the adjacent mirrors. The Dewar support (Fig. 3) is inserted into the vertical grooves vacated by the cuvette holder. It is constructed in brass, has a square aperture for light transmission, and bears near the top a large clip (not shown in Fig. 3) which is used to hold the Dewar flask. The part-silvered Dewar vessel was manufactured to order by Day–Impex Ltd., Station Works, Station Road, Earls Colne, Essex. The bottom 4 cm are silvered, such that, when in position in the spectrophotometer, the silvering stops at approximately the level of the openings in the standard cuvette holder. The brass cuvette holder (Fig. 3C) differs from some earlier designs in having facilities for correct loca-tion of the sometimes irregular Dewar stop (using the four top screws), vertical and rotational adjustment of the cuvettes with respect to the Dewar (using the vertical threaded post and lock-nuts), and the opportu-nity for removing the two cuvettes individually, thus allowing the record-

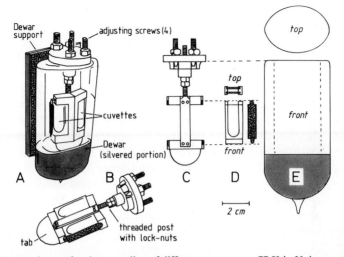

Fig. 3. An attachment for the recording of difference spectra at 77 K in Unicam split-beam spectrophotometers. A and B are sketches of the complete apparatus, and cuvette holder with cuvettes, respectively, and are not to scale. C is a scale diagram of the cuvette holder viewed from the back (i.e., the face adjacent *in situ* to the photomultiplier). D shows one of the two removable cuvettes from three faces. The cross-hatched structures in the top view are the Perspex windows. E is the Dewar vessel in top and front views, and shows the extent of silvering. The scale applies to diagrams C, D, and E only.

ing of differences between various pairs of samples. The angle between the two cuvettes equals that between the positions for 1 × 1-cm cuvettes in the standard cuvette holder. Two kinds of cuvettes have been used. The first are constructed entirely from Perspex and are cemented together to provide liquid-tight cuvettes. The path length is only 2 mm. The small light path hastens freezing, facilitates examination of the very opaque and light-scattering samples, and is generally more than compensated for by the enhancement of peak heights at 77 K. In the second design (Fig. 3D), a brass framework supports two Perspex windows that fit tightly into grooves in the brass frame, spaced so as to give a path length of 2 mm. Although not always as liquid tight, these cuvettes are more robust, are less likely to crack on repeated immersion in liquid nitrogen and aid rapid freezing of samples squirted into them (see Chance and Schoener, 1966).

In use, the Dewar is mounted in its holder in the light path and filled with liquid nitrogen to the level of the silvered margin. A moderate stream of compressed air is directed over the lower, outer surface of the Dewar via a catheter tube (led into the sample compartment through the openings designed for water circulation) and serves to demist the cold glass surfaces. The cuvettes are generally positioned in their holder and filled from hypodermic syringes with needles. Using tongs, the complete assembly is immersed in a separate, large Dewar flask filled with liquid nitrogen until all bubbling stops. Some care is necessary at this stage to prevent the cuvettes from sliding laterally from the holder. The cooled assembly is then located in the partly silvered Dewar with the orientation of the cuvettes matching that in the standard cuvette holder. The brass semicircular tab at the bottom of the cuvette holder (Fig. 3A–C) dips into the liquid, whilst the cuvettes are raised just clear of the surface. When the "correct" rotational and vertical position of the cuvettes has been established, and the adjusting screws at the top of the holder have been set to hold it snugly on top of the Dewar (all adjustments being made to give the flattest baseline), the apparatus has proved to be reliable and simple in use (Fig. 4). For slight modifications of the above procedures, designed to permit rapid freeze trapping of redox states, the reader is referred to Chance and Schoener (1966) and Gibson *et al.* (1980).

(*d*) *Interpretation of spectra.* The spectral characteristics of the cytochromes of many bacteria have been described and cannot be reiterated in detail here (but see Sections II and IV; also Jurtshuk *et al.,* 1975; Haddock and Jones, 1977; Bragg, 1980; Poole, 1983, Ingledew and Poole, 1984; for cytochromes in eukaryotic microorganisms, see Lloyd, 1974). Figure 4 presents a guide to the visible absorbance bands of cytochromes commonly found in microbial systems and identifiable in difference spectra. Extinction coefficients are given in Table I. It is important to note that most of the extinction coefficients cited have not been measured in microbial systems, or at best have been determined for one organism and the values extrapolated to cytochromes of the same type in other species.

2. Spectra of ligand-bound forms

The ability of certain cytochromes to bind small ligands such as CO and cyanide has been widely exploited and, in the case of cytochrome components of the respiratory chain, the ability of a cytochrome to react with CO has been taken as evidence for that component acting as a functional or potential oxidase. Other haemoproteins, however, namely haemoglobin, cytochrome P-450, "cytochrome P-420," and cytochrome *c* peroxidase, also bind CO, and other ligands may confuse spectral analysis (Lemberg and Barrett, 1973). Although the ability of a cytochrome to bind CO is indicative of an oxidase function, the pigment cannot be considered competent to function as such unless (1) the absorption bands are identifiable in photochemical action spectra (for methods, see Castor and

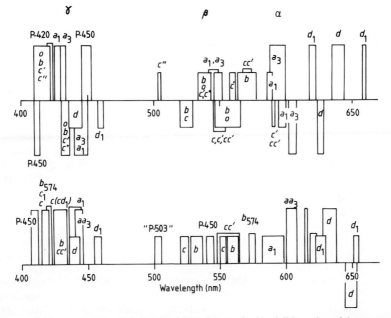

Fig. 4. A guide to the absorption bands of cytochromes in the visible region of the spectrum. The lower section shows the expected variation in positions of the reduced bands of the cytochromes shown at room temperature. At 77 K bands would be expected to shift 1–4 nm to the blue. Although the band positions shown are those for reduced *minus* oxidised difference spectra, the variations of band positions shown should also allow identification from absolute spectra. In the upper section are shown the approximate positions of the peaks (above the wavelength scale) and troughs (below) of the indicated cytochromes seen in CO + reduced *minus* reduced difference spectra. Depiction of band intensity is intended to show only striking variations in the relative intensity of α and γ bands for a given pigment. In the lower section, the unlabelled bands at 615 and 620 nm are, respectively, the 615-nm pigment of *Leptospira* and cytochrome a_{620} of *Tetrahymena*. For references and more details, see especially Lemberg and Barrett (1973), Lloyd (1974), and Poole (1983).

Chance, 1959; Rosenthal and Cooper, 1967; Hoffman *et al.*, 1980; Lloyd and Scott, 1983), (2) it can be demonstrated to be kinetically competent to support observed respiration rates (e.g., Haddock *et al.*, 1976), or (3) its reaction with O_2 can be demonstrated directly, such as in low-temperature ligand-exchange reactions (Poole *et al.*, 1979a). More detailed descriptions of the ligand-binding behaviour of haemoproteins are given by Lemberg and Barrett (1973), Wilson and Erecinska (1978), Wikstrom *et al.* (1981), and Poole (1983).

(*a*) *Carbon monoxide.* CO binds to the reduced form of cytochrome oxidases and certain other cytochromes, acting as a competitive inhibitor

Cytochrome	Spectral conditions	Wavelength (nm)	Absorption (mM cm^{-1})	Source	Reference
aa_3	Reduced minus oxidised	605–630[a]	16	Animal mitochondria	Chance (1957)
		445–465	91	Animal mitochondria	Chance (1957)
		445–455	60	Animal mitochondria	Chance (1957)
		605–630	11.7	*Paracoccus denitrificans*	Ludwig and Schatz (1980)
a_3	Reduced + CO minus reduced	430–444	91	Various, including yeast, *Bacillus subtilis*	Chance (1957); Smith (1955)
		590–605	10.1	Keilin–Hartree preparation	Vanneste and Vanneste (1965)
		428.5–455	148	Keilin–Hartree	Vanneste and Vanneste (1965)
		592–608	3.5	*P. denitrificans*	Ludwig and Schatz (1980)
	Photolytic method	589	11–12	Muscle and yeast	Chance (1953)
a_1	Reduced minus oxidised	427	120	*A. pasteurianum*	Chance (1953)
	Reduced + CO minus reduced	427–440	60	*A. pasteurianum*	Smith (1955)
	Photolytic method	589	12	*A. pasteurianum*	Chance (1953)
	Reduced minus oxidised	594–610	8.5	Suggested value	Meyer and Jones (1973b)
b	Reduced minus oxidised	562–575	22	By analogies with other haemoproteins	Chance (1957)
	Reduced minus oxidised	430–410	160	By analogies with other haemoproteins	Chance (1957)
	Reduced minus oxidised	563–577	25.6	Beef heart mitochondria	van Gelder (1978)
	Absolute reduced spectrum	557.5	16	*E. coli*	Deeb and Hager (1964)
	Reduced minus oxidised	560–575	17.5	*E. coli*	Jones and Redfearn (1966)
b haem (from *bd* complex)	Reduced minus oxidised	562–580	10.8[b]	*E. coli*	Miller and Gennis (1983)
$c(c + c_1)$[c]	Reduced minus oxidised	550–540	19.1	Animal mitochondria	Chance (1957)
c	Reduced minus oxidised	550–540	14.3	Beef heart mitochondria	van Gelder (1978)

TABLE I—*continued*

Cytochrome	Spectral conditions	Wavelength (nm)	Absorption (mM cm^{-1})	Source	Reference
c_1	Reduced minus oxidised	552–540	20.9	Beef heart mitochondria	van Gelder (1978)
c_1 (+c)	Reduced minus oxidised	553–540	19.1	Animal mitochondria	Chance (1957)
d	Reduced minus oxidised	630–615	8.5	By analogy with *P. aeruginosa* cd_1	Jones and Redfearn (1966)
	Reduced minus oxidised	630–655[d]	19	*E. coli/A. vinelandii*	Haddock *et al.* (1976)
d (of bd complex)	Reduced minus oxidised	628–608	13.6[e]	*E. coli*	Miller and Gennis (1983)
cd_1 complex[f]	Absolute reduced spectrum	418(c)	182	*P. aeruginosa*	Horio *et al.* (1961)
		554(c)	30.2	*P. aeruginosa*	
		625(d_1)	36	*P. aeruginosa*	
	Absolute oxidised spectrum	411(c)	282	*P. aeruginosa*	Silvestrini *et al.* (1979)
o	Photodissociation spectrum[g]	417	80–90	Various bacteria	Chance (1961)
	Reduced + CO minus reduced	417–432	170	Used for *Acetobacter*	Daniel (1970)
	Reduced minus oxidised	430–475[h]	91	*Vitreoscilla*	Tyree and Webster (1978)
		553–571[h]	8	*Vitreoscilla*	Tyree and Webster (1978)
	Absolute oxidised spectrum	538	10.6	*Vitreoscilla*	Choc *et al.* (1982)
		575[i]	6.9	*Vitreoscilla*	Choc *et al.* (1982)
		407	88.3	*Vitreoscilla*	Choc *et al.* (1982)
	Absolute reduced spectrum	420	192	*Vitreoscilla*	Choc *et al.* (1982)
		540	14.5	*Vitreoscilla*	Choc *et al.* (1982)
		564	13.5	*Vitreoscilla*	Choc *et al.* (1982)

[a] Major contribution is from cytochrome *a*.

[b] Pyridine haemochromogen.

[c] For details of the many bacterial *c*-type cytochromes, including those with more than one haem, see Lemberg and Barrett (1973).

[d] Probably unreliable, because of the variability in the amount of the oxygenated form of cytochrome *d* at about 655 nm (Poole *et al.* 1983a).

[e] Calculated by us from Miller and Gennis (1983) assuming 1 haem *d* per M_r 100,000.

[f] Nitrite reductase.

[g] Isosbestic points in the spectrum were about 404, 423, and 450 nm and could be used as reference wavelengths in the absence of interfering components.

[h] Calculated by us from published absolute spectra.

[i] Shoulder.

with respect to oxygen. It generally forms a reduced cytochrome–CO compound with characteristic absorption maxima (Fig. 4). The CO complexes of various bacterial cytochrome oxidases are photodissociable and, with few exceptions, the rate of CO reassociation is relatively slow. As expected for a ligand that has a higher affinity for the reduced form of the enzyme than for the oxidised form, the half-reduction potential of CO-binding cytochromes is raised upon addition of CO.

The most common practical application of these important properties is the recording of reduced-plus-CO *minus* reduced difference spectra ["CO difference spectra" (Chance, 1957)]. Here, the difference of absorbance between two reduced samples of the preparation is recorded using a split-beam or dual-wavelength scanning spectrophotometer to give a baseline. This is not always a trivial matter and is sometimes facilitated by prior reduction of a single sample with substrate or dithionite before the sample is carefully split between two matched, dry, clean cuvettes, taking care to avoid aeration during transfer. One suspension (conventionally called the "front" or "sample" suspension in a split-beam experiment) is bubbled with a fine stream of CO for a minute or so. If the top of the cuvette is smeared or sprayed with silicone antifoam, problems with excessive bubbling and frothing of concentrated protein are circumvented. The spectrum representing the difference between the samples is again recorded and, after baseline correction, if necessary, may reveal a complex spectrum comprising peaks and troughs. For example, in the classical case of cytochrome a_3, the CO-binding component of cytochrome oxidase in mitochondria and certain bacteria, a peak in the difference spectrum at about 430 nm results from the characteristic absorbance maximum of the CO complex (in the "front" cuvette) whilst a trough at 445 nm arises from the corresponding loss of absorbance of the reduced form in this cuvette, i.e., the peak absorbance of the CO-reactive form prior to bubbling with CO. The other cytochrome oxidases of bacteria also bind CO and give characteristic absorbance changes (see Table I and Fig. 4). Exploitation of the light reversibility of CO binding in photodissociation spectra is described separately below.

(b) *Cyanide*.   The reactions of cyanide with cytochromes are more complex than those of CO and less well defined even in the case of the mitochondrial oxidase (Nicholls, 1983). Here, cyanide reacts with both oxidised and reduced forms; the affinity for the oxidised form is high, but the reaction rate is slow, whereas the opposite is true for the reduced enzyme. Cyanide reactivity is most simply investigated in reduced-plus-cyanide *minus* reduced difference spectra analogous to those described for CO above and has been used to identify cyanide-reactive cytochromes of the aa_3 type and other putative oxidases in bacteria (Poole, 1981). Other features of the reaction of cyanide with the former may be exploited, however. Thus, because the reduced-CN^-–cytochrome a_3 complex is easily auto-oxidised, whereas the oxidised-CN^-–cytochrome a_3 complex is not readily reduced, the spectra of cytochromes a and a_3 can be recorded individually (for an application and references, see Salmon and Poole, 1980).

(c) *Azide*. In contrast to cyanide, azide reacts quickly with oxidised oxidases of the aa_3 type, but gives only small spectral changes, including a slight, blue shift of the Soret band. At higher azide concentrations, the Soret band is slightly red shifted.

(d) *Nitrogen oxyanions and NO*. Inhibition of mitochondrial respiration by hydroxylamine results from the production of NO. Reduced cytochrome oxidase exhibits a high affinity for NO and reacts with it readily in both mitochondrial (Wikstrom *et al.*, 1981) and bacterial systems (*e.g.* Poole, 1981). However, NO also reacts rapidly with O_2 and so the inhibition is transient unless a NO-generating system is used. For simple binding studies, NO can be readily generated by sodium nitrite and a reducing agent such as dithionite.

In bacterial systems, particular interest attaches to the reactions of oxidases with NO and related ligands because of the role of such small molecules as intermediates in denitrification (Payne *et al.*, 1980). There are detailed accounts of the binding of NO to the cytochrome cd_1 (nitrite reductase) of *Pseudomonas aeruginosa* (for references, see Poole, 1983). Recently, we have shown that addition to *E. coli* membranes of nitrate, nitrite, NO, or trioxodinitrate generates the same spectral changes in cytochrome d (Hubbard *et al.*, 1983), presumably due to the formation of a nitrosyl complex, suggesting that caution should be applied in using nitrogen oxyanions with bacteria exhibiting partial reactions of denitrification.

(e) *Fluoride, isonitriles, and sulphide*. Fluoride causes only small blue shifts in the absorbance maxima of mitochondrial cytochrome oxidase. Isonitriles cause a shift in the α band of reduced cytochrome c oxidase from 605 to 600 nm and from 444 to 439 nm. Sulphide reacts readily with oxidised cytochrome c oxidase, giving optical spectral signals similar to those induced by cyanide. None of these ligands has been widely used for microbial systems. Further information may be found in Wilson and Erecinska (1978) and Wikstrom *et al.* (1981).

3. Photodissociation spectra

Advantage was taken of the ability of light to dissociate the CO complex of diverse bacterial oxidases in the early work of Chance who thus obtained the photodissociation difference spectra of these pigments. Such spectra take the form of an inverted CO difference spectrum, being the difference in absorbance between a CO-liganded, reduced sample that has been photolysed and an identical non-irradiated sample. Two factors facilitate the recording of such spectra. First, although the recombination of CO with the reduced cytochrome is relatively slow (e.g., K "on" = $7 \times 10^4 \ M^{-1} \ s^{-1}$ at 25°C for mitochondrial aa_3) by comparison with either the photodissociation event or the reaction of O_2 with the reduced enzyme, the spectrum is most conveniently recorded at low temperatures (typically -100 to -196°C) thus allowing leisurely spectrophotometric observation and the use of light sources of modest intensity such as a tungsten arc lamp or the focused beam from a slide projector. At 77 K, the rate of CO recombination with most oxidases is usually negligible (see, however, Poole et al., 1982). Secondly, when the extent of dissociation may be small, the difference spectrum is best recorded in the dual-wavelength scanning model in which a single sample is used as reference and, after photolysis in situ, rescanned to give the difference spectrum. One of us (R.K.P.), however, has obtained some success with the simple low-temperature (77 K) split-beam accessory for a Unicam SP1700/SP800 (described previously). The cuvette assembly, initially containing two CO-supplemented, reduced samples was removed from the Dewar and one cuvette shielded while the sample cuvette was irradiated with a camera flash gun, during which time the cuvette did not warm significantly from 77 K. In the more convenient Johnson Foundation design for a dual-wavelength scanning instrument, the sample is contained in a flat cuvette which is cooled by a flow of cold nitrogen gas (see next section). Light guides (fibre optics) couple the sample compartment to both the chopper and photomultiplier tube. The guide is also bifurcated, allowing the photolysing beam to irradiate the same flat face of the sample as that illuminated by the chopped measuring and reference beams (Fig. 5).

This apparatus also forms the basis for recording the binding of O_2 to reduced cytochrome oxidases using the ligand exchange and low-temperature trapping techniques described in detail elsewhere (Chance et al., 1975b; Chance, 1978). In essence, O_2 (as air-saturated buffer) is added at -25°C to a CO-saturated preparation which is kept fluid during the addition by inclusion of a cryosolvent (e.g., 30% v/v ethylene glycol). At these temperatures, and in the dark, O_2 does not displace CO at a significant rate and the O_2-supplemented, but CO-bound oxidase is then trapped by

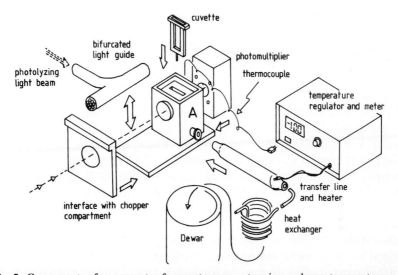

Fig. 5. Components of an apparatus for spectroscopy at various subzero temperatures and photolytic activation of cytochrome oxidase reactions. The main sample compartment (A) has an insulating lining of chipboard and Styrofoam and a locating hole for the pin on the cuvette bottom. A short light guide, abutting onto the back face of the installed cuvette, transmits light to the photomultiplier, whilst a bifurcated guide couples the sample compartment to the input measuring and reference beams from the chopper. The limit of this guide is used to transmit light from the photolysing light source (e.g., tungsten arc or laser), via fibres mixed randomly with those in the afferent guide, to the sample. Low temperatures are achieved by passing N_2 gas (typically 10 litre min^{-1}) through a coiled copper heat exchanger (internal diameter 7 mm; length 0.5 m) immersed in liquid N_2 in a large Dewar. The cold gas passes through an EPR transfer line (Wilmad Glass, Fluorochem Ltd., Glossop, Derbyshire, England) and is reheated to a temperature selected on the control and measuring unit. The temperature at the side of the cuvette is sensed by a thermocouple. For further details, see Chance *et al.* (1975a). (Redrawn and modified from a sketch kindly made available by the Biomedical Instrumentation Shops of the Johnson Research Foundation.)

further lowering the temperature to about −80°C. Photolysis at such temperatures (or lower) dissociates the CO–oxidase complex and is followed by O_2 binding and reaction. For examples of applications to microbial systems, see Poole *et al.* (1979a, b, 1983a).

4. Spectrophotometry at other subzero temperatures

In media such as those containing mannitol or sucrose, which do not exhibit phase transitions in the experimental range of temperatures, the absorption peak heights and peak areas are nearly linear functions of temperature between −40°C and liquid N_2 temperatures (Wilson, 1967).

Furthermore, if the sample is warmed within this temperature range, and cooled again to the original temperature, the resultant spectrum is indistinguishable from that recorded originally (Wilson, 1967; Vincent et al., 1982). At temperatures higher than $-40°C$, the absorbance intensity decreases abruptly and refreezing does not fully restore the enhancement (Wilson, 1967).

For kinetic studies at subzero temperatures, it is sometimes necessary to use temperatures other than 77 K. Temperatures between about 0 and $-40°C$ can be maintained simply by circulating cooling water with ethylene glycol around a cuvette containing the ethylene glycol-supplemented sample. For temperatures between about -40 and $140°C$, as required for kinetic studies of ligand binding to microbial cytochrome oxidases and of subsequent electron transfer (see previous section) the procedure developed by Chance and co-workers for their "triple trapping" studies of mitochondrial oxidases is appropriate (Fig. 5). Here, the sample is maintained at any chosen temperature within this range by a steady flow of nitrogen gas, cooled by passing through a copper coil immersed in liquid N_2, and reheated by a thermostatically controlled resistor. Measurements of the sample temperature with a calibrated copper–constantan thermocouple next to the cuvette show that temperatures are maintained to better than $0.5°C$ by the thermostat. This system is simple, reliable, and relatively inexpensive. The techniques of cryoenzymology are outside the scope of this chapter but are described in detail by Douzou (1977).

Temperatures much lower than 77 K are occasionally required. Thus, in a study of the recombination of photodissociated CO with cytochrome d in E. coli, temperatures as low as 4.2 K were achieved during split-beam spectrophotometry using a commercial liquid He cryostat (Poole et al., 1982). A double Dewar system for use at temperatures down to 4.2 K in a split-beam instrument has also been described by Hagihara and Iizuka (1971). These methods have been more widely used in studies of the ligand-binding kinetics of haemoglobin and myoglobin.

5. Numerical (higher derivative) analyses of complex absorption spectra

Higher derivative analysis provides a powerful ancillary technique to aid in the resolution of complex spectra, such as those recorded at 77 K of intact microbial cells and subcellular fractions. The method exploits the narrower half-bandwidths, typically four-fold for a Lorentzian band, of components obtained by higher derivative analysis. Thus, bands that would otherwise be too close together for resolution are resolved (Butler and Hopkins, 1970a). Derivatives are obtained by computing a difference spectrum between the original curve, $A(\lambda)$, and the same curve shifted by a small, finite wavelength interval, $A(\lambda + \Delta\lambda)$, with the difference value being assigned the wavelength corresponding to the midpoint of $\Delta\lambda$, the so-called differencing or differentiating interval. Consider the absorbance values, taken at equal increments along the wavelength axis, to be numbered n, sequentially. Thus, the first derivative A^1 is given by

$$A^1(n + w/2) = A(w + n) - A(n)$$

where w is the differencing interval (chosen to be as large as possible to maximise the signal but not so large that spectral detail is lost; see below) and numbers in parentheses indicate the positions (number of increments along the wavelength axis) of the absorbance or derivative values. Higher derivatives (A^{II}, A^{III}, A^{IV}, etc.) are obtained by a similar differentiation of the preceding derivative curve (Butler and Hopkins, 1970b). In such cases, an appreciable improvement in signal-to-noise ratio is obtained when the four differencing intervals are very similar but not equal (for an example, see Fig. 6). It should be noted that second, fourth, and sixth derivative spectra give minima, maxima, and minima, respectively, at wavelength positions very close to, or coincident with, the peak of the original spectrum. First and third derivative spectra have the characteristic "derivative" shape with a crossover point close to the original peak position. Thus the fourth-order spectrum is the first in the series to give a peak close to the original position and represents a useful compromise, offering substantially improved resolution yet tolerable noise levels and facility of computing. First- and second-order derivative spectra may be used to advantage, however (e.g., Kühn and Gottschalk, 1983).

The major difficulties that arise (Butler and Hopkins, 1970a) are (1) slight shifts of the band positions in the higher derivative spectra from their true positions, (2) the generation of spurious derivative bands by two appropriately spaced bands (Morrey, 1968), (3) the suppression of peak heights in the derivative spectra by overlap with "wings" or troughs arising from adjacent bands, (4) the marked discrimination in favour (i.e., enhancement) of narrow bands (Butler, 1979), and (5) limitations of detectability of the higher derivative bands by noise in the original data (Butler, 1979).

Despite these problems, the methods can yield invaluable information when the user is fully aware of the potential artefacts and the resolution limits (Shrager, 1983). Examples of applications to bacteria are to *E. coli* (Scott and Poole, 1982), and diverse bacteria (Shipp, 1972), the thermophilic bacterium PS3 (Poole *et al.*, 1983b), and *Spherotilus natans* (Fig. 6).

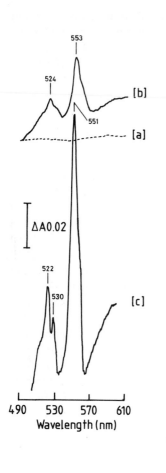

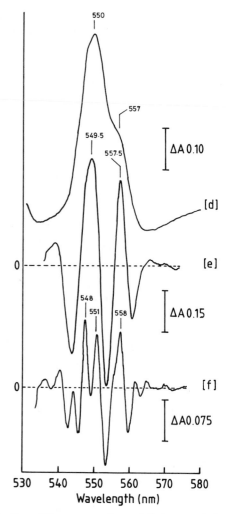

Fig. 6. Examples of the use of low-temperature difference spectra and their numerical analysis in resolution of multiple cytochrome bands. The suspension of *Spherotilus natans* membranes contained in 10 m*M* phosphate, buffer pH 7.4, and 0.5 *M* mannitol (final protein concentration 5 mg ml⁻¹). Spectrum (a) is the "native"–"native" baseline at 290 K. In (b) contents of the sample cuvette were reduced with 10 m*M* ascorbate plus 1 m*M* TMPD and those of the reference cuvette oxidised with a grain of K_3 (FeCN)$_6$ before the spectrum was run at 290 K. Spectrum (c) was recorded after the same cuvettes had been cooled to, and scanned at, 77 K. In (d) five spectra of the type in (c) were summed and replotted. Fourth-order finite difference analyses were performed using differencing intervals of 3.4, 3.0, 2.8, and 2.6 nm (e) or 2.2, 1.8, 1.6, and 1.4 nm (f). In (a) to (c) the scan rate was 2 nm s⁻¹, whilst for (d) it was 0.2 nm s⁻¹. The spectral bandwidth was 0.75 nm in all cases. (Unpublished results of R. K. Poole, M. V. Jones, and R. I. Scott.)

More recently, higher derivative analysis has been incorporated into an integrated approach to the resolution of absorbance bands in the complex spectra of various bacterial membranes (van Wielink *et al.*, 1982). Here the number of component peaks in a complex, composite band is assessed from potentiometric titrations at 77 K, curve-fitting routines, and confirmation in some cases by numerical analyses. It is encouraging to note that in the case of PS3 membranes (Poole *et al.*, 1983b), numerical analyses obtained using either the Butler and Hopkins (1970a) algorithm or the method of Savitsky and Golay (1964) were in good agreement with the potentiometric studies and spectroscopy at 77 K.

Higher derivative analyses have generally been performed by on-line computation and sampling of absorbance values at small wavelength increments (e.g., 0.125–0.25 nm; Butler and Hopkins, 1970a). Our own early work (Salmon and Poole, 1980 and references above) involved interrogation of the analogue output from the spectrophotometer by a digital voltmeter and data logger unit which transferred absorbance values to a paper-tape punch, the paper tape was analysed by a main-frame computer and the numerical analysis plotted on an *X–Y* recorder under computer control. On-line data sampling and computation would now be more widely available.

Whatever hardware is used, it is important to maximise the signal-to-noise ratio (Butler, 1979). To this end, successive spectra may be recorded and summed or averaged; the ratio is then improved by $n/\sqrt{n}$, where n is the number of scans, so that six scans will increase the ratio 2.45-fold. In addition, the Butler and Hopkins (1970a) algorithm has noise-reducing properties when the differencing intervals are similar but not identical (see above).

6. Combination of other measurements with spectrophotometry

Being a non-invasive technique, spectrophotometry is well suited for supplementation with measurement of other experimental variables. Although most of the earlier devices were designed to be accommodated in the sample compartments of commercial spectrophotometers, the ready availability and ease of use of light guides and fibre optics make feasible the analysis of cytochromes, using either reflectance or transmission methods, in situations remote from the primary light beam(s). Examples of easily constructed devices are those for simultaneous measurements of O_2 consumption (Ribbons *et al.*, 1968; Hamilton *et al.*, 1979; Horiuchi *et al.*, 1982), redox potential (Dutton, 1978; Chance *et al.*, 1982), steady state O_2 tensions in the "open" electrode system (Degn *et al.*, 1980), and bacterial luminescence as an indicator of O_2 concentration (Poole *et al.,* 1979a). Samples in standard EPR tubes may be studied optically using either transmission (Hansen *et al.,* 1970) or reflectance methods (Palmer, 1967). Angular rotation within the spectrophotometer of a membrane preparation in which there is substantial orientation of the membrane planes (oriented multilayers) can provide information on the arrangement of the cytochrome haems relative to the membrane (Erecinska, 1982).

7. *Extraction, identification and determination of haems*

Although visible absorption spectroscopy of cytochromes in intact cells, in membranes, or in the purified state is frequently informative about the haem type present (i.e., *a, b, c,* or *d*), in other cases the prosthetic group must be extracted and analysed. The so-called "cytochrome a_1" in bacteria is a good example; many other haemoproteins, not containing haem *a* (e.g., catalase, peroxidase, sirohaem), may be mistakenly identified as cytochrome a_1 (Poole, 1983). The classical method is determination of the pyridine haemochromes in which two molecules of the base are coordinated to the ferroporphyrin.

Haems may generally be extracted from microbial cells by acidified organic solvents. Note that haem *c* and the haem of certain peroxidases are covalently bound to protein and not split off by these reagents. The most commonly used solvent mixtures are ether–acetic acid, ethyl acetate–acetic acid, or acetone containing 0.015 *M* HCl (Fuhrhop and Smith, 1975). The detailed methods given by Smith and Caughey (1978) for haem *a* are suitable for bacteria by substituting a known (wet) weight of cells for beef heart (B.S. Baines and R.K. Poole, unpublished). Examples of other reports that provide adequate experimental detail are those for chlorin (the haem of cytochrome *d;* Barrett, 1956), haems, especially *a,* from *Corynebacterium* (Rawlinson and Hale, 1949), sirohaem from *E. coli* (Murphy *et al.,* 1973), and haems from yeast (Barrett, 1961), unusually refractory to the usual procedures.

Pyridine haemochrome spectra are measured in aqueous alkaline pyridine solutions after reduction, usually with sodium dithionite. Final concentrations of 0.075 *M* NaOH and 2.1 *M* pyridine are suitable.

The characteristics of absolute spectra of the pyridine haemochromes of various haems are tabulated by Fuhrhop and Smith (1975). Reduced (with dithionite, 0.7 mg ml^{-1}) *minus* oxidised (with 0.05 m*M* $K_3Fe(CN)_6$) difference spectra, however, permit the determination of haem in the presence of considerable amounts of porphyrins, as well as decreasing the errors from non-specific absorption. Extinction coefficients and the characteristic and maxima and minima for such spectra are also given in the above reference.

C. Non-spectrophotometric techniques and their applications

1. Staining procedures

(a) *The oxidase test.* This test measures the ability of the terminal cytochrome system to catalyse the aerobic synthesis of indophenol blue from a mixture of the non-physiological substrates N,N-dimethyl-p-phenylenediamine (DMPD) and α-naphthol (Fig. 7). The original version of the test was designed for bacterial colonies on agar plates and has proved over the years to be an extremely simple but useful chemotaxonomic tool (Gordon and McLeod, 1928; Kovacs, 1956). It involves flooding the plates for 15 s with a mixture of DMPD and α-naphthol, then draining the substrates away and waiting for the colonies to become stained by the nascent indophenol blue; the appearance of a deep blue colouration within approximately 30 s is usually taken to mean a positive reaction and hence to signify the presence of a membrane-bound, high-redox-potential cytochrome c linked to an active cytochrome oxidase (usually cytochrome oxidase aa_3 or o; possibly also a_1). In its more recent versions, the test has been applied to suspensions of bacteria and has been endowed with quantitative properties; i.e., the speed with which a positive reaction is attained can be used as a measure of the activity of the terminal cytochrome system (Jurtshuk and Liu, 1983); the test can also be carried out using TMPD (N,N,N',N'-tetramethyl-p-phenylenediamine) in place of DMPD.

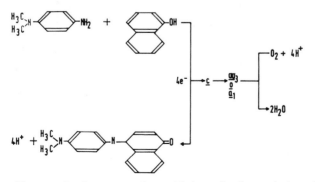

Fig. 7. The oxidase test for the presence in aerobic bacteria of a respiratory chain which contains cytochrome c linked to an active cytochrome oxidase.

Although the effective absence of either cytochrome c or cytochrome oxidase is enough to ensure a negative reaction to the oxidase test, in practice under aerobic conditions it usually reflects the absence of cytochrome c (either via a lack of synthesis or by its extraction from the respiratory membrane *in vitro*). The test has therefore been used with some success to screen for cytochrome c deficient mutants of various organisms, including the aerobic chemoheterotrophs *Azotobacter vinelandii, Paracoccus denitrificans,* and *Pseudomonas* AM1, and the facultative phototroph *Rhodopseudomonas capsulata* (Marrs and Guest, 1973; Anthony, 1975; Hoffman *et al.* 1980). It has additionally, but less widely, been used to screen for cytochrome aa_3 deficient mutants of *P. denitrificans* (Willison and Haddock, 1981) and cytochrome o deficient mutants of *R. capsulata* (La Monica and Marrs, 1976).

Interestingly, it has recently been claimed that TMPD will selectively donate electrons to cytochrome oxidase d in *E. coli,* albeit slowly (Kranz *et al.,* 1982); this property has been used to select a cytochrome oxidase d-deficient mutant of this organism.

(b) The peroxidase test.   This test is based upon the inherent peroxidase activity of most partially denatured haemoproteins (including cytochromes, myoglobin and haemoglobin, as well as peroxidases themselves), i.e.,

$$2[X \cdot Fe^{2+}] + 2H^+ + H_2O_2 \longrightarrow 2H_2O + 2[X \cdot Fe^{3+}]$$

This activity can be linked to the oxidation of an artificial substrate such as benzidine (Culliford and Nickolls, 1964), o-tolidine (Reid and Ingledew, 1980) or 3,3',5,5'-tetramethylbenzidine (TMBZ) (Thomas *et al.,* 1976) to yield a greenish-blue product, thus allowing the detection and location of haemoprotein bands on SDS–polyacrylamide gels. TMBZ is now the preferred substrate since it is a far less potent carcinogen than benzidine or o-tolidine, as well as being a much more sensitive reagent; it also causes much less background staining and forms a product that is stable for several months.

SDS–polyacrylamide gels of membrane samples denatured in the absence of mercaptans such as β-mercaptoethanol are immersed in a freshly prepared mixture of methanolic TMBZ and acetate buffer, pH 5.0, and left in the dark for 1–2 h. Hydrogen peroxide is then added and the haemoprotein bands become fully stained within 30 min. If required, the stain can subsequently be removed by soaking the gels in sodium sulphite and then washing them several times with 30% (v/v) isopropanol. The destained gels can then be restained for protein using conventional procedures.

The method has been used successfully to detect haemoprotein bands in SDS–polyacrylamide gels of membranes from several bacterial sources (see, for example, Miller and Wood, 1983).

2. Fluorescence procedures

Cytochrome c can be detected on SDS–polyacrylamide gels of membranes or purified respiratory complexes by its ability to fluoresce when exposed to ultraviolet light at 360 or 366 nm (Katan, 1976). This probably reflects the loss of the iron atom under these dissociating conditions such that the non-fluorescent mesohaem is converted into the fluorescent mesoporphyrin. The method is specific for c-type cytochromes if electrophoresis is carried out in the presence of mercaptans, since the latter markedly stimulate the removal of the non-covalently bound haems of the a-, b-, and d-type cytochromes but not the covalently bound mesohaem of the c-type cytochromes. The gels are viewed against a dark background at right angles to the source of ultraviolet light, and exhibit an orange-red fluorescence which can be successfully photographed provided that a guard filter is placed between the ultraviolet light source and the gel to cut out stray visible light from the lamp (e.g., Kodak–Wratten 18A), and also between the gel and the camera to cut out background ultraviolet light (e.g., Kodak–Wratten 16). The lower detectable limit for cytochrome c is reported to be 0.2 μg (<20 pmol).

This method has enabled the detection of c-type cytochromes in SDS–polyacrylamide gels of respiratory or photosynthetic membranes, and of partially purified cytochrome oxidase complexes (see, for example, Wood, 1980).

3. Primary sequence and X-ray diffraction studies

These studies have until recently been restricted to cytochromes that can be purified to homogeneity. Thus the primary sequences of almost 50 c-type cytochromes from bacterial sources, together with cytochrome b from *E. coli,* have been determined using conventional amino acid sequencing techniques, and X-ray diffraction analyses have revealed the tertiary structures of several c-type cytochromes (see for example Ambler, 1976, 1980; Dickerson *et al.,* 1976). The recent advent of recombinant DNA technology should soon allow the primary sequences of less readily purified cytochromes to be determined via nucleotide sequencing of cloned DNA. The methods employed in these various structural analyses are outside the scope of this article.

IV. Cytochrome function during respiration

The ability of the cytochromes to exist in ferro (Fe^{2+}) and ferri (Fe^{3+}) forms obviously enables them to participate in oxidation–reduction or redox reactions, the most important of which constitute the membrane-bound processes of respiration and photosynthetic electron transfer. In both of these systems, reducing equivalents in the form of hydrogen atoms, hydride ions, or electrons are transferred from a donor to a higher redox potential acceptor via sequential redox reactions, with the concomitant ejection of H^+ across the energy coupling membrane (the cytoplasmic membrane of prokaryotes). This vectorial H^+ movement thus conserves the free energy released by the oxidation–reduction reactions as a trans- or intramembrane proton-motive force (Δp or $\Delta\bar{\mu}H^+/F$). This is variably composed of a difference in pH across the membrane (ΔpH) and an electrical potential difference or membrane potential ($\Delta\psi$), according to the ambient external pH and/or to the concomitant activity of various solute transport systems such that

$$\Delta p = \Delta\psi - Z \cdot \Delta pH$$

where $Z = 2.303\ RT/F$ (Mitchell, 1966). The magnitude of the Δp is largely dependent on the intrinsic properties of the donor and acceptor couples, the nature of the redox system, and the integrity of the energy coupling membrane with respect to H^+ leakage. Provided the Δp exceeds a threshold value, proton retranslocation occurs via the $BF_0 \cdot BF_1$ ATP phosphohydrolase complex with the concomitant synthesis of ATP. The proton-motive force also drives various other energy-dependent membrane functions including reversed electron transfer, solute transport, and, in some organisms, motility, sensory perception, and pyrophosphate synthesis (Haddock and Jones, 1977; Garland, 1977; Downie et al. 1979; Fillingame, 1980; Ferguson and Sorgato, 1982; Taylor, 1983).

A. Cytochromes b and c

Following growth under aerobic conditions, the membrane-bound respiratory chains of chemoheterotrophic, chemolithotropic, and facultatively phototrophic bacteria classically contain several cytochromes b, two or more cytochromes c, and at least one cytochrome oxidase. The cytochromes b function close to the quinone (either ubiquinone or menaquinone according to species) and, by analogy with more intensively studied mitochondrial respiratory chains, are probably most commonly involved in a complex, protonmotive quinone cycle which serves to transfer electrons from the iron–sulphur centres of the primary dehydrogenases (e.g., NADH, succinate or lactate dehydrogenases) located on the cytoplasmic side of the membrane, to cytochrome c on the outside of the membrane; this outwardly directed electron transfer is accompanied by proton ejection (energy coupling site 2; $\rightarrow H^+/O = 4$, $\rightarrow$charge/O $= -2$). All bacterial respiratory chains examined so far appear to contain cytochrome b, although in some chemolithotrophs oxidising substrates of particularly high redox potential (e.g., *Thiobacillus ferrooxidans* oxidising Fe^{2+}, and *Nitrobacter* sp. oxidising nitrite) it participates in energy-dependent reversed electron transfer to NAD^+ rather than forward electron transfer to oxygen. Cytochrome b is generally very firmly embedded in the respiratory membrane and hence is difficult to purify. However, cytochromes b_{556} and b_{562} from *E. coli* have been purified to homogeneity (Itagaki and Hager, 1966, 1968; Kita et al., 1978). Each is composed of only one type of subunit (M_r 17.5 K and 62–68 K, respectively) and, interestingly, the apoprotein of cytochrome b_{562} exhibits some "vestigial homology" with cytochrome c' (Weber et al., 1981). Cytochromes b of similar molecular weight to cytochrome b_{556} have been identified in several other organisms.

Cytochromes b also function in various types of anaerobic respiratory chains including those of the sulphate reducers *Desulphovibrio* and *Desulphotomaculum* (Le Gall et al., 1979), fumarate reducers such as *Vibrio succinogenes* (Kröger, 1978), and methanogens such as *Methanosarcina barkeri* which are able to use methylated substrates (Kühn and Gottschalk, 1983). The cytochromes b in these organisms generally exhibit particularly low E_m values, in line with their physiological roles. Some obligately fermentative anaerobes of the genus *Clostridium* also contain cytochromes b, but their roles have not yet been determined.

The function of cytochrome c is generally to accept electrons from the quinone–cytochrome b system and transfer them to the cytochrome oxidase(s) or to terminal reductases. Two cytochromes c have been identified in the respiratory chains of many species of bacteria including *A. vinelandii, P. denitrificans, Pseudomonas AM1, Methylophilus methylotrophus, Thermus thermophilus,* and *Nitrobacter europaea* (see, for example, Swank and Burris, 1969; Cross and Anthony, 1980). They tend to have significantly different molecular weights (within the range M_r 8500–38,000) and isoelectric points, and to be bound with different affinities to the outer surface of the cytoplasmic membrane. Particularly large amounts of c-type cytochromes are present in some relatively specialised respiratory chains, such as those of Gram-negative methylotrophic and sulphur- or ammonia-oxidising bacteria, where these cytochromes additionally accept electrons directly from specialised primary dehydrogenases or cytochrome c reductases located on the periplasmic side of the membrane (e.g., methanol dehydrogenase, methylamine dehydrogenase, hydroxylamine cytochrome c reductase) (see, for example, Burton et $al.$, 1983; Olson and Hooper, 1983). Cytochromes c are thus absolutely essential for aerobic respiration and associated energy conservation under these conditions. Since they are located on the outside of the respiratory membrane, and the site of oxygen reduction by the cytochrome oxidases is tenuously accepted as being on the cytoplasmic side of the membrane, terminal electron transfer is inwardly directed and is minimally associated with the consumption of $2H^+$ on the cytoplasmic surface (energy coupling site 3; $\rightarrow H^+/O = 0$, $\rightarrow$charge$/O = -2$).

In contrast to the bacteria discussed above, many chemoheterotrophs lack significant concentrations of a high redox potential, c-type cytochrome (e.g., $E.$ $coli$ and other enterobacteria, $Acinetobacter$ $calcoaceticus$, $Pseudomonas$ $syringae$, $Bacillus$ $pumilis$, and several other mesophilic strains of $Bacillus$) (Jones and Meyer, 1982; Jurtshuk and Liu, 1983; see also Section V,A). It is likely that these organisms transfer electrons directly from the quinone–cytochromes b system to the cytochrome oxidase(s), probably without the involvement of a protonmotive quinone cycle, and hence do not conserve energy at site 3. These organisms thus exhibit low stoichiometries of respiration-linked proton translocation during the oxidation of endogenous NAD(P)H, and are characterised by low molar growth yields with respect to carbon substrate and oxygen (Jones et $al.$, 1975, 1977, 1978; Jones, 1977).

A similar situation also pertains during anaerobic respiration to lower redox potential acceptors such as nitrate or fumarate, both of which are reduced by appropriate reductases located on the cytoplasmic side of the membrane; in neither case is cytochrome c involved, even when present in the respiratory chain. Paradoxically, although anaerobic respiration to nitrite in various denitrifying bacteria occurs via cytochromes c, the associated energy conservation is lower than might be expected because the terminal nitrite reductase (cytochrome cd_1) is located on the same side of the membrane as the c-type cytochromes. The terminal stage of nitrite reduction is therefore accompanied by scalar, rather than vectorial, electron transfer and hence by proton consumption on the "wrong" side (i.e., the outside) of the membrane. Cytochrome cd_1 (M_r 120,000) has been purified from several species of denitrifying bacteria and is composed of two identical subunits, each of which contains one molecule of haems c and d_1 (Kuronen and Ellfolk, 1972; Gudat et $al.$, 1973).

The anaerobic oxidation of hydrogen by sulphate-reducing bacteria such as $Desulphovibrio$ sp. also involves a c-type cytochrome, in this particular case the multihaem, low redox potential cytochrome c_3 (E_m − 300 mV; M_r 13,000) which transfers electrons from periplasmically located hydrogenases to menaquinone or cytochrome *b* within the respiratory membrane.

B. Cytochrome oxidases

The functions of the cytochrome oxidases (cytochromes aa_3, *o, d,* and, in some species of bacteria, a_1) is to transfer electrons from either a high redox potential cytochrome *c* or the quinone–cytochrome *b* region to molecular oxygen with the concomitant formation of water. Unlike cytochrome oxidases aa_3, *o,* and a_1, which can function in both types of system, cytochrome oxidase *d* appears to be incapable of accepting electrons from cytochrome *c* during aerobic respiration and hence is not associated with energy conservation at site 3. Most bacterial respiratory chains are terminated by more than one cytochrome oxidase (e.g., aa_3, *o* or a_1, *o* or *d, o,* a_1), the concentration and activity of the individual oxidases varying as a complex function of the ambient concentrations of oxygen and/or various inhibitors. Respiratory systems terminated by single oxidases are extremely rare; such linear systems are present in *E. coli* and other enterobacteria cultured under conditions of excess oxygen (*o*), and in *Brochothrix thermosphacta* (*Microbacterium thermosphactum*) (*aa_3*). Neither cytochrome oxidases a_1 or *d* occur alone (Meyer and Jones, 1973a; see also Section V,A).

Cytochrome oxidase aa_3 has been purified from several species of chemoheterotrophic or facultatively phototrophic bacteria including *P. denitrificans, B. subtilis, N. agilis, Thiobacillus novellus, Rhodopseudomonas sphaeroides,* and the thermophiles PS3 (probably *Bacillus stearothermophilus*) and *Thermus thermophilus* (for reviews see Ludwig, 1980; Poole, 1983; Ingledew and Poole, 1984). The bacterial enzyme, like its mitochondrial counterpart, contains two molecules each of haem *a* and copper per molecule and is probably a transmembrane protein. It is composed, however, of only two or three subunits, rather than the seven or more present in the mitochondrial enzyme; these subunits probably correspond to subunits I to III which are coded for by the mitochondrial DNA. A general picture is emerging that subunit I (M_r 45,000–58,000) probably contains one haem centre (a_3) and possibly one copper (Cu_B). By analogy with mitochondrial aa_3, subunit II (M_r 28,000–38,000) would be expected to carry the other redox centres (*a* and Cu_A); however, in the thermophile oxidases this subunit also carries a molecule of haem *c* such that these enzymes may be regarded as caa_3 complexes. Subunit III (M_r 21,000) has been detected in several preparations and, again by analogy with the mitochondrial enzyme, might be expected to endow the oxidase with the ability to pump H^+ across the membrane (in addition to its capacity to catalyse inwardly directed electron transfer and hence to consume $2H^+$ per atom of oxygen reduced on the cytoplasmic side of the membrane). Indeed, whole cells of the moderate thermophile *B. stearothermophilus* exhibit $\rightarrow H^+/O$ and $\rightarrow$charge/O quotients commensurate with the oxidase acting as a combination of an electron-transferring arm ($\rightarrow H^+/O = 0$, $\rightarrow$charge/O $= -2$) and a proton pump ($\rightarrow H^+/O = 2$, $\rightarrow$charge/O $= -2$) (Chicken *et al.*, 1981); the purified oxidase also pumps H^+ when reconstituted into proteoliposomes. It should be noted, however, that an ability to pump H^+ has also been reported for whole cells of *P. denitrificans* and for the reconstituted oxidases from *P. denitrificans* and *T. thermophilus*, both of which purify as two-subunit enzymes. In contrast, whole cells of *M. methylotrophus* do not exhibit H^+ pumping via cytochrome oxidase aa_3 (Dawson and Jones, 1981) and the three-subunit oxidase from *R. sphaeroides* appears to be incapable of proton translocation when reconstituted into proteoliposomes (Gennis *et al.*, 1982). The proton-pumping abilities of bacterial cytochrome oxidases aa_3 thus remain the subject of some controversy.

Cytochrome oxidase *o*, like cytochrome oxidase aa_3, has been purified or partially purified from several species of bacteria, including the chemoheterotrophs *E. coli*, *Vitreoscilla* (*Beggiatoa*) sp., *P. aeruginosa*, *A. vinelandii*, *M. methylotrophus*, and thermophile PS3, and the facultative phototrophs *Rhodopseudomonas palustris* and *Rhodopseudomonas capsulata* (for reviews see Jurtshuk *et al.*, 1975; Poole, 1983; Ingledew and Poole, 1984; also Carver and Jones, 1983). The various preparations exhibit considerable diversity in terms of molecular weight, subunit composition, and haem content. Subunit I (most frequent M_r 27,000–29,000, but ranging from 13,000 to 65,000) binds one molecule of haem *b;* two such subunits make up the *Vitreoscilla* oxidase, the two haems *b* exhibiting different E_m values. Subunit I is accompanied in the *A. vinelandii*, *P. aeruginosa*, *M. methylotrophus* and *R. palustris* enzymes by at least one other subunit (M_r 12,200 to 21,000) which binds haem *c;* in these organisms the oxidase thus purifies as a *co* complex, analogous to the *caa_3* complex of thermophilic bacteria, which readily oxidises the artificial substrate ascorbate–TMPD. Cytochrome oxidase *o* is generally fairly sensitive to cyanide, but exhibits wide variations in its sensitivity to carbon monoxide; not unexpectedly, the oxidase in the carbon monoxide-oxidising bacterium *Pseudomonas carboxydovorans* is almost completely resistant to this inhibitor (Meyer and Schlegel, 1983).

Cytochrome oxidase *d* exhibits a number of properties which set it firmly apart from the other cytochrome oxidases, the most striking of which is its marked insensitivity to inhibition by cyanide (see Henry, 1981). Thus cytochrome oxidase *d* is adaptively synthesised by several organisms including *E. coli, A. vinelandii, Chromobacterium violaceum,* and *Achromobacter* during growth in the presence of exogenous cyanide and/or under conditions which promote cyanogenesis. In addition cytochrome oxidase *d* exhibits a significantly higher affinity for oxygen than the other cytochrome oxidases, and this is often reflected in a dramatically increased synthesis of this oxidase (and also of its cyanide-sensitive companion, cytochrome a_1) during oxygen-limited growth. Cytochrome *d* is also present in several species of anaerobic bacteria, including the sulphate reducer *Desulphotomaculum* and various fermentative bacteria, where it probably has an important role as an oxygen scavenger. In this context it is interesting to note that cytochrome oxidase *d* is the major functional oxidase in *Azotobacter* spp., which reduce dinitrogen to ammonia via the action of an oxygen sensitive nitrogenase; in these organisms cytochrome oxidase *d* is maximally synthesised during growth under highly aerated conditions and, by terminating a non-energy conserving branch of the terminal respiratory chain, is closely associated with the phenomenon of respiratory protection (Yates and Jones, 1974; Robson and Postgate, 1980).

Cytochrome oxidase *d* has been partially purified from *E. coli, K. aerogenes,* and *Photobacterium phosphoreum* (see Poole, 1983). It contains two molecules each of haems *d* and *b,* which in *E. coli* are distributed between two subunits (M_r 55,000 and 44,000), but no copper. Preliminary reports suggest that the cytochrome oxidase b_{588} · *d* complex from *E. coli* is able to catalyse respiration-linked H^+ ejection when reconstituted into proteoliposomes (Kranz *et al.,* 1982), but whole-cell proton translocation studies and growth yield measurements do not support this view (see Jones *et al.,* 1978).

Relatively little is known about cytochrome oxidase a_1 compared with the other bacterial oxidases (see Jurtshuk *et al.,* 1975; Ingledew, 1978; Poole, 1983). Like cytochrome oxidase aa_3, it is firmly membrane bound and contains two molecules of haem *a* with significantly different E_m values; again no copper is present. Although a combination of photodissociation spectra, photochemical action spectra, and/or kinetic analysis have indicated an oxidase role for cytochrome a_1 in *T. ferrooxidans* and various species of *Acetobacter,* conflicting evidence has been presented for cytochrome a_1 in various other organisms including *E. coli* and *A. vinelandii.* Interestingly, cytochrome a_1 has a non-oxidase role in the respiratory system of the chemolithotroph *Nitrobacter* spp. where it transfers electrons from nitrite to cytochrome *c,* possibly in collaboration with a molybdoprotein (see Aleem, 1977).

The mechanisms by which the various bacterial cytochrome oxidases trap, and subsequently reduce, molecular oxygen have been the subject of extensive investigations. Significant differences are apparent, in particular the ability of cytochrome oxidases o and d (but not aa_3 or a_1) to form stable oxygenated complexes, but further discussion of these properties is beyond the scope of this chapter (for a recent review see Poole, 1983).

V. Taxonomic significance

In recent years, comparative analyses of cytochromes have been used increasingly as a taxonomic tool. Interspecies comparisons have generally been based upon differences either in cytochrome patterns or in cytochrome structures. The former approach is based upon spectrophotometric analyses of whole cells or membranes, principally by measuring reduced *minus* oxidised or reduced + CO *minus* reduced difference spectra (see Sections III,B,1 and 2); a large number of such analyses have been reported over the past 60 years, but unfortunately the quality of the data is extremely variable since many of the reports are incomplete, some spectra have been misinterpreted and the majority of the papers refer only to room temperature studies with their attendant limitations. In contrast, the structural approach has been applied to far fewer species and has been almost entirely restricted to easily purified types of cytochrome c; this method compares these cytochromes on the basis of their primary and tertiary structures (see Section III,C,3), both of which have proved to be particularly informative.

A. Cytochrome patterns

The bacteria are composed of two kingdoms, the eubacteria and the archaebacteria. For the purposes of this discussion, however, they will be subdivided into several broad groupings on the basis of their metabolic behaviour and/or their reaction to the Gram stain: (1) chemoheterotrophs (Gram-positive and Gram-negative eubacteria, archaebacteria), (2) chemolithotrophs (Gram-negative eubacteria, archaebacteria), and (3) phototrophs (Gram-negative eubacteria). A number of general conclusions can be drawn with respect to the cytochrome patterns exhibited by these various groups (Smith, 1968; Meyer and Jones, 1973b; Jones, 1977, 1980; Jones and Meyer, 1982).

Gram-positive chemoheterotrophs that are capable of obtaining energy via aerobic respiration (e.g., the Micrococcaceae, Actinoplanaceae, Streptomycetaceae, most of the Bacilliaceae, and many other families), generally exhibit cytochrome patterns of the $bcaa_3o$ type, i.e., their cytochrome systems superficially resemble those of eukaryote mitochondria with the addition of cytochrome oxidase o. A few species of these bacteria contain only one cytochrome oxidase (e.g., *Brochothrix thermosphacta* contains only aa_3) and a significant number of species lack cytochrome c. Interestingly, it has recently been shown that cytochrome oxidase d is present in some micrococci and coryneforms (see Schleifer and Stackebrandt, 1983). Indeed, although *Arthrobacter globiformis* exhibits the expected $bcaa_3o$ pattern during the logarithmic phase of batch growth (and during carbon-limited continuous culture), this changes to a $bcaa_3oa_1d$ pattern when growth is limited by oxygen; this change is accompanied by the organism losing the ability to accept the Gram stain. The cytochrome systems of Gram-negative chemoheterotrophs (e.g., the Azotobacteriaceae, Vibrionaceae, Enterobacteriaceae, Pseudomonadaceae, and other families) exhibit much less homogeneous patterns during aerobic growth. Most of these organisms cannot synthesise cytochrome oxidase aa_3 and consequently exhibit a cytochrome $bcoa_1d$ pattern, from which cytochrome c is often entirely absent or virtually so. Exceptionally, some species of *Pseudomonas* are able to synthesise cytochrome oxidase aa_3, as also are most methylotrophs, *P. denitrificans,* and the genera *Halobacterium* and *Rhizobium;* these organisms thus exhibit a cytochrome $bcaa_3o$ pattern. *P. fluorescens*, *P. cichorri*, and *P. aptata* do not synthesise a- or d-type cytochromes, and hence exhibit extremely simple bco or bo patterns with their attendant loss of energy coupling at site 3 and/or flexibility of electron transfer at low oxygen concentrations.

When energy is conserved as a result of anaerobic respiration, the chemoheterotrophs generally lack cytochrome oxidases or synthesise considerably decreased amounts. A few obligate anaerobes (e.g., *Propionibacterium* and *Desulphotomaculum*) synthesise low levels of cytochrome oxidases a_1 and/or d, which probably enables them to scavenge molecular oxygen efficiently and thus to maintain a highly anaerobic environment. Usually, however, the cytochrome systems of organisms catalysing anaerobic respiration are characterised by extremely simple patterns of the b or bc types. In the sulphate reducers and methanogens, both of these types of cytochromes have extremely low redox potentials, whereas in *P. denitrificans* and the denitrifying pseudomonads they exhibit normal redox potentials and are accompanied during growth on nitrate or nitrite by cytochrome cd_1.

The chemolithotrophs (e.g., the Nitrobacteraceae, Thiobacillaceae, and some species of *Pseudomonas* or other genera) generally exhibit a cytochrome pattern of the $bcaa_3o$ type; cytochrome c is almost always present and is intimately involved in electron transfer during the oxidation of the high redox potential substrates favoured by many of these organisms. Exceptionally, *Thiobacillus ferro-oxidans* and *Nitrobacter* spp.

contain cytochrome a_1, the former as an oxidase and the latter probably to catalyse the initial oxidation of nitrite. Denitrifying chemolithotrophs (e.g., *T. denitrificans*) of course contain cytochrome cd_1.

The phototrophs (the families Rhodospirillaceae, Chromatiaceae, Chlorobiaceae, Chloroflexaceae, and Cyanobacteriaceae) all contain *b*- and *c*-type cytochromes as part of their cyclic electron transfer systems; various other cytochromes *c* and flavocytochromes *c* are present in the green photosynthetic bacteria where they serve to transfer electrons from sulphide or thiosulphate into the cyclic system. Facultative phototrophs of the genera *Rhodospirillum* and *Rhodopseudomonas* synthesise a conventional respiratory chain during aerobic growth in the dark, the cytochrome system exhibiting a $bcaa_3o$ pattern of the type shown by most Gram-positive heterotrophs and also by the Gram-negative chemoheterotrophs mentioned above; indeed, ribosomal RNA typing has confirmed that quite close phylogenetic relationships exist between these photosynthetic and non-photosynthetic Gram-negative organisms (Stackebrandt and Woese, 1981).

B. Cytochrome structures

The primary sequences of almost forty species of cytochrome *c* from bacteria have now been reported (see Ambler, 1976, 1980; Ambler *et al.*, 1979). The majority of these are type I cytochromes *c* either from facultative phototrophs such as *Rhodospirillum* or *Rhodopseudomonas*, from the obligate phototroph *Chlorobium*, or from a wide variety of chemoheterotrophs including *P. denitrificans*, *A. vinelandii*, *Micrococcus* sp., and several species of *Pseudomonas*. These type I cytochromes can be subdivided into three major groups according to the number of amino acid residues present: (1) long (L) cytochromes with 112–134 residues (i.e., cytochromes c_2 from several species of *Rhodospirillum* and *Rhodopseudomonas*, *P. denitrificans* c_{550}), (2) medium (M) cytochromes with 100–105 residues (i.e., cytochromes c_2 from several species of *Rhodospirillum*, *Rhodopseudomonas*, and also *Rhodomicrobium vanniellii*; mitochondrial *c* also falls into this group), and (3) short (S) cytochromes with 82–86 amino acid residues (i.e., cytochromes c_{551} from *A. vinelandii*, *Rhodospirillum tenue*, *Rhodopseudomonas gelatinosa*, and several species of *Pseudomonas*, and cytochrome c_{555} from *Chlorobium thiosulphatophilum*). It is strikingly clear from these data that some species of *Rhodospirillum* are more closely related, for example, to *P. denitrificans* and to some species of *Rhodopseudomonas* than they are to other species of their own genus. Since there is good evidence that phylogenetic relationships between these organisms have not become blurred by the lateral transfer of genes (Dickerson, 1980; Woese *et al.*, 1980), it seems likely that taxonomic divisions based solely on morphological and metabolic properties may require some re-evaluation. Indeed, it would appear that some Gram-negative chemoheterotrophs are very closely related to certain phototrophs and probably evolved from photosynthetic ancestors via the loss of their light harvesting apparatus.

These taxonomic views are strongly supported by the X-ray studies which have been carried out on four species of cytochrome c from bacteria: *R. rubrum* c_2 (L), *P. denitrificans* c_{550} (L), *P. aeruginosa* c_{551} (S) and *Chlorobium thiosulphatophilum* c_{555} (S) (Dickerson *et al.*, 1976; Timkovitch and Dickerson, 1976; Korszun and Salemme, 1977; Almassey and Dickerson, 1978; Dickerson, 1980). Thus, the structures of the first two cytochromes differ significantly only by the presence of an enlarged loop of approximately 20 additional amino acid residues close to the N-terminal end of the *P. denitrificans* c_{550}, and the other two cytochromes exhibit almost identical structures.

Acknowledgments

R.K.P. thanks the SERC, the Royal Society, and the Nuffield Foundation for financial support, and Professor P. B. Garland for permission to draw on information collected by him for a Biochemical Society Refresher Course. Figure 5 was obtained by R.K.P. in collaboration with Dr. M. V. Jones and Dr. R. I. Scott. C.W.J. thanks the SERC and ICI for financial support. Both authors are indebted to Mrs. Amelia Dunning for her good-humoured and efficient processing of the manuscript.

References

Aleem, M. I. H. (1977). *In* "Microbial Energetics" (B. A. Haddock and W. A. Hamilton, Eds.), Soc. Gen. Microbial. Symp. 27, pp. 351–381. Cambridge University Press, Cambridge.

Almassy, R. J., and Dickerson, R. E. (1978). *Proc. Natl. Acad. Sci. U.S.A.* **75**, 2674–2678.

Ambler, R. P. (1976). *In* "Handbook of Biochemistry and Molecular Biology" (G. D. Tasman, Ed.), 3rd ed., Vol. 3, pp. 292–307. CRC Press, Cleveland.

Ambler, R. P. (1980). *In* "From Cyclotrons to Cytochromes" (N. O. Kaplan and A. Robinson, eds.), pp. 263–280. Academic Press, New York.

Ambler, R. P., Meyer, T. E., and Kamen, M. D. (1979). *Nature (London)* **278**, 661–662.

Anthony, C. (1975). *Biochem. J.* **146**, 289–298.

Bashford, C. L., Barlow, C. H., Chance, B., Haselgrove, J., and Sorge, J. (1982). *Am. J. Physiol.* **242**, C265–C271.

Barrett, J. (1956). *Biochem. J.* **64**, 626–639.

Barrett, J. (1961). *Biochim. Biophys. Acta* **54**, 580–582.

Bragg, P. D. (1980). *In* "Diversity of Bacterial Respiratory Systems" (C. J. Knowles, Ed.), Vol. 1, pp. 115–136, CRC Press, Boca Raton, Florida.

Burton, S. M., Byrom, D., Carver, M. A., Jones, G. D. D., and Jones, C. W. (1983). *FEMS Microbial. Lett.* **17**, 185–190.

Butler, W. L. (1979). *In* "Methods in Enzymology," Vol. 56, pp. 501–515. Academic Press, New York.

Butler, W. L., and Hopkins, D. W. (1970a). *Photochem. Photobiol.* **12**, 439–450.

Butler, W. L., and Hopkins, D. W. (1970b). *Photochem. Photobiol.* **12**, 451–456.

Carver, M. A., and Jones, C. W. (1983). *FEBS Letts.* **155**, 187–191.

Castor, L. N., and Chance, B. (1959). *J. Biol. Chem.* **234**, 1587–1592.

Chance, B. (1953). *J. Biol. Chem.* **202**, 407–416.

Chance, B. (1957). *In* "Methods in Enzymology," Vol. 4, pp. 273–329. Academic Press, New York.

Chance, B. (1961). *In* "Haematin Enzymes" (J. E. Falk, R. Lemberg, and R. K. Morton, Eds.), pp. 433–435. Pergamon, Oxford.

Chance, B. (1966). *J. Franklin Inst.* **282**, 349–356.

Chance, B. (1978). *In* "Methods in Enzymology" (S. Fleischer and L. Packer, Eds.), Vol. 54, pp. 102–111. Academic Press, New York.

Chance, B., and Graham, N. (1971). *Rev. Sci. Instrum.* **42**, 941–945.

Chance, B., and Schoener, B. (1966). *J. Biol. Chem.* **241**, 4567–4573.

Chance, B., Smith, L., and Castor, L. N. (1953). *Biochim. Biophys. Acta* **12**, 289–298.

Chance, B., Graham, N., and Legallais, V. (1975a). *Anal. Biochem.* **67**, 552–579.

Chance, B., Legallais, V., Sorge, J., and Graham, N. (1975b). *Anal. Biochem.* **66**, 498–514.

Chance, B., Moore, J., Powers, L., and Ching, Y. (1982). *Anal. Biochem.* **124**, 239–247.

Chicken, E., Spode, J. A., and Jones, C. W. (1981). *FEMS Microbiol. Lett.* **11**, 181–185.

Choc, M. G., Webster, D. A., and Caughey, W. S. (1982). *J. Biol. Chem.* **257**, 865–869.

Claisse, M. L., Pere-Aubert, G. A., Clavilier, L. P., and Slonimski, P. (1970). *Eur. J. Biochem.* **16**, 430–438.

Cross, A. R., and Anthony, C. (1980). *Biochem. J.* **192**, 421–427.

Culliford, B. J., and Nickolls, L. C. (1964). *J. Forensic Sci.* **9**, 175.

Daniel, R. M. (1970). *Biochim. Biophys. Acta* **216**, 328–341.

Dawson, M. J., and Jones, C. W. (1981). *Biochem. J.* **194**, 915–924.

Deeb, S., and Hager, L. P. (1964). *J. Biol. Chem.* **239**, 1024–1031.

Degn, H., Lundsgaard, J. S., Petersen, L. C., and Ormicki, A. (1980). *Methods Biochem. Anal.* **26**, 47–77.

Dickerson, R. E. (1980). *Nature (London)* **283**, 210–212.

Dickerson, R. E., Timkovitch, R., and Almassey, R. J. (1976). *J. Mol. Biol.* **100**, 473–491.

Downie, J. A., Gibson, F., and Cox, G. B. (1979). *Annu. Rev. Biochem.* **48**, 103–131.

Douzou, P. (1977). "Cryobiochemistry." Academic Press, New York.

Dutton, P. L. (1978). *In* "Methods in Enzymology" (S. Fleischer and L. Packer, Eds.), Vol. 54, pp. 411–435. Academic Press, New York.

Edwards, C., and Lloyd, D. (1973). *J. Gen. Microbiol.* **79**, 275–284.

Elliott, W. B., and Tanski, W. (1962). *Anal. Chem.* **34**, 1672–1673.

Erecinska, M. (1982). *In* "Membranes and Transport" (A. N. Martonosi, Ed.), Vol. 1, pp. 397–404. Plenum, New York.

Estabrook, R. W. (1956). *J. Biol. Chem.* **223**, 781–794.

Estabrook, R. W. (1961). *In* "Haematin Enzymes" (J. E. Falk, R. Lemberg, and R. K. Morton, Eds.), pp. 436–460. Pergamon, Oxford.

Ferguson, S. J., and Sorgato, M. C. (1982). *Annu. Rev. Biochem.* **51**, 155–184.

Fillingame, R. H. (1980). *Annu. Rev. Biochem.* **49,** 1079–1113.

Fleischer, S., and Packer, L. (Eds.) (1978). "Methods in Enzymology," Vol. 54. Academic Press, New York.

Fujita, T. (1966). *J. Biochem. (Tokyo)* **60,** 329–334.

Fuhrhop, J-H., and Smith, K. M. (1975). *In* "Porphyrins and Metalloporphyrins" (K. M. Smith, Ed.), pp. 757–869. Elsevier, Amsterdam.

Garland, P. B. (1977). *In* "Microbial Energetics" (B. A. Haddock, and W. A. Hamilton, Eds.), Soc. Gen. Microbiol. Symp. 27, pp. 1–21. Cambridge Univ. Press, Cambridge.

Garland, P. B., Littleford, S. J., and Haddock, B. A. (1976). *Biochem. J.* **154,** 277–284.

Gennis, R. B., Casey, R. P., Azzi, A., and Ludwig, B. (1982). *Eur. J. Biochem.* **125,** 189–195.

Gibson, J. F., Hadfield, S. G., Hughes, M. N., and Poole, R. K. (1980). *J. Gen. Microbiol.* **116,** 99–110.

Gordon, J., and McLeod, J. W. (1928). *J. Pathol. Bacteriol.* **31,** 185–190.

Gudat, J. C., Singh, J., and Wharton, D. C. (1973). *Biochim. Biophys. Acta* **292,** 376–390.

Haddock, B. A., and Jones, C. W. (1977). *Bacteriol. Rev.* **41,** 47–99.

Haddock, B. A., Downie, J. A., and Garland, P. B. (1976). *Biochem. J.* **154,** 285–294.

Hager, L. P., and Deeb, S. S. (1967). *In* "Methods in Enzymology," Vol. 10, pp. 367–372. Academic Press, New York.

Hagihara, B., and Iizuka, T. (1971). *J. Biochem.* **69,** 355–362.

Hamilton, R., Maguire, D., and McCabe, M. (1979). *Anal. Biochem.* **93,** 386–389.

Hansen, R. E., van Gelder, B. F., and Beinert, H. (1970). *Anal. Biochem.* **35,** 287–292.

Hartridge, H. (1912). *J. Physiol.* **44,** 1–21.

Henry, M. F. (1981). *In* "Cyanide in Biology" (B. Vennesland, E. E. Conn, C. J. Knowles, J. Westley, and F. Wissing, Eds.), pp. 415–436. Academic Press, New York.

Hoffman, P. S., Irwin, R. M., Carreira, L. A., Morgan, T. V., Ensley, B. D., and DerVartanian, D. V. (1980). *Eur. J. Biochem.* **105,** 177–185.

Hoffman, P. S., Morgan, T. V., and DerVartanian, D. V. (1980). *Eur. J. Biochem.* **110,** 349–354.

Holloway, M. R., and White, H. A. (1975). *Biochem. J.* **149,** 221–231.

Horio, T., Higashi, T., Yamanaka, T., Matsuhara, H., and Okunuki, K. (1961). *J. Biol. Chem.* **236,** 944–951.

Horiuchi, K., Kaneko, S., and Asai, H. (1982). *Int. J. Biochem.* **14,** 359–362.

Hubbard, J. A. M., Hughes, M. N., and Poole, R. K. (1983). *FEBS Lett.* **164,** 241–243.

Ingledew, W. J. (1978). *In* "Functions of Alternative Terminal Oxidases" (H. Degn, D. Lloyd, and G. C. Hill, Eds.), pp. 79–87. Pergamon, Oxford.

Ingledew, W. J., and Cobley, J. G. (1980). *Biochim. Biophys. Acta* **590,** 141–158.

Ingledew, W. J., and Poole, R. K. (1984). *Microbiol. Rev.* **48,** 222–271.

Itakagi, E., and Hager, L. P. (1966). *J. Biol. Chem.* **241,** 3687–3695.

Itakagi, E., and Hager, L. P. (1968). *Biochem. Biophys. Res. Commun.* **32,** 1013–1019.

Jones, C. W. (1977). *In* "Microbial Energetics" (B. A. Haddock, and W. A. Hamilton, Eds.), Soc. Gen. Microbiol. Symp. 27, pp. 23–59. Cambridge Univ. Press, Cambridge.

Jones, C. W. (1980). *In* "Microbial Classification and Identification" (M. Goodfellow and R. G. Board, Eds.), pp. 127–138. Academic Press, New York.

Jones, C. W., and Meyer, D. J. (1982). *In* "Handbook of Microbiology" (H. Lechevalier, Ed.), Vol. 4, pp. 583–598. CRC Press, Boca Raton, Florida.

Jones, C. W., and Redfearn, E. R. (1966). *Biochim. Biophys. Acta* **113,** 467–481.

Jones, C. W., Brice, J. M., Downs, A. J. and Drozd, J. W. (1975). *Eur. J. Biochem.* **52,** 265–271.

Jones, C. W., Brice, J. M., and Edwards, C. (1977). *Arch. Microbiol.* **115**, 85–93.
Jones, C. W., Brice, J. M., and Edwards, C. (1978). *In* "Functions of Alternative Terminal Oxidases" (H. Degn, D. Lloyd, and G. C. Hill, Eds.), pp. 89–97. Pergamon, Oxford.
Jones, R. W., and Garland, P. B. (1977). *Biochem. J.* **164**, 195–211.
Jurtshuk, P., and Liu, J. K. (1983). *Int. J. Syst. Bacteriol.* **33**, 887–891.
Jurtshuk, P., Mueller, T. J., and Acord, W. C. (1975). *CRC Crit. Rev. Microbiol.* **3**, 399–468.
Katan, M. B. (1976). *Anal. Biochem.* **74**, 132–137.
Keilin, D. (1925). *Proc. R. Soc. London, Ser. B* **98**, 312–319.
Keilin, D. (1927). *C.R. Soc. Biol. (Paris)* **96**, SP39–68.
Keilin, D. (1929). *Proc. R. Soc. London, Ser. B* **104**, 206–252.
Keilin, D., and Hartree, E. F. (1949). *Nature (London)* **164**, 254–259.
Keilin, D., and Hartree, E. F. (1950). *Nature (London)* **165**, 504–505.
Kita, K., Yamato, I., and Anraku, Y. (1978). *J. Biol. Chem.* **253**, 8910–8915.
Koch, A. L. (1970). *Anal. Biochem.* **38**, 252–259.
Korszun, S. R., and Salemme, F. R. (1977). *Proc. Natl. Acad. Sci. U.S.A.* **74**, 5244–5247.
Kovacs, N. (1956). *Nature (London)* **178**, 703.
Kranz, R. G., Lorence, R. M., Miller, M., Green, G. N., and Gennis, R. B. (1982). *In* "Second European Bioenergetics Conference," pp. 637–638. LBTM-CNRS, Villeurbanne, France.
Kröger, A. (1978). *Biochim. Biophys. Acta* **505**, 129–145.
Kubowitz, F., and Haas, E. (1932). *Biochem. Z.* **255**, 247–277.
Kühn, W., and Gottschalk, G. (1983). *Eur. J. Biochem.* **135**, 89–94.
Kuronen, T., and Ellfolk, N. (1972). *Biochim. Biophys. Acta* **275**, 308–318.
Labbe, P., and Chaix, P. (1971). *Anal. Biochem.* **39**, 322–326.
La Monica, R. F., and Marrs, B. L. (1976). *Biochim. Biophys. Acta* **423**, 431–439.
Le Gall, J., DerVartanian, D. V., and Peck, H. D. (1979). *Curr. Top. Bioenerg.* **9**, 237–265.
Lemberg, R., and Barrett, J. (1973). "The Cytochromes." Academic Press, New York.
Lemberg, R., Morell, D. B., Newton, N., and O'Hagan, J. E. (1961). *Proc. R. Soc. London Ser. B* **155**, 339–355.
Liu, C. Y., and Webster, D. A. (1974). *J. Biol. Chem.* **249**, 4261–4266.
Lloyd, D. (1974). "The Mitochondria of Microorganisms." Academic Press, New York.
Lloyd, D., and Scott, R. I. (1983). *Anal. Biochem.* **128**, 21–25.
Ludwig, B. (1980). *Biochim. Biophys. Acta* **594**, 177–189.
Ludwig, B., and Schatz, G. (1980). *Proc. Natl. Acad. Sci. U.S.A.* **77**, 196–200.
MacMunn, C. A. (1886). *Philos. Trans. R. Soc. London* **177**, 167–298.
Marrs, B., and Guest, H. (1973). *J. Bacteriol.* **114**, 1045–1051.
Meyer, D. J., and Jones, C. W. (1973a). *Int. J. Syst. Bacteriol.* **23**, 459–467.
Meyer, D. J., and Jones, C. W. (1973b). *FEBS Lett.* **33**, 101–105.
Meyer, O., and Schlegel, H. G. (1983). *Annu. Rev. Microbiol.* **37**, 277–310.
Miller, D. J., and Wood, P. M. (1983). *FEMS Microbiol. Lett.* **20**, 323–326.
Miller, M. J., and Gennis, R. B. (1983). *J. Biol. Chem.* **258**, 9159–9165.
Mitchell, P. (1966). *Biol. Rev. Cambridge Philos. Soc.* **41**, 445–502.
Morrey, J. R. (1968). *Anal. Chem.* **40**, 905–914.
Murphy, M. J., Siegel, L. M., Kamin, H., and Rosenthal, D. (1973). *J. Biol. Chem.* **248**, 2801–2814.
Nicholls, P. (1983). *Trends Biochem. Sci.* **8**, 353.
Nuner, J. H., and Payne, W. J. (1973). *Anal. Biochem.* **52**, 355–362.
Olson, T., and Hooper, A. B. (1983). *FEMS Microbiol. Lett.* **19**, 47–50.
Orri, Y., and Morita, M. (1977). *J. Biochem.* **81**, 163–168.

Palmer, G. (1967). *In* "Methods in Enzymology," Vol. 10, pp. 594–609. Academic Press, New York.

Payne, W. J., Rowe, J. J., and Sherr, B. F. (1980). *In* "Nitrogen Fixation" (W. E. Newton and W. H. Orme-Johnson, Eds.), Vol. 1, pp. 29–42. University Park Press, Baltimore.

Poole, R. K. (1981). *FEBS Lett.* **133**, 255–259.

Poole, R. K. (1983). *Biochim. Biophys. Acta* **726**, 205–243.

Poole, R. K., and Chance, B. (1981). *J. Gen. Microbiol.* **126**, 277–287.

Poole, R. K., Lloyd, D., and Chance, B. (1979a). *Biochem. J.* **184**, 555–563.

Poole, R. K., Waring, A. J., and Chance, B. (1979b). *Biochem. J.* **184**, 379–389.

Poole, R. K., Sivaram, A., Salmon, I., and Chance, B. (1982). *FEBS Lett.* **141**, 237–241.

Poole, R. K., Kumar, C., Salmon, I., and Chance, B. (1983a). *J. Gen. Microbiol.* **129**, 1335–1344.

Poole, R. K., van Wielink, J. E., Baines, B. S., Reijnders, W. N. M., Salmon, I., and Oltmann, L. F. (1983b). *J. Gen. Microbiol.* **129**, 2163–2173.

Rawlinson, W. A., and Hale, J. H. (1949). *Biochem. J.* **45**, 247–255.

Reid, G. A., and Ingledew, W. J. (1979). *Biochem. J.* **182**, 465–472.

Reid, G. A., and Ingledew, W. J. (1980). *FEBS Lett.* **109**, 1–4.

Ribbons, D. W., Hewett, A. J. W., and Smith, F. A. (1968). *Biotechnol. Bioeng.* **10**, 238–242.

Robson, R. L., and Postgate, J. R. (1980). *Annu. Rev. Microbiol.* **34**, 183–207.

Rosenthal, O., and Cooper, D. Y. (1967). *In* "Methods in Enzymology," Vol. 10, pp. 616–629. Academic Press, New York.

Salmon, I., and Poole, R. K. (1980). *J. Gen. Microbiol.* **117**, 315–326.

Savitsky, A., and Golay, J. E. (1964). *Anal. Chem.* **36**, 1627–1639.

Schleifer, K. H., and Stackebrandt, E. (1983). *Annu. Rev. Microbiol.* **37**, 143–187.

Scott, R. I., and Poole, R. K. (1982). *J. Gen. Microbiol.* **128**, 1685–1696.

Shipp, W. S. (1972). *Arch. Biochem. Biophys.* **150**, 482–488.

Shipp, W. S., Piotrowski, M., and Friedman, A. E. (1972). *Arch. Biochem. Biophys.* **150**, 473–481.

Shrager, R. I. (1983). *Photochem. Photobiol.* **38**, 615–617.

Silvestrini, M. C., Colosimo, A., Brunori, M., Walsh, T. A., Barber, D., and Greenwood, C. (1979). *Biochem. J.* **183**, 701–709.

Smith, L. (1955). *In* "Methods in Enzymology," Vol. 2, pp. 732–740. Academic Press, New York.

Smith, L. (1968). *In* "Biological Oxidations" (T. P. Singer, Ed.), pp. 55–122. Wiley Interscience, New York.

Smith, M. L., and Caughey, W. S. (1978). *In* "Methods in Enzymology," Vol. 52, pp. 421–436. Academic Press, New York.

Stackebrandt, E., and Woese, C. R. (1981). *In* "Molecular and Cellular Aspects of Microbiol Evolution" (M. J. Carlile, J. F. Collins, and B. E. B. Moseley, Eds.), Soc. Gen. Microbiol. Symp. 32, pp. 1–31. Cambridge Univ. Press, Cambridge.

Swank, T. R., and Burris, H. R. (1969). *Biochim. Biophys. Acta* **180**, 473–483.

Taylor, B. L. (1983). *Annu. Rev. Microbiol.* **37**, 551–573.

Thomas, P. E., Ryan, D., and Levin, W. (1976). *Anal. Biochem.* **75**, 168–176.

Timkovitch, R., and Dickerson, R. E. (1976). *J. Biol. Chem.* **251**, 4033–4046.

Tyree, B., and Webster, D. A. (1978). *J. Biol. Chem.* **253**, 7635–7637.

van Gelder, S. R. (1978). *In* "Methods in Enzymology," Vol. 53, pp. 126–128. Academic Press, New York.

Vanneste, W. H., and Vanneste, M-T. (1965). *Biochem. Biophys. Res. Comm.* **19**, 182–186.

C. W. JONES AND R. K. POOLE

van Verseveld, H. W., and Stouthamer, A. H. (1978). *Arch. Microbiol.* **118,** 13–20.
van Wielink, J. E., Oltmann, L. F., Leeuwerik, F. J., de Hollander, J. A., and Stouthamer, A. H. (1982). *Biochim. Biophys. Acta* **681,** 177–190.
Vincent, J-C., Kumar, C., and Chance, B. (1982). *Anal. Biochem.* **126,** 86–93.
von Jagow, G., Weiss, H., and Klingenberg, M. (1973). *Eur. J. Biochem.* **33,** 140–157.
Warburg, O., and Negelein, E. (1929). *Biochem. Z.* **214,** 64–100.
Warburg, O., Negelein, E., and Haas, E. (1933). *Biochem. Z.* **266,** 1–8.
Weber, P. C., Salemme, F. R., Mathews, F. S., and Bethge, P. H. (1981). *J. Biol. Chem.* **256,** 7702–7704.
White, D. C., and Sinclair, P. R. (1970). *Adv. Microbiol. Physiol.* **5,** 173–211.
Wikstrom, M. K. F. (1971). *Biochim. Biophys. Acta* **253,** 332–345.
Wikstrom, M., Krab, K., and Saraste, M. (1981). "Cytochrome Oxidase. A Synthesis." Academic Press, New York.
Willison, J. C., and Haddock, B. A. (1981). *FEMS Microbiol. Lett.* **10,** 53–57.
Wilson, D. F. (1967). *Archiv. Biochem. Biophys.* **121,** 757–768.
Wilson, D. F. (1978). *In* "Methods in Enzymology," Vol. 54, pp. 1–3. Academic Press, New York.
Wilson, D. F., and Erecinska, M. (1978). In "Methods in Enzymology," Vol. 53, pp. 191–201. Academic Press, New York.
Wood, P. M. (1980). *Biochem. J.* **189,** 385–391.
Woese, C. R., Gibson, J., and Fox, G. E. (1980). *Nature (London)* **283,** 212–214.
Yamagutchi, S. (1934). *Acta Phytochim.* **8,** 157–172.
Yaoi, H., and Tamiya, H. (1928). *Proc. Imp. Acad. (Tokyo)* **4,** 436–439.
Yates, M. G., and Jones, C. W. (1974). *Adv. Microbiol. Physiol.* **11,** 97–135.

11

Analysis of Isoprenoid Quinones

M. D. COLLINS

*Department of Microbiology, National Institute for Research in Dairying,
Shinfield, Reading, England*

I. Introduction

The respiratory quinones represent an important group of isoprenoid lipids which occur in the cytoplasmic membrane of most prokaryotes. Two major structural groups of bacterial isoprenoid quinones can be recognised: the naphthoquinones and benzoquinones. Naphthoquinones can be divided further into two major types: phylloquinones and menaquinones (Fig. 1).

Phylloquinone (formerly vitamin K_1) was first isolated by Dam and associates in 1939 from alfalfa, and shown by MacCorquodale *et al.* (1939b) to be 2-methyl-3-phytyl-1,4-naphthoquinone (Fig. 1a). Phylloquinone is normally associated with the green tissues of plants and its presence within the prokaryotes seems to be limited to the cyanobacteria (see Collins and Jones, 1981a). Menaquinone (formerly vitamin K_2) was first isolated from bacterially putrified fish meal in 1939 (MacCorquodale *et al.*, 1939a; McKee *et al.*, 1939) and was thought to be 2-methyl-3-farnesyl-farnesyl-1,4-naphthoquinone (McKee *et al.*, 1939; Binkley *et al.*, 1940). It was not, however, until 1958 that the compound from fish meal was shown to have a farnesyl–geranyl–geranyl side chain (i.e., containing 7 isoprene units, MK-7 in present nomenclature), with the related compound MK-6 present in only minor amounts (Isler *et al.*, 1958). Today,

METHODS IN MICROBIOLOGY
VOLUME 18

a

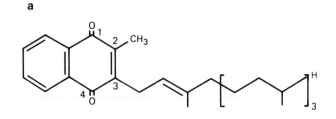

b

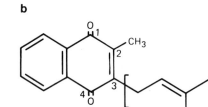

c

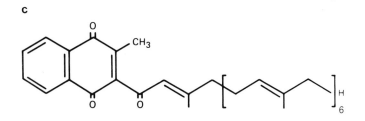

d

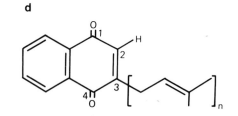

e

Fig. 1. Structures of (a) phylloquinone, (b) menaquinone, (c) chlorobiumquinone, (d) de-methylmenaquinone, and (e) methionaquinone.

naturally occurring menaquinones form a large class of molecules in which length of the C-3 multiprenyl side chain varies from 1 up to 15 isoprene units (i.e., 5–75 carbon atoms) (Collins and JOnes, 1981a). In addition to variations in the length of the side chain, varying degrees of saturation or hydrogenation of the C-3 multiprenyl side chain of menaquinones have been reported. Gale et al. (1963) reported that the principal menaquinone isolated from *Mycobacterium phlei* possessed a C_{45} side chain with one isoprene unit saturated [abbreviated MK-9(H_2)] whereas the major menaquinone from *Corynebacterium diphtheriae* was shown to be MK-8(H_2) (Scholes and King, 1964, 1965). Subsequent studies have shown that such dihydrogenated menaquinones are widespread amongst corynebacteria and mycobacteria, whereas even more highly saturated (tetra-, hexa-, and octahydrogenated) menaquinones are common in certain actinomycetes (see Collins and Jones, 1981a for review). Other modifications of the menaquinone series include the replacement of the methyl group at C-2 with a hydrogen atom, as in demethylmenaquinones (Fig. 1d). To date, demethylmenaquinones with polyprenyl side chains varying in length from 1 up to 10 isoprene units have been described. Interestingly, although demethylmenaquinones are found in many Gram-negative eubacteria, they have so far been found in only a single Gram-positive species (*Streptococcus faecalis*) and are apparently absent from archaebacteria (see Collins and Jones, 1981a). An unusual menaquinone, chlorobiumquinone, is produced by certain green photosynthetic bacteria (Frydman and Rapaport, 1963; Redfearn and Powls, 1968). Whereas initially chlorobiumquinone was thought to be a modified MK-7 in which the first methylene of the normal polyisoprenyl side chain was absent (Frydman and Rapaport, 1963), subsequent work by Powls et al. (1968) has shown that chlorobiumquinone is in fact 1'-oxomenaquinone-7 (Fig. 1c). Chlorobiumquinone is the only example of a bacterial isoprenoid quinone containing a side-chain carbonyl group.

Ishii et al. (1983) reported the presence of a novel sulphur-containing "menaquinone" (designated methionaquinone) in a thermophilic hydrogen bacterium. Methionaquinone was shown to be 2-methylthio-3-VI,VII-tetrahydroheptaprenyl-1,4-naphthoquinone (Ishii et al., 1983) (Fig. 1e). Collins and Langworthy (1983) recently reported the isolation of a new quinone, designated thermoplasmaquinone, from the thermophilic, acidophilic archaebacterium *Thermoplasma acidophilum*. Thermoplasmaquinone possessed an ultraviolet absorption spectrum similar to those of the menaquinone series. Although the detailed structure of thermoplasmaquinone was not established, mass spectral analyses indicated the quinone had seven isoprene units and a molecular formula $C_{47}H_{66}O_2$ (Collins and Langworthy, 1983). Carlone and Anet (1983) and Collins et al.

(1984) reported the isolation of a novel methyl-substituted menaquinone-6 from *Campylobacter* species. This methyl-substituted menaquinone had similar UV and mass spectral (except containing 6 isoprene units) characteristics to thermoplasmaquinone. Carlone and Anet (1983) on the basis of proton nuclear magnetic resonance (^{1}H NMR) analyses established that the methyl substituent of this modified MK-6 from *Campylobacter* was in the 5 or 8 position of the ring (i.e., 2,[5 or 8]-dimethyl-3-farnesyl-farnesyl-1,4-naphthoquinone). ^{1}H NMR studies (see Section II,C and Fig. 20) have shown that thermoplasmaquinone-7 also contains a methyl group in the 5 or 8 position of the ring and corresponds to 2,[5 or 8]-dimethyl-3-farnesyl-geranyl-geranyl-1,4-naphthoquinone. The naphthoquinones from *Thermoplasma acidophilum* and *Campylobacter* species thus represent a new series of methyl-substituted menaquinones. 2,[5 or 8]-dimethyl-3-farnesyl-farnesyl-1,4-naphthoquinone (thermoplasmaquinone-6) has also recently been isolated from the rumen anaerobe *Wolinella succinogenes* (Collins and Fernandez, 1984).

The second major class of bacterial isoprenoid quinones are the benzoquinones, of which there are two main types, the ubiquinones and plastoquinones (Fig. 2). The discovery of ubiquinones (or coenzyme Q) was the result of independent studies by Morton and associates in the United Kingdom and Crane and colleagues in the United States. Ubiquinones contain a 2,3-dimethoxy-5-methyl-1, 4-benzoquinone nucleus with a multiprenyl side chain in position 6 (Fig. 2a) (see Crane, 1961; Isler *et al.*, 1961 for early literature). To date, ubiquinones with multiprenyl side chains varying in length from 1 up to 15 isoprene units have been de-

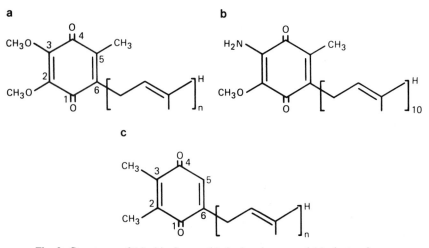

Fig. 2. Structures of (a) ubiquinone, (b) rhodoquinone, and (c) plastoquinone.

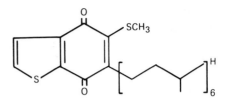

Fig. 3. Structure of caldariellaquinone.

scribed. It is worth noting that unlike menaquinones, ubiquinones with partially saturated side chains do not occur in bacteria. Within the prokaryotes, ubiquinones have a more restricted distribution than menaquinones (see Collins and Jones, 1981a). Although ubiquinones are present in many Gram-negative eubacteria, they have not as yet been reported in Gram-positive eubacteria nor archaebacteria. An unusual "ubiquinone," designated rhodoquinone, was isolated from *Rhodospirillum rubrum* by Glover and Threlfall (1962). Moore and Folkers (1966) showed that rhodoquinone was an amino quinone related to ubiquinone-10, with the methoxyl group *para* to the side chain replaced by an NH_2 group (Fig. 2b). The other major bacterial benzoquinone is plastoquinone (Fig. 2c). Plastoquinone was originally isolated by Kofler (1946) from alfalfa but was not identified. Only after the discovery of ubiquinone was the compound found once more by Crane (1959) and shown to be a 2,3-dimethylbenzoquinone with a C-5 side chain containing nine isoprene units (Kofler *et al.*, 1959; Erickson *et al.*, 1959). Although plastoquinone is found in the photosynthetic tissues of higher plants and red, brown, and green algae, its distribution within the prokaryotes is apparently restricted to the cyanobacteria (Collins and Jones, 1981a).

De Rosa *et al.* (1977) reported the presence of a novel sulphur-containing quinone, designated caldariellaquinone, in the thermophilic and acidophilic bacterium *"Caldariella acidophilum."* Caldariellaquinone was shown to be 6-(3,7,11,15,19,23-hexamethyl-tetracosyl)-5-methylthiobenzo-[b]-thiopen-4,7-quinone (Fig. 3) (De Rosa *et al.*, 1977). Caldariellaquinone has recently been shown to be present in all three species of the genus *Sulfolobus* (Collins and Langworthy, 1983).

II. Analysis of isoprenoid quinones

A. Extraction and purification

In choosing the methods to be employed for the extraction and purification of menaquinones, ubiquinones, and related quinones one must take into account the susceptibility of these compounds to degradation. Isoprenoid quinones are quite rapidly photo-oxidised in the presence of oxygen and strong light. They are also particularly susceptible to alkaline conditions (the last limitation rules out alkaline saponification). Thus, it is preferable to conduct extraction and subsequent purification procedures fairly rapidly, avoiding extremes of pH and strong light. It is not, however, necessary to work in dim light or a nitrogen atmosphere. Isoprenoid quinones are "free" lipids and can be readily extracted from bacterial cells using normal lipid solvents, such as petroleum ether, chloroform, acetone, diethyl ether, etc. Adequate extraction of these components can be achieved with any one of these solvents or a mixture of two. Extraction procedures normally yield a complex mixture of lipids containing small amounts of nonlipid material from which the quinones can be readily purified by simple chromatographic procedures.

The extraction method most commonly employed by the author is direct extraction of dry cells with chloroform/methanol (see Collins *et al.*, 1977). Approximately 50–100 mg of lyophilised cells are extracted with a small volume (~25–50 ml) of chloroform/methanol (2 : 1, v/v) for approximately 2 h using a magnetic stirrer. The cell/solvent mixture is passed through a filter funnel (Whatman filter paper) to remove cell debris, collected in a flask, and evaporated to dryness under reduced pressure at ~40°C on a rotary evaporator. The lipid extract is resuspended in a small volume of chloroform/methanol and applied (with a Pasteur pipette or syringe) as a uniform streak (~3–5 cm long) to a silica gel F_{254} TLC plate (e.g., Kieselgel $60F_{254}$, Merck Art. 5735; plastic-backed cut to 10×10 cm) (Fig. 4). The TLC plate is developed in hexane diethyl ether (85 : 15, v/v) and purified quinones revealed (as dark brown/purple bands on a green fluorescent background) by *brief* irradiation with ultraviolet light (254 nm). The quinones are scraped from the plate with a spatula and eluted (through a sintered glass filter) with a suitable solvent (e.g., chloroform). The purified quinones are finally evaporated to dryness under a stream of nitrogen gas (oxygen free). Using the above chromatographic system menaquinones and ubiquinones occur at $R_f \sim 0.7$ and 0.3, respectively (Collins *et al.*, 1977). Vitamin K_1 (Sigma Art. V 3501) and Coenzyme Q_{10} (Sigma Art. C 9629) may be applied to the TLC plate as respective standards. Thermoplasmaquinone (methylmenaquinones) and caldariellaquinone occur at $R_f \sim 0.8$ and 0.6, respectively, in the above system. The procedure gives good quantitative recovery of the quinones from lyophilised cells. However, it should be noted that the presence of water can cause a reduction in the yield of quinones. Purified quinones awaiting analysis should therefore preferably be kept free of moisture and refrigerated.

Alternatively, quinones can be extracted directly from wet cells using

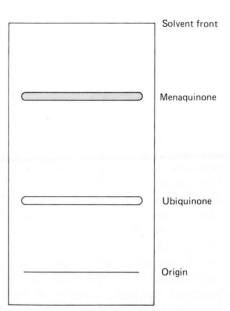

Solvent front

Menaquinone

Ubiquinone

Origin

Fig. 4. Thin-layer chromatographic separation of menaquinones and ubiquinones (solvent hexane diethyl ether, 85 : 15, v/v).

acetone (see Kroppenstedt, 1982a). Approximately 1–2 g wet cells are suspended in 25 ml cold acetone (cell pellets should first be homogenised) and broken by ultrasonification (1–2 min). Cell debris is removed by centrifugation and the lipid extract reduced to ~0.5 ml on a rotary evaporator (~40°C). The extract is then filtered (e.g., Hamilton syringe filter), applied to a TLC plate, and purified as described above.

B. Chromatographic analysis

1. Reverse-phase partition thin-layer chromatography

The composition of natural mixtures of bacterial quinones can be investigated using partition chromatography. Separation of compounds by partition chromatography is generally on the basis of relative solubilities, which in the case of homologous series such as ubiquinones and menaquinones is determined by the length and degree of hydrogenation of the multiprenyl side chain. In the case of ubiquinones and menaquinones, which are highly hydrophobic, separations are achieved using a non-polar (hydrophobic) stationary phase and polar developing solvents (referred to as reverse-phase partition chromatography). Although reverse-phase partition-paper chromatography has in the past been extensively used for the separation of quinone mixtures (for example, see Sommer and Kofler, 1966) this method has now been superseded by reverse-phase partition thin-layer chromatography (RPTLC). The latter method achieves better separation of prenologues in a shorter time and facilitates the purification of larger amounts of quinone. RPTLC utilises a stationary phase of non-polar, non-volatile hydrocarbon, such as paraffin, silicone, or hexadecane and polar developing mixtures such as acetone/water or alcohol/water mixtures. The adsorbants most commonly employed include silica gel G, kieselguhr, or cellulose. The TLC plates are normally impregnated with non-polar phase by immersing the activated plate into a solution of the hydrocarbon (e.g., 5% paraffin in petroleum ether or diethyl ether) and allowing the solvent to evaporate. Alternatively, plates may be coated with non-polar phase by developing the plate in a closed chamber containing a solution of %5 hydrocarbon in petroleum ether. After development the volatile solvent is allowed to evaporate. It should be noted, however, that impregnation of the adsorbant with hydrocarbon is time consuming and the method is further hampered by lack of reproducibility and loading problems. Recently, ready-made high-performance reverse-phase plates in which the non-polar phase (C_{18}) is chemically bonded to the adsorbant have been employed for ubiquinone and menaquinone analyses (Collins *et al.*, 1980a; Collins and Jones, 1981b). The use of ready-made reverse-phase plates avoids the tedious process of achieving satisfactory impregnation with hydrocarbons. Furthermore, the resolution, reproducibility, and developing times ($\sim$15 min) are superior to those of conventional plates. The high loading capacity of the ready-made plates together with the presence of a chemically bonded non-polar phase (which alleviates further chromatographic steps to remove contaminating paraffin) also facilitates preparative work: the gel is simply scraped from the plate and quinones eluted with solvent.

(a) RPTLC of ubiquinones. Ubiquinone samples are suspended in a small volume of acetone and spotted onto a Merck HPTLC RP18F$_{254}$ (Art. 13724) reverse-phase plate (10 × 10 cm) with a fine pipette or glass capillary ($\sim$1 cm from bottom of plate). The plate is developed in a standard chromatography tank containing acetone acetonitrile (80 : 20, v/v) (Collins and Jones, 1981b). After development, separated components are revealed by irradiation with ultraviolet light (254 nm): quinones appear as dark spots on a blue fluorescent background.

The above system allows clear separation of Q-1 through to Q-15. Identification of components can be achieved by comparing R_f values with known ubiquinones. Standard ubiquinones (Q-6 through to Q-10) may be obtained from Sigma Chemical Company. Since bacterial ubiquinones

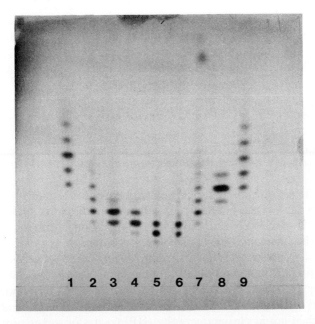

Fig. 5. Reverse-phase partition TLC of ubiquinones: (1) standard mixture Q-6 through to Q-10; (2) *"Legionella morrisei"*; (3) *L. pneumophila;* (4) *L. micdadei;* (5) *Legionella* sp. (strain Gilbart, F268); (6) *L. feelei;* (7) *"L. morrisei"* (fluorescent strain); (8) *L. oakridgensis;* (9) Q-6 through to Q-10 (Merck RP18-F_{254} plate 10 × 10 cm; solvent acetone acetonitrile, 80 : 20, v/v).

vary only in the length of the C-6 multiprenyl side chain this method is sufficient for the qualitative determination of ubiquinone mixtures (see Collins and Jones, 1981b). The separation of ubiquinone prenologues by RPTLC is illustrated in Fig. 5.

(b) RPTLC of menaquinones. Menaquinone samples are suspended in a small volume of acetone and spotted (~1 cm from bottom) onto a Merck HPTLC RP18F_{254} (Art. 13724) reverse-phase plate (10 × 10 cm). The plate is developed in a standard chromatography tank containing acetone water (99 : 1, v/v) (Collins *et al.,* 1980a) and separated components observed under ultraviolet light (254 nm). Menaquinones appear as dark spots on a blue fluorescent background.

The above method allows clear separation of unsaturated mena-quinones (MK-1 through to MK-14). The chromatographic system also facilitates the separation of menaquinone components with the same number of isoprene units but differing degrees of saturation [e.g., MK-9, MK-9(H_2), MK-9(H_4), MK-9(H_6), MK-9(H_8)]. Saturation of one double bond causes a negative shift in R_f value approximately equivalent to 0.8 of the effect of adding one isoprene unit. Although the described procedure facilitates the qualitative determination of simple unsaturated menaquinones (for example, as found in Gram-negative and many Gram-positive taxa), if partially saturated menaquinones are present (for example, as found in most coryneform and actinomycete taxa), this method by itself is insufficient for their characterisation.

2. Reverse-phase partition high-performance liquid chromatography

The thin-layer chromatographic techniques described in the preceding section provide only qualitative or semi-quantitative (relying upon the intensity of spots on chromatograms) data. Although this is sufficient for organisms which have only one (or less commonly two) major component, in the case of certain taxa (e.g., legionellae, actinomycetes etc.) which contain complex mixtures of prenologues valuable taxonomic information may be lost. Until recently, the vast majority of quinone studies generated only qualitative/semi-quantitative data. The advent of high-performance liquid chromatography (HPLC) has however provided taxonomists with a rapid and sensitive means of obtaining quantitative quinone profiles. The separation of quinone mixtures by HPLC is normally by the reverse-phase partition (RPHPLC) mode (alternatively, a silver ion exchanger may be employed; see Section B,3) using octadecylsilane (ODS) as stationary phase with polar eluents. As with RPTLC, separation by this method is determined by the lipophilic character of the quinones, which depends mainly on the length and degree of saturation of the multiprenyl side chain.

(a) RPHPLC of ubiquinones.   Ubiquinone samples (injection solvent 1-chlorobutane) are separated isocratically with either a Spherisorb ODS (5 μm) (Laboratory Data Control) or LiChrosorb RP-18 (5 μm) (Merck) prepacked column (250 $\times$ 4.6 mm i.d.). Ubiquinones are monitored by their absorption at 270 nm using a UV detector and eluted with methanol/1-chlorobutane (100 : 10, by vol) at 1.5 ml min^{-1}. The elution times and area of each peak may be determined by a computer integrator. Although the above system provides excellent resolution of commonly encountered bacterial ubiquinones, the level of non-polar phase may be increased (e.g., methanol/1-chlorobutane, 100 : 20, v/v) to facilitate the elution of exceptionally long ubiquinone prenologues (for example, Q-13, Q-14, etc., as found in some legionellae; see Collins and Gilbart, 1983). This not only speeds up elution times of the very large prenologues but prevents spreading of the peaks. The separation of some natural mixtures of bacterial ubiquinones by RPHPLC is illustrated in Fig. 6.

As with RPTLC identification of the vast majority of components can be achieved by comparing retention times with known ubiquinones (Q-6 through to Q-10 may be obtained from Signa Chemical Co.). Since large ubiquinones (>Q-10) are not commercially available, preparations of known structure from bacterial strains may be used as standards. Since the relationship between common logarithm of netto retention time (t_R) and the number of isoprene units is linear, the length of the side chain of any unknown ubiquinone can be determined graphically. A straight-line graph is obtained by plotting netto retention times of standard ubiquinone homologues against their number of isoprene units on semilogarithmic paper. As can be seen in Fig. 7, the t_R of any unknown component can then be used to determine the number of isoprene units in the side chain. Alternatively, the length of the side chain of an unknown ubiquinone may be calculated by the following equation when a known Q-n_s is present:

$$n_x = n_s + K[\log t_R(n_x) - \log t_R(n_s)]$$

where n_s = number of isoprene units in standard, n_x = number of isoprene units in unknown ubiquinone, $t_R(n_s)$ = netto retention time of standard

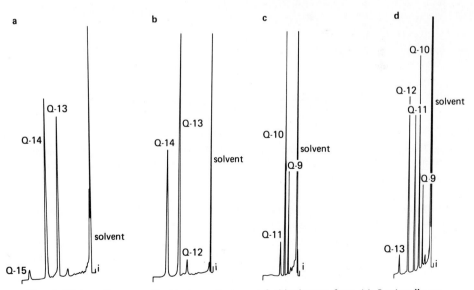

Fig. 6. High-performance liquid chromatograms of ubiquinones from (a) *Legionella* sp. (strain Gilbart, F268); (b) *L. feelei;* (c) *L. oakridgensis;* (d) *L. sp.* 1308. Spherisorb ODS (5 μm) column using methanol/1chlorobutane (100:20, v/v) as mobile phase (1.5 ml min^{-1}).

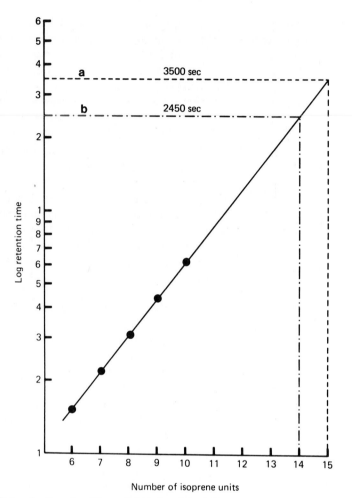

Fig. 7. Determination of multiprenyl chain length by RPHPLC by plotting logarithm of netto retention times of ubiquinone homologues (a) Q-15 and (b) Q-14 against the number of isoprene units.

ubiquinone (Q-n_s), $t_R(n_x)$ = netto retention time of unknown ubiquinone, K = constant (determined for particular chromatographic conditions). The K value can be calculated for a particular chromatographic system from the retention times of known standard ubiquinones. Using the data from Fig. 7 as an illustration $K \simeq 6.683$. Thus if t_R of an unknown ubiquinone is 3500 s, substituting this figure together with the t_R of a standard ubiquinone (e.g., Q-10, $t_R = 620$) in the above equation gives a value of $n = 15.02$ for the unknown component (i.e., Q-15).

(b) *RPHPLC of menaquinones*. Menaquinones (injection solvent 1-chlorobutane) may be separated isocratically on either a Spherisorb ODS (5 μm) (Laboratory Data Control) or LiChrosorb RP-18 (5 μm) (Merck) pre-packed column (250 × 4.6 mm i.d.) using methanol/1-chlorobutane (100 : 10, v/v) as mobile phase (1.5 ml min^{-1}) (Collins *et al.*, 1982a). Alternatively, acetonitrile/tetrahydrofuran (70 : 30, v/v) (Kroppenstedt, 1982b) or methanol/isopropyl ether (3 : 1, v/v) (Tamaoko *et al.*, 1983) at 1 ml min^{-1} may be used as eluents. Menaquinones are monitored at 270 nm using a UV detector and quantitation achieved with a computer integrator. Although the above chromatographic systems give good separation of most commonly encountered menaquinones, the relative proportions of the solvents of the mobile phase may be altered to facilitate certain separations. For example, improved separation of very small unsaturated (e.g., MK-1 to MK-6) or hydrogenated [e.g., MK-6 (H$_2$), MK-5 (H$_2$), etc.] menaquinones can be achieved by using pure methanol as eluent. Conversely, the elution of very large menaquinones (e.g., MK-13, MK-14, MK-15, etc.) can be facilitated by increasing the proportion of non-polar solvent (e.g., 1-chlorobutane, isopropyl ether) in the mobile phase.

As with RPTLC, identification of menaquinones is done by comparing retention times with menaquinones of known structure. Since menaquinones are not commercially available [with the exception of vitamin K$_1$, MK-4 (H$_6$)], menaquinone preparations from bacterial strains of established structure (see Collins and Jones, 1981a, for survey of structures) may be used as standards. The identification of unknown menaquinones can also be assisted by plotting netto retention times (t_R) of menaquinone homologues against their number of isoprene units on semilogarithmic paper. As the relationship between common logarithm of netto retention times and the number of isoprene units is almost linear in equally hydrogenated menaquinones, five parallel lines are formed for the five homologous menaquinone series (i.e., MK-n (H$_0$, H$_2$, H$_4$, H$_6$, H$_8$) (see Fig. 8 and Kroppenstedt, 1984). The separation of some natural mixtures of bacterial menaquinones by RPHPLC is illustrated in Figs. 9–11.

It should be noted that if reliable quantitative fingerprints are to be obtained by RPHPLC it is essential to perform analyses on freshly prepared quinone samples. The presence of water in the sample (during storage) or exposure to strong light can result in the formation of artefacts (which may elute close to short quinones and the solvent peak). Partially hydrogenated quinones are particularly susceptible to degradation. It is thus preferable to perform analyses as soon as possible after completion of sample purification (i.e., same day or as soon as possible thereafter). Quinone samples should be freeze-dried or sealed under nitrogen for long-term storage.

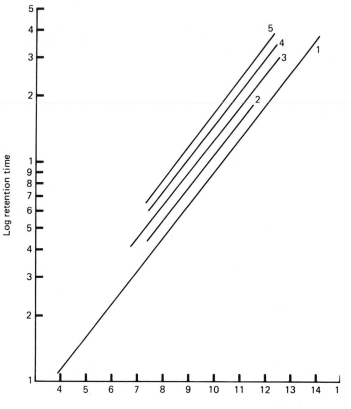

Fig. 8. Plots of logarithm of netto retention times of homologous series of menaquinones against the number of isoprene units. (1) MK $-$ n(H$_0$); (2) MK $-$ n(H$_2$); (3) MK $-$ n(H$_4$); (4) MK $-$ n(H$_6$); (5) MK $-$ n(H$_8$).

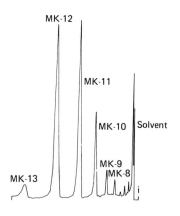

Fig. 9. High-performance liquid chromatogram of unsaturated menaquinones from *Aureobacterium liquefaciens*. Spherisorb ODS (5 μm) column using methanol/1chlorobutane (100 : 10, v/v) as mobile phase (1.5 ml min^{-1}).

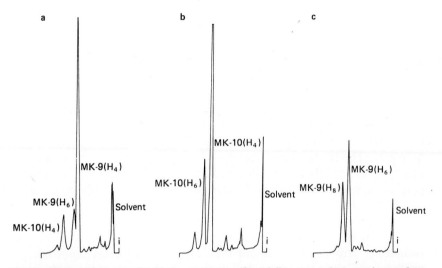

Fig. 10. High-performance liquid chromatogram of partially saturated menaquinones from (a) *Actinoplanes philippinensis;* (b) *Micromonospora purpurea;* (c) *Sporichthya polymorpha.* Spherisorb ODS (5 μm) column using methanol/1chlorobutane (100 : 10, v/v as mobile phase (1.5 ml min^{-1}).

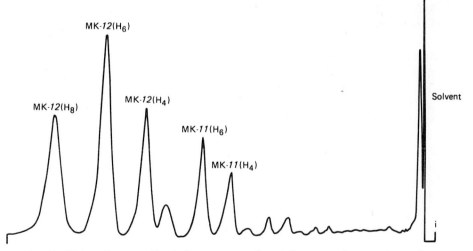

Fig. 11. High-performance liquid chromatogram of partially saturated menaquinones from *Micromonospora echinospora* subsp. *pallida.* Spherisorb ODS (5 μm) column using methanol/1chlorobutane (100 : 10, v/v) as mobile phase (1.5 ml min^{-1}).

3. Silver-ion-modified chromatography

The majority of organisms synthesise menaquinones of the simple unsaturated types (see Collins and Jones, 1981a, for review), the structures of which can be determined readily by reverse-phase partition chromatography. However, a number of taxa (e.g., certain actinomycetes and coryneforms) produce menaquinones with partially saturated side chains. Reverse-phase partition chromatography by itself is insufficient for the unequivocal structural determination of these hydrogenated menaquinones. The latter should therefore be used in conjunction with either physicochemical methods (such as mass spectrometry) or chromatographic procedures on which the menaquinones display different separation behaviour.

The capacity of silver ions (Ag^+) to form loose complexes with electrons of olefins has been known for a long time. The formation of such complexes provides a means for the separation of menaquinone components according to the length and degree of saturation of the multiprenyl side chain. Silver ion chromatographic analysis of menaquinones is normally performed using thin-layer methods (e.g., Dunphy et al., 1971). Alternatively separations can be obtained on a liquid chromatograph using silver loaded ion exchangers (Kroppenstedt, 1982a,b).

(a) Silver-ion TLC of menaquinones.   Plates (20 × 20 cm) may be prepared in the normal way from a slurry of silica gel G containing 3–5% $AgNO_3$ (dissolve $AgNO_3$ in distilled water before addition of adsorbent) (Dunphy et al., 1971). For rhodamine 6G impregnated silver layers (to aid visualisation) 1 ml of a 2% rhodamine 6G solution is mixed with the aqueous $AgNO_3$ before addition of the adsorbent. Plates are allowed to dry in the air for 10 min followed by heating in an oven at 100°C for 40 min. Alternatively, precoated silica gel plates (e.g., Camlab Polygram Sil G/UV$_{254}$ or Merck kieselgel 60F$_{254}$ Art. 5735) may be impregnated with Ag^+ by immersing (~1 min) in a solution of 2–3% $AgNO_3$ in aqueous acetone (dissolve $AgNO_3$ in 10 ml H_2O and add 90 ml acetone) (Collins, 1984). Plates are allowed to dry at room temperature (10 min) and heated in an oven at 80°C for ~15 min. To prevent darkening, the plates should be stored in a dark, moisture-free container until required. By using one of a variety of solvent systems [e.g., methyl ethyl ketone/hexane (15 : 85, v/v) or methanol/benzene (5 : 95, v/v) Beau et al., 1966; Dunphy et al., 1971] it is possible to achieve the separation of a variety of menaquinone types. As with RPTLC, small unsaturated menaquinones run ahead of longer unsaturated menaquinones. However, in contrast to RPTLC saturation of a double bond causes a *positive* shift in R_f value (approximately equivalent to 1.5 to 2 times the effect of one isoprene unit) (see Dunphy *et al.*, 1971). Thus by comparing the behaviour (amplitute and direction of change) on both Ag$^+$ TLC and RPTLC it is possible to elucidate the structure of unknown hydrogenated menaquinones.

(*b*) *Silver ion HPLC of menaquinones.* The composition of menaquinone mixtures may also be investigated by Ag$^+$ HPLC (Kroppenstedt, 1982a,b). Menaquinones (injection solvent ethanol) are separated on a stainless-steel column (250 × 4.6 mm i.d.) custom packed (Chrompack Nederland, Middleburg, Netherlands) with silver-loaded Nucleosil 10SA (Macherey, Nagel and Co., Düren, Federal Republic of Germany). The column is maintained at 65°C by use of a precision thermostat and menaquinones eluted with methanol (1.5 ml min^{-1}). Menaquinones are monitored at 270 nm and elution times and peak areas determined by a computer integrator. Due to the strong nature of the olefinic–Ag$^+$ bonds, the above system can be used only for menaquinones with nine or less double bonds. For the separation of very long unsaturated menaquinones (e.g., MK-9 to MK-12) shorter columns (125 × 4.6 mm i.d.) packed with Nucleosil 5SA may be employed.

Using Ag$^+$ HPLC separation of menaquinone components is determined primarily by the length and degree of unsaturation of the C-3 multiprenyl side chain. Menaquinones with fewer double bonds are eluted before those with a larger number. Similarly, as expected from Ag$^+$ TLC data, menaquinones with the same number of double bonds but longer isoprenoid chain length are eluted earlier [e.g., MK-9(H$_2$) is eluted before MK-8 even though both contain eight double bonds]. The above Ag$^+$ HPLC system also facilitates the separation of cis and trans isomers (*cis*-MKs elute before *trans*-MKs) and stereoisomers of hydrogenated menaquinones (i.e., partially hydrogenated menaquinones which differ only in the *position* of saturation of the C-3 multiprenyl side chain) (Kroppenstedt, 1982a,b, 1984; Fischer *et al.*, 1983)[1]. The separation of menaquinone mixtures by Ag$^+$ HPLC is illustrated in Fig. 12.

---

[1] The detailed structure (points of saturation) of several hydrogenated menaquinones have now been established by mass spectrometry and NMR [e.g., MK-9 (II-H$_2$) in *Mycobacterium phlei* (Azerad and Cyrot-Pelletier, 1973); MK-9 (II,III,IX-H$_6$) and MK-9 (II,III,VIII,IX-H$_8$) in *Streptomyces olivaceus* (Batrakov and Bergelson, 1978); MK-9 (II,III,VIII-H$_6$) in *Actinomadura madurae* (Yamada *et al.*, 1982a); MK-9 (II,III,VIII,IX-H$_8$) in *Streptomyces albus* (Yamada *et al.*, 1982b)]. Assignment of unknown menaquinones may be achieved by comparing retention times with stereoisomers of established structure.

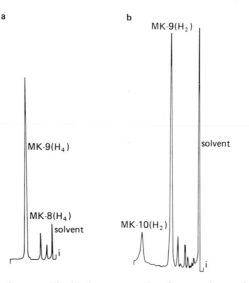

Fig. 12. Ag⁺ high-performance liquid chromatography of menaquinones from a, *Oerskovia turbata;* b, *Corynebacterium bovis*. Silver-loaded Nucleosil 5SA (125 × 4.6 mm i.d.) column using methanol.

(c) Silver ion HPLC of ubiquinones. Ubiquinones (injection solvent ethanol) can be separated on a stainless steel column (125 × 4.6 mm i.d.) packed with silver-loaded Nucleosil 5SA (Macherey, Nagel and Co., Düren, Federal Republic of Germany). The column is maintained at a constant temperature by use of a precision thermostat and quinones eluted with methanol (1–2 ml min⁻¹). Ubiquinones are monitored at 270 nm and quantitation achieved with a computer integrator. As with menaquinones, the column temperature and flow rate can be altered to facilitate particular separations. For the resolution of most commonly encountered bacterial ubiquinones (i.e., Q-6 to Q-10) a column temperature of 65°C and flow rate of 1.5 ml min⁻¹ is suitable. Improved separation of very small prenologues (Q-6) can be achieved by reducing the column temperature (e.g., 50°C) or flow rate. For the separation of ubiquinones with more than 10 isoprene units a shorter column may be employed.

Identification of unknown ubiquinones can be achieved by comparing retention times with ubiquinones of known structure. Alternatively, as with RPHPLC assignments can be determined graphically (by plotting netto R_t vs number of isoprene units, see Fig. 13) or using the equation given in Section II,B,2,a. It is worth re-emphasising that since partially saturated ubiquinones are not found in bacteria (see Collins and Jones, 1981a) RPHPLC is by itself normally sufficient for ubiquinone determina-

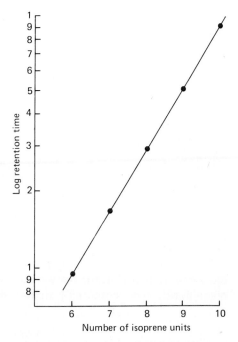

Fig. 13. Plots of logarithm of netto retention times (Ag⁺HPLC) of homologous series of ubiquinones against the number of isoprene units.

tion. The described Ag⁺ HPLC system, however, may be used as a simple alternative to mass spectrometry to check the structural assignments made by RPHPLC of ubiquinones for which no standards are commercially available (e.g., Q-5, Q-4, etc.). The separation of some ubiquinone mixtures by Ag⁺ HPLC is illustrated in Fig. 14.

C. Physicochemical analysis

1. Ultraviolet spectroscopy

The class or category to which an unknown isoprenoid quinone belongs can be established readily using ultraviolet spectroscopy. Details of the absorption characteristics of the major types of bacterial quinones are given in Table I. Menaquinones and phylloquinone exhibit very characteristic absorption spectra with five absorption maxima (λ_{max}) at 242, 248, 260, 269, and 326 nm and one point of inflection at 238 nm (Fig. 15). Absorption bands at 242, 248, and 238 (inf) nm are due to benzenoid contributions while bands at 260 and 269 are due to quinone contributions. The ultraviolet spectrum of demethylmenaquinones is distinct from that

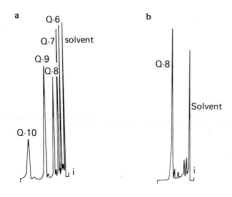

Fig. 14. Ag⁺ high-performance liquid chromatography of (a) standard ubiquinone mixture Q-6 through to Q-10 and (b) ubiquinones from *Vitreoscilla stercoraria*. Silver-loaded Nucleosil 5SA (125 × 4.6 mm i.d.) column using methanol (a, 65°C and b, 60°C) as eluant (1.5 ml min⁻¹).

of menaquinones, with λ_{max} at 242, 248, 254, 263, 325, and 238 (inf) nm (Table I). The replacement of the CH_3 in position 2 of menaquinones by SCH_3, as in methionaquinone, causes a marked change in the spectrum with three absorption maxima produced at 238, 262, and 310 nm (Ishii *et al.*, 1983). In contrast, the methyl-substituted menaquinones (thermoplasmaquinones) have essentially similar ultraviolet characteristics to mena-

TABLE I

Ultraviolet absorption characteristics of menaquinones, ubiquinones, and related compounds

Quinone	Solvent	γ_{max} (nm)
Menaquinone	Iso-octane	242, 248, 260, 269, 326, 238(inf)[a]
Phylloquinone	Iso-octane	242, 248, 260, 269, 326, 238(inf)
Demethylmenaquinone	Iso-octane	242, 248, 254, 263, 326, 238(inf)
Chlorobiumquinone	Ethanol	254, 265(inf)
Thermoplasmaquinone[b]	Iso-octane	242, 248, 259, 269, 337
Ubiquinone	Ethanol	275
Rhodoquinone	Cyclohexane	251, 280, 320, 500(inf)
	Ethanol	253, 283, 320, 500(inf)
Plastoquinone	Iso-octane	254, 262
Caldariellaquinone	Methanol[c]	241, 283, 333
	Iso-octane[d]	237, 272, 278, 322
Methionaquinone[e]	Ethanol	238, 262, 310

[a] inf, point of inflection.

[b] Thermoplasmaquinone-7(2,[5 or 8]-dimethyl-3-heptaprenyl-1,4-naphthoquinone) from *Thermoplasma acidophilum* (Collins and Langworthy, 1983).

[c] De Rosa *et al.* (1977).

[d] Collins and Langworthy (1983).

[e] 2-methylthio-3-VI,VII-tetrahydroheptaprenyl-1,4-naphthoquinone (Ishii *et al.*, 1983).

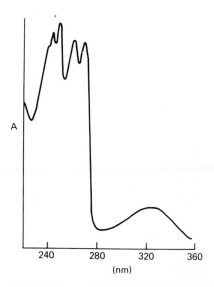

Fig. 15. Ultraviolet spectrum of MK-7 (isooctane).

240   280   320   360

(nm)

quinones except the absorption band at 326 nm shifts about 10 nm to longer wavelengths (Collins and Langworthy, 1983). Chlorobiumquinone (1'-oxo-MK-7) exhibits absorption characteristics quite distinct from those of other bacterial naphthoquinones with λ_{max} at 254 nm and a point of inflection at 265 nm (Powls *et al.*, 1968).

Ubiquinones produce a very simple ultraviolet spectrum with a single λ_{max} at 270 to 275 nm (Fig. 16). The replacement of the methoxyl group in position 3 of ubiquinone by an amino group, as in rhodoquinone, causes a marked change in the spectrum with three absorption maxima at 251, 280, and 320 nm and one point of inflection at 500 nm (Glover and Threlfall, 1962). Plastoquinone can be readily distinguished from other bacterial

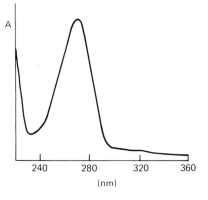

Fig. 16. Ultraviolet spectrum of Q-10 (isooctane).

240   280   320   360

(nm)

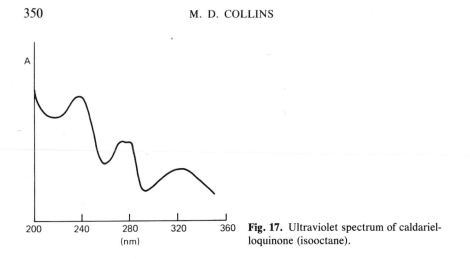

Fig. 17. Ultraviolet spectrum of caldariel-loquinone (isooctane).

benzoquinones in producing two absorption maxima at 254 and 262 nm (Table I).

The unusual sulphur-containing quinone, caldariellaquinone, from the thermophilic and acidophilic archaebacteria *Sulfolobus* exhibits absorption characteristics quite distinct from those of other described respiratory quinones. Caldariellaquinone has been reported to possess three absorption maxima at 241, 283, and 333 nm (methanol) (De Rosa *et al.,* 1977). Collins and Langworthy (1983) however have shown that the broad absorption at 283 nm in fact corresponds to two fused peaks [λ_{max} (isooctone) 237, 272, 278, and 322 nm] (Fig. 17).

2. Mass spectrometry

The most powerful technique for the structural determination of isoprenoid quinones is mass spectrometry. This technique not only provides very accurate molecular weights of unidentified quinones but also structural information such as the nature of the ring system and length and degree of saturation of the multiprenyl side chain. Mass spectrometry is also at present the only reliable means of determining the exact points of saturation of partially hydrogenated menaquinone side chains (See Section II,B).

When subjected to mass spectrometry, ubiquinones produce a base peak at *m/z* 235, with a second intense peak at *m/z* 197. The peak at *m/z* 235 can be assigned to that of ion a (Fig. 18) which is highly stabilised by virtue of extensive charge delocalisation. The peak at *m/z* 197 is generated from the corresponding hydroquinone and can formally be represented as ion b (Fig. 18). The base peak in the mass spectrum of plastoquinones

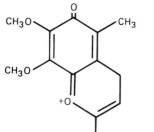

a, m/z 235

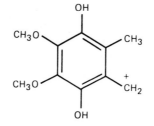

b, m/z 197

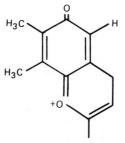

c, m/z 189

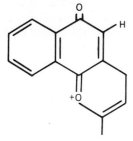

d, m/z 151

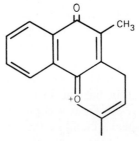

e, m/z 225

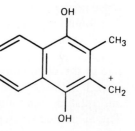

f, m/z 187

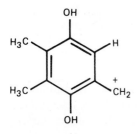

g, m/z 211

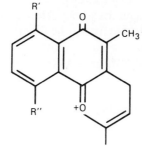

h, m/z 239

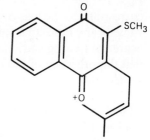

i, m/z 257

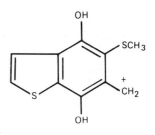

Fig. 18. Mass spectral nuclear fragments (R' and R'' correspond to H and CH₃, not necessarily respectively).

j, m/z 225

occurs at m/z 189, with a second intense peak at m/z 151, and are similarly derived from the 2,3-dimethyl-1,4-benzoquinone nucleus (ions c and d, Fig. 18). Menaquinones (including phylloquinone) exhibit a fragmentation pattern in which m/z 225 (ion e corresponding to ion a of ubiquinones, Fig. 18) is the most prominant peak. Peaks of lower intensity at m/z 187 (ion f corresponding to ion b of ubiquinones) and sometimes at m/z 186 are also observed (Fig. 18). In contrast, demethylmenaquinones and thermoplasmaquinones (2,[5 or 8]-dimethylmultiprenyl-1,4-naphthoquinones) have corresponding nuclear fragments at m/z 211 (ion g, Fig. 18) and 172, and at m/z 239 (ion h, Fig. 18) and 201 (Collins and Langworthy, 1983; Carlone and Anet, 1983), respectively. Methylthionaphthoquinone produces intense peaks at m/z 257 and 211 in the low mass region (Ishii et al., 1983). The fragment peak at m/z 257 is probably derived from the methylthio-1,4-naphthoquinone nucleus (analogous to ion e of menaquinones) and can be assigned to that of ion i (Fig. 18) whereas the peak at m/z 211 is possibly due to the elimination of $S=CH_2$ (produced by hydrogen transfer) from ion i. It is thus evident that the base peaks in ubiquinones, menaquinones, and related compounds result from a common mode of fragmentation. The m/z values of these peaks facilitate the identification of the various isoprenoid quinone classes. Strong peaks corresponding to molecular ions (M^+) are observed in the high mass region of all isoprenoid quinones (see Figs. 19 and 20). Peaks of lower intensity are also observed at M^+-15 and correspond to the loss of a methyl group from the molecular ions. The m/z value of the molecular ions facilitates the determination of the length and degree of hydrogenation of the multiprenyl side chains. It is worth noting that the presence of intense M^+ peaks also facilitates accurate molecular weight determination. Although high-resolution measurements of M^+ peaks are not normally necessary, the latter are essential for the elucidation of the molecular formulae of any new quinones (for example, see Ishii et al., 1983; Collins and Langworthy, 1983). High-resolution measurements are also useful for distinguishing between compounds which have similar low-resolution unit M^+ values but differing molecular formulae. For example, thermoplasmaquinone-7 (2, [5 or 8]-dimethyl-3-heptaprenyl-1,4-naphthoquinone) and chlorobiumquinone (1'-oxomenaquinone-7) both produce a molecular ion at m/z 662 (Powls et al., 1968; Collins and Langworthy, 1983). The M^+ of thermoplasmaquinone-7, however, has an accurate mass of 662.5063 (molecular formula $C_{47}H_{66}O_2$) whereas the M^+ of chlorobiumquinone has a mass 662.4699 (molecular formula $C_{46}H_{62}O_3$). A series of low-intensity peaks are observed in the mass spectra of isoprenoid quinones due to fragmentation of the side chain. Indeed, the pattern of fragmentation of the multiprenyl substituent of menaquinones and related compounds is characteristic of the cracking

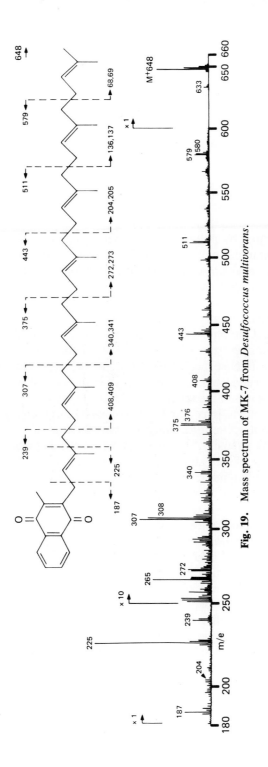

Fig. 19. Mass spectrum of MK-7 from *Desulfococcus multivorans*.

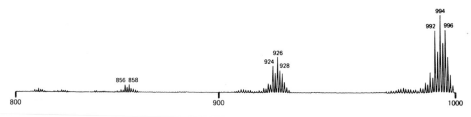

Fig. 20. Peaks corresponding to molecular ions in the mass spectrum of menaquinones from *Micromonospora echinospora* subsp. *pallida*.

pattern exhibited by polyisoprenoid chains in general. Fragmentation of the side chain under electron impact involves solely diallylic bonds. Using the mass spectrum of MK-7 from *Desulfococcus multivorans* as an illustration (Fig. 19) fragmentation of the side chain involves the loss of a terminal isoprenyl unit (m/z 579; 648–69) followed by five successive losses of 68 mass units (m/z 511, 443, 375, 307, 239).

The unusual sulphur-containing terpenoid, caldariellaquinone, produces a mass spectrum quite distinct from those of other respiratory quinones. An intense peak (derived from the corresponding hydroquinone) is observed at m/z 225 (ion j, Fig. 18) and corresponds to ions b and f of ubiquinones and menaquinones, respectively (Fig. 18). Peaks structurally analogous to ions a and e (base peaks) of ubiquinones and menaquinones do not occur due to the presence of a saturated side chain. Peaks of comparable intensity, however, occur at m/z 192 and 193 formed by the elimination of SH and S, respectively, from ion j. As with menaquinones and ubiquinones strong peaks corresponding to molecular ions are present in the high-mass region; peaks of similar intensity, however, are observed at M^+-32, formed by the elimination of S from the molecular ion (Collins and Langworthy, 1983). It is worth noting that the characteristic side-chain fragmentation pattern of menaquinones and related compounds (Fig. 18) is not observed in the caldariellaquinone due to the presence of a C_{30} saturated isoprenoid side chain.

Certain bacteria (for example, coryneforms, actinomycetes, etc.) are rich in partially hydrogenated menaquinones (see Collins and Jones, 1981a). As noted earlier, fragmentation of the multiprenyl side chain involves solely diallylic bonds. Saturation of olefinic bonds within the side chain causes a marked alteration in the fragmentation pattern. Although this alteration in the cracking pattern may provide an "insight" into the position of saturation of relatively simple hydrogenated menaquinones (for example, MK-9(II-H$_2$), see Azerad and Cyrot-Pelletier, 1973) it is not normally possible to determine the exact points of saturation directly by mass spectrometry. Menaquinones can, however, be converted to chromenyl acetates and subjected to ozonolysis. Reduction of the ozonides of the corresponding aldehydes, followed by mass spectrometry can, however, provide information on the exact points of saturation of the side chain. Using this procedure the structures of several hydrogenated menaquinones have now been established [e.g., MK-9(II-H$_2$), Azerad and Cyrot-Pelletier, 1973; MK-9(II,III,IX-H$_6$) and MK-9(II,III,VIII,IX-H$_8$), Batrakov et al., 1976; Batrakov and Bergelson, 1978; MK-9(II,III,VIII-H$_6$), Yamada et al., 1982a; MK-5(V-H$_2$), Collins and Widdel, 1984]. The position of saturation in the recently described tetrahydrogenated methylthionaphthoquinone has been shown to be in sixth and seventh units of the isoprenoid side chain [i.e., MTK-7(VI,VII-H$_4$)] (Ishii et al., 1983).

Although mass spectrometry is the most precise and sensitive tool for the structural determination of isoprenoid quinones, it must be emphasised that this technique does *not* provide accurate quantitative information. The intensity of peaks corresponding to molecular ions may vary considerably and does not always accurately reflect the relative proportions of prenologues in a mixture. This problem becomes particularly acute when either large involatile isoprenologues (e.g., MK-11, MK-12, MK-13, MK-14 etc.) and/or complex mixtures of hydrogenated [e.g., MK-11(H$_4$), MK-11(H$_6$), MK-11(H$_8$), MK-12(H$_4$), MK-12(H$_6$), MK-12(H$_8$); Fig. 20] quinones are present (see Collins and Kroppenstedt, 1983; Collins and Gilbart, 1983; Collins et al., 1983a). Mass spectrometry should therefore preferably be used in conjunction with high-performance liquid chromatographic techniques. Certain quinones such as ubiquinones also have a tendency to form $(M + 2)^+$ (hydroquinones). These $(M + 2)^+$ peaks can be confused with molecular ions of ubiquinones containing a partially saturated side chain. The absence of partially hydrogenated ubiquinones can, however, be established by either reverse-phase TLC or HPLC.

3. Nuclear magnetic resonance spectroscopy

Nuclear magnetic resonance (NMR) spectroscopy is a powerful tool in the structural determination of isoprenoid quinones. This technique yields information about the chemical environment of hydrogen atoms, the number of hydrogen atoms in each kind of environment, and the structure of groups adjacent to each hydrogen atom. Spectra are normally recorded in deuterated chloroform or carbon tetrachloride with tetramethylsilane (TMS) as an internal standard.

Menaquinones and related 2,3- disubstituted naphthoquinones (e.g., chlorobiumquinone, methionaquinone, etc.) exhibit characteristic complex absorption at δ 7.5 to 8.1[2] due to the presence of four adjacent ring protons (see Table II). Broad absorption at δ 5.0 to 5.2 is attributable to olefinic protons whereas the presence of a doublet at δ 3.3 to 3.4 is due to a methylene group adjacent to the ring (*N.B.* the latter is characteristic of all isoprenoid quinones containing a double bond in the first C_5 unit). The methyl substituents at the C-2 position of menaquinone and phylloquinone produce a singlet at δ 2.1 to 2.2. Menaquinones display strong bands in the δ 1.5 to 2.0 region due to methylene and methyl groups adjacent to double bonds. The methylenes appear around δ 1.9 to 2.0 and the olefinic methyls from δ 1.5 to 1.8. The latter show a complexity not seen in phylloquinone. In phylloquinone and partially saturated menaquinones signals occur at δ 1.2 to 1.3 for methylene groups and at δ 0.8 to 0.9 for protons of methyl groups on saturated carbon atoms. The number of isoprenoid units and degree of hydrogenation of the side chain can be determined by comparison of the proton ratios of the various functional units. Demethylmenaquinones produce essentially similar spectra to menaquinones except that a signal occurs at ca. δ 6.7 due to quinoid hydrogen whilst the singlet δ 2.1 to 2.2 due to C-2 methyl is absent (see Table II). The spectrum of methionaquinone is similar to that of hydrogenated menaquinones except for the ring —CH_3 resonance. A singlet is observed at δ 2.6 due to the presence of SCH_3 (ring) whereas the singlet at δ 2.1 to 2.2 attributable to C-2 methyl is absent (see Ishii *et al.*, 1983, and Table II).

Thermoplasmaquinones (methyl-substituted menaquinones) exhibit characteristic NMR spectra (Carlone and Anet, 1983). The proton spectrum of thermoplasmaquinone-7 (abbreviated TPQ-7) from *Thermoplasma acidophilum* is illustrated in Fig. 21. Complex absorption in the δ 7.4 to 8.0 region is due to the presence of three aromatic protons. The splitting pattern of these aromatic protons indicates the additional methyl group is in a peri position on the aromatic ring (C-5 or C-8). The methyl group causes a shielding of the three aromatic protons with the *ortho* proton (H_3) the most shielded (Fig. 21). The doublet at δ 8.0 (J=7.5 Hz) corresponds to a peri proton (i.e., H_1 in position C-5 and/or C-8) (Fig. 21) whereas the doublet at δ 7.45 (J=7.5 Hz) and triplet at δ 7.52 (J=7.5 Hz) correspond to aromatic protons at C-6 and C-7 (not necessarily respectively[3]). The singlet at δ 2.73 is due to the presence of a methyl group at

---

[2] Chemical-shift parameter δ is dimensionless and is expressed in parts per million (ppm). Alternatively, when TMS is used as a reference, shifts are often reported as τ value, where $\tau = 10.00 - \delta$.

[3] If the methyl group is at C-8 then the triplet (δ 7.52) and doublet (δ 7.45) correspond to aromatic H at C-6 and C-7, respectively; if the methyl is at C-5 then the triplet (δ 7.52) and doublet (δ 7.45) correspond to aromatic H at C-7 and C-6, respectively.

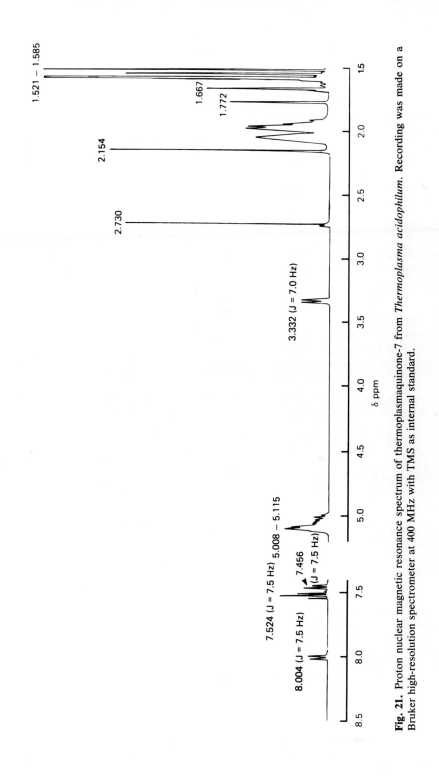

Fig. 21. Proton nuclear magnetic resonance spectrum of thermoplasmaquinone-7 from *Thermoplasma acidophilum*. Recording was made on a Bruker high-resolution spectrometer at 400 MHz with TMS as internal standard.

TABLE II

Chemical shifts (δ) and splitting patterns for ^{1}H NMR spectra of menaquinones and related 1,4-naphthoquinones[a]

Assignment	Phyllo-quinone	Unsaturated menaquinone	Partially saturated menaquinone	Demethylmen-aquinone
Aromatic hydrogens	7.6–8.1(m)[f]	7.6–8.2(m)	7.6–8.2(m)	7.6–8.1(m)
Quinonoid hydrogen (C-2)	—	—	—	6.7(t)[g]
Olefinic hydrogen adjacent to carbonyl	—	—	—	—
Olefinic hydrogens	5.0–5.2(t)[h]	5.0–5.2(m)	5.0–5.2(m)	5.0–5.2(m)
Allylic methylene at C-1 (next to ring)	3.3–3.4(d)	3.3–3.4(d)	3.3–3.4(d)	3.3(d)
SCH$_3$ (ring)	—	—	—	—
Benzenoid methyl (C-5 or C-8)	—	—	—	—
Quinonoid methyl (C-2)	2.1–2.2(s)	2.1–2.2(s)	2.1–2.2(s)	—
Allylic methylenes	1.9–2.0(m)	1.9–2.1(m)	1.9–2.1(m)	1.9–2.1(m)
trans-Methyl 1st isoprene unit	1.8	1.8	1.8	1.8
cis end of chain methyl	—	1.65–1.7	1.65–1.7[k]	1.65–1.7
trans internal methyl groups	—	1.52–1.59(bs)	1.52–1.59(bs)	1.52–1.59(bs)
Aliphatic methy-lenes and methine groups	1.1–1.3	—	1.1–1.3	—
Alliphatic methyl groups (i.e., methyl groups on saturated carbon atoms)	0.8–0.9	—	0.8–0.9	—

[a] In CDCl$_3$ unless otherwise stated.
[b] Data from Powls et al. (1968).
[c] Data from Carlone and Anet (1983); solvent acetone-d_6.
[d] Data from TPQ-7 of Thermoplasma acidophilum (Fig. 21).
[e] Data from Ishii et al. (1983); solvent not cited.
[f] s, singlet; d, doublet; t, triplet; m, multiplet, b, broad.

TABLE II—*continued*

	Thermoplasmaquinones		
Chlorobiumquinone[b]	TPQ-6[c]	TPQ-7[d]	Methionaquinone[e]
7.73–8.06(m)	7.593(d)	7.456(d)	7.79–8.08(m)
	7.645(t)	7.524(t)	
	7.970(d)	8.004(d)	
—	—	—	—
6.15(s)	—	—	—
4.92(m)	5.016–5.12(m)	5.008–5.115(m)	5.06(m)
—	3.356(d)[i]	3.332(d)[i]	3.6(d)
—	—	—	2.64(s)
—	j	2.730(s)	—
2.28(s)	2.165(s)	2.154(s)	—
1.99–2.08(m)	1.889–2.025(m)	1.894–2.044(m)	1.97(m)
2.22	1.794(bs)	1.772(bs)	1.81
1.66	1.640	1.667	—
1.58(s)	1.534–1.581(bs)	1.521–1.585(bs)	1.57
—	—	—	1.28
—	—	—	0.87

[g] Due to long-range splitting.
[h] Due to single olefinic proton.
[i] $J = 7$ Hz.
[j] Signal due to benzenoid-CH_3 obscured by solvent lines (Carlone and Anet, 1983).
[k] Signal absent if terminal isoprene unit is saturated (see Collins and Widdel, 1984).

TABLE III

Chemical shifts and splitting patterns for ^{1}H NMR spectra
of ubiquinones

Assignment	δ values[a]
Olefinic hydrogens	5.0–5.15(b)
Methoxyl groups (C-2,C-3)	3.9–4.1(s)
Allylic methylene adjacent to ring (C-1′)	3.18–3.21(d)
Allylic methylenes and ring methyl (C-5)	1.98–2.03
trans-Methyl 1st isoprene unit	1.75–1.77(s)
cis end of chain methyl	1.67–1.70(s)
trans internal methyl groups	1.56–1.62(bs)

[a] As in Table II.

C-5 or C-8 of the ring (see Fig. 20). This latter peak is characteristic of thermoplasmaquinones, and serves to distinguish these molecules from menaquinones which lack this signal (Collins and Fernandez, 1984). It is worth noting that in the ^{1}H NMR spectrum of TPQ-6 from *Campylobacter* (Carlone and Anet, 1983) this signal was obscured by solvent lines (CH_3COCHD_2). The signal due to the benzenoid methyl (C-5 or C-8) is, however, readily visible if $CDCl_3$ is employed as solvent (see Fig. 21). The remaining signals in the spectrum of thermoplasmaquinone-7 are essentially similar to those of menaquinones and correspond to the C-2 ring methyl (δ 2.15, singlet) and contributions from the C-3 multiprenyl side chain [i.e., δ 5.0 to 5.12, olefinic protons; δ 3.33 ($J = 7$ Hz), —CH_2— adjacent to ring; δ 1.91 to 2.04, —CH_2— allylic; δ 1.77, *trans*-CH_3 next to ring; δ 1.67, *cis*-CH_3 end of chain; δ 1.52 to 1.59, *trans*-CH_3] (Fig. 21).

The NMR spectra of ubiquinones differ from those of menaquinones in lacking bands due to aromatic ring protons (δ 7.4 to 8.1 in menaquinones). In addition, ubiquinones exhibit intense signals (δ 3.9 to 4.0, singlet) due to the presence of two methoxy groups. The remainder of the signals in the spectra of ubiquinones are similar to those of menaquinones and related compounds and are due to the presence of a multiprenyl side chain (see Sommer and Kofler, 1966; Table III). Plastoquinones similarly lack the complex absorption due to adjacent ring protons although a signal due to single proton at the C-5 of the ring is observed at δ 6.5 (triplet due to long-range splitting). Broad signals from the two ring methyls occur at about δ 2.0 but these may overlap with bands due to side-chain methylenes (see Sommer and Kofler, 1966).

The ^{1}H NMR spectrum of caldariellaquinone is quite distinct from those of other respiratory quinones. Two doublets centered at δ 7.58 and 7.51 ($J = 5.0$ Hz) derived from ring protons and a singlet at δ 2.62 due to

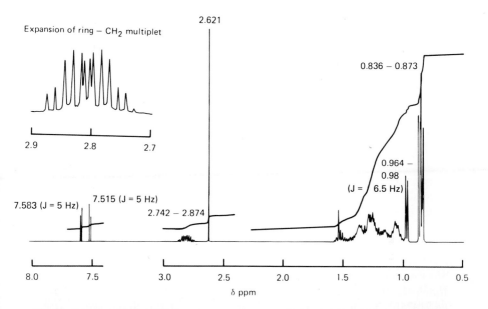

Fig. 22. Proton nuclear magnetic resonance spectrum of caldariellaquinone from *Sulfolobus acidocaldarius*. Recording was made on a Bruker high-resolution spectrometer at 400 MHz with TMS as internal standard.

the presence of SCH$_3$ are observed (see Fig. 22). A multiplet centred at δ 2.8 ($J = 5.5$ Hz) is due to the methylene group adjacent to the ring whereas a broad absorption centred at δ 1.2 corresponds to methylene and methine protons. Signals from the protons of methyl groups on saturated carbon atoms appear in the region δ 0.8 to 1.0 (Fig. 22). The doublet at δ 0.97 ($J = 6.5$ Hz) is due to the methyl group of the first isoprene unit whereas overlapping doublets at δ 0.84 and 0.86 correspond to the remaining six methyl groups. It is worth emphasising that bands present in menaquinones and related compounds in the δ 1.5 to 2.0 region due to methylene and methyl groups adjacent to double bonds are not observed in caldariellaquinone due to the presence of a C$_{30}$ saturated side chain.

III. Taxonomic significance

Isoprenoid quinones are located in the cytoplasmic membranes of bacteria and play an important role in electron transport and active transport (see Kröger, 1977; Taber, 1980). Although the physiological importance of these compounds has been appreciated for a long time, it is only during the past few years that their importance as taxonomic criteria has been fully realised. It became evident during the early studies of Bishop, Crane, Page, and others (Lester and Crane, 1959; Page *et al.*, 1960; Bishop *et al.*, 1962; Gale *et al.*, 1964) that different bacteria not only synthesised different quinone types (e.g., MK, Q) but that the number of isoprene units in the multiprenyl side chain often differed among taxa. The first systematic demonstration of the value of isoprenoid quinones in microbial taxonomy was by Jeffries and associates during an examination of the menaquinone composition of some aerobic Gram-positive cocci (Jeffries *et al.*, 1967a,b,c). Since these pioneering studies, a larger number of systematic studies, most notably those of Yamada and colleagues (e.g., Yamada *et al.*, 1969, 1976a,b, 1977a,b,c, 1982a,b,c) and Collins and associates (e.g., Collins *et al.*, 1977, 1979a,b, 1980b, 1981, 1982a,b, 1983a,b,c; Collins and Jones, 1979a,b, 1980) have firmly established the value of isoprenoid quinones as taxonomic criteria. Isoprenoid quinone analyses have proved to be of considerable value in the taxonomy of many groups of Gram-positive [i.e., *Bacillus* (Watanuki and Aida, 1972; Collins and Jones, 1979a), *Listeria* (Collins *et al.*, 1979b), *Micrococcus* (Yamada *et al.*, 1976b), *Staphylococcus* (Yamada *et al.*, 1976a; Collins, 1981), *Streptococcus* (Collins and Jones, 1979b), *Thermoactinomyces* (Collins *et al.*, 1982b), actinomycetes and coryneforms (Collins *et al.*, 1977, 1979a, 1980b, 1982a, 1983a,b,c; Yamada *et al.*, 1976a, 1977a,b; Fischer *et al.*, 1983)] and Gram-negative [i.e., acetic acid bacteria (Yamada *et al.*, 1969), *Bacteroides* (Shah and Collins, 1980), *Campylobacter* (Carlone and Anet, 1983; Collins *et al.*, 1984), *Flavobacterium* (Callies and Mannheim, 1978), *Haemophilus* (Hollander and Mannheim, 1975), *Legionella* (Karr *et al.*, 1982; Collins and Gilbart, 1983), *Pasteurella* (Mannheim *et al.*, 1978), *Pseudomonas* (Yamada *et al.*, 1982c), phototrophic bacteria (Collins and Jones, 1981a), *Thiobacillus* (Katayama-Fujimura *et al.*, 1982)] eubacteria. Preliminary studies indicate that isoprenoid quinones may also be of value in the taxonomy of the archaebacteria (Collins *et al.*, 1981; Collins and Langworthy, 1983). The distribution of isoprenoid quinones in the prokaryotes, and their value as taxonomic criteria has been extensively reviewed by Collins and Jones (1981a).

It is worth re-emphasising that it is not only the distribution of the different classes of respiratory quinones (MK, Q, DMK) which is taxonomically significant. The inherent value of menaquinones, ubiquinones, and related compounds as taxonomic criteria lies basically in the variations in the length of their multiprenyl side chains (Collins and Jones, 1981a). To date, menaquinones and ubiquinones with side chains containing 5–14 and 7–14 isoprene units, respectively, have been found as principal components in bacteria. In the case of the menaquinone series, differing degrees of hydrogenation or saturation (1 to 5 double bonds saturated) of the multiprenyl side chain are also of taxonomic importance. Partially saturated menaquinones with side chains varying in length from 5 to 12 isoprene units have been reported as major components in bacteria (e.g., see Collins and Jones, 1981a; Collins *et al.*, 1983c). Preliminary studies also indicate the "position" of saturation in the multiprenyl side chain of hydrogenated menaquinones may be of taxonomic significance. For example, the hexa- and octahydrogenated menaquinones-9 in *Actinomadura* and *Streptomyces* have been shown to differ in their points of saturation (see Batrakov *et al.*, 1976; Batrakov and Bergelson, 1978; Yamada *et al.*, 1982a,b; Fischer *et al.*, 1983). Similarly, *Nocardia autotrophica* and *N. brasiliensis* have been reported to possess two "different" tetrahydrogenated menaquinones-8 (Kroppenstedt, 1984). Interestingly, although variations in the length (and, in the case of menaquinones, degree and position of hydrogenation) of the multiprenyl side chain are of taxonomic value, the biological significance of this variation is not known.

The primary function of the side chain is probably to bind the quinone in the membrane and/or to enzyme complexes during certain enzymatic reactions. For instance, a specific interaction between protein and ubiquinone has been demonstrated in succinate–ubiquinone reductase (Yu and Yu, 1982). It is therefore possible that the structure of the side chain may be critical in interactions/binding the quinone to protein/enzyme complexes (Collins and Gilbart, 1983; Collins, *et al.*, 1983c). Although the physiological significance of the side-chain variations is not as yet understood, this does not detract from the value of these molecules as taxonomic criteria. The wide range of side-chain structures exhibited by different taxonomic groups, together with their relative uniformity and stability within strains of the same taxon, indicate that the nature of the polyprenyl side tail is a genetic trait. Indeed, this structural variation between different taxonomic groups, and the relative ease with which isoprenoid quinones can now be isolated and characterised, make these molecules ideal chemotaxonomic markers.

References

Azerad, R., and Cyrot-Pelletier, M.-O. (1973). *Biochimie* **55**, 591–603.
Batrakov, S. G., and Bergelson, L. D. (1978). *Chem. Phys. Lipids.* **21**, 1–29.
Batrakov, S. G., Panosyan, A. G., Rosynov, B. V., Konova, I. V., and Bergelson, L. D. (1976). *Bioorg. Khim.* **2**, 1538–1546.
Beau, P. S., Azerad, R., and Lederer, E. (1966). *Bull. Soc. Chim. Biol.* **48**, 569–581.
Binkley, S. B., McKee, R. W., Thayer, S. A., and Doisy, E. A. (1940). *J. Biol. Chem.* **133**, 721–729.

Bishop, D. H. L., Pandya, K. P., and King, H. K. (1962). *Biochem. J.* **83**, 606–614.

Callies, E., and Mannheim, W. (1978). *Int. J. Syst. Bacteriol.* **28**, 14–19.

Carlone, G. M., and Anet, F. A. L. (1983). *J. Gen. Microbiol.* **129**, 3385–3395.

Collins, M. D. (1981). *FEMS Microbiol. Lett.* **12**, 83–85.

Collins, M. D. (1984). *In* "The Use of Chemotaxonomic Methods for Bacteria" (M. Goodfellow and D. E. Minnikin, Eds.) No. 20 in the Society for Applied Bacteriology Technical Series, No. 20. Academic Press, New York.

Collins, M. D., and Fernandez, F. (1984). *FEMS Microbiol. Lett.* **22**, 273–276.

Collins, M. D., and Gilbart, J. (1983). *FEMS Microbiol. Lett.* **16**, 251–255.

Collins, M. D., and Jones, D. (1979a). *J. Appl. Bacteriol.* **47**, 293–297.

Collins, M. D., and Jones, D. (1979b). *J. Gen. Microbiol.* **114**, 27–33.

Collins, M. D., and Jones, D. (1980). *J. Appl. Bacteriol.* **48**, 459–470.

Collins, M. D., and Jones, D. (1981a). *Microbiol. Rev.* **45**, 316–354.

Collins, M. D., and Jones, D. (1981b). *J. Appl. Bacteriol.* **51**, 129–134.

Collins, M. D., and Kroppenstedt, R. M. (1983). *Syst. Appl. Microbiol.* **4**, 95–104.

Collins, M. D., and Langworthy, T. A. (1983). *Syst. Appl. Microbiol.* **4**, 295–304.

Collins, M. D., and Widdel, F. (1984). *System. Appl. Microbiol.* **5**, 281–286.

Collins, M. D., Pirouz, T., Goodfellow, M., and Minnikin, D. E. (1977). *J. Gen. Microbiol.* **100**, 221–230.

Collins, M. D., Goodfellow, M., and Minnikin, D. E. (1979a). *J. Gen. Microbiol.* **110**, 127–136.

Collins, M. D., Jones, D., Goodfellow, M., and Minnikin, D. E. (1979b). *J. Gen. Microbiol.* **111**, 453–457.

Collins, M. D., Shah, H. N., and Minnikin, D. E. (1980a). *J. Appl. Bacteriol.* **48**, 277–282.

Collins, M. D., Goodfellow, M., and Minnikin, D. E. (1980b). *J. Gen. Microbiol.* **118**, 29–37.

Collins, M. D., Ross, H. N. M., Tindall, B. J., and Grant, W. D. (1981). *J. Appl. Bacteriol.* **50**, 559–565.

Collins, M. D., McCarthy, A. J., and Cross, T. (1982a). *Zentralbl. Bakteriol. Parasitenkd. Infektionkr. Hyg Abt. 1: Orig.* **C3**, 358–363.

Collins, M. D., MacKillop, G. C., and Cross, T. (1982b). *FEMS Microbiol. Lett.* **13**, 151–153.

Collins, M. D., Jones, D., and Kroppenstedt, R. M. (1983a). *Syst. Appl. Microbiol.* **4**, 65–78.

Collins, M. D., Jones, D., Keddie, R. M., Kroppenstedt, R. M., and Schleifer, K. H. (1983b). *Syst. Appl. Microbiol.* **4**, 236–252.

Collins, M. D., Faulkner, M., and Keddie, R. M. (1983c). *System. Appl. Microbiol.* **5**, 20–29.

Collins, M. D., Costas, M., and Owen, R. J. (1984). *Arch. Microbiol.* **137**, 168–170.

Crane, F. L. (1959). *Plant Physiol.* **34**, 546–551.

Crane, F. L. (1961). *In* "Ciba Foundation Symposium on Quinones in Electron Transport" (G. E. W. Wolstenholme and C. M. O'Connor, Eds.), pp. 36–75. Churchill, London.

De Rosa, M., De Rosa, S., Gambacorta, A., Minale, L., Thomson, R. H., and Worthington, R. D. (1977). *J. Chem. Soc. Perkin Trans.* I, 653–657.

Dunphy, P. J., Phillips, P. G., and Brodie, A. F. (1971). *J. Lipid Res.* **12**, 442–449.

Erickson, R. E., Shunk, C. H., Trenner, N. R., Arison, B. H., and Folkers, K. (1959). *J. Am. Chem. Soc.* **81**, 4999.

Fischer, A., Kroppenstedt, R. M., and Stackebrandt, E. (1983). *J. Gen. Microbiol.* **129**, 3433–3446.

Frydman, B., and Rapaport, H. (1963). *J. Am. Chem. Soc.* **85**, 823–825.

Gale, P. H., Arison, B. H., Trenner, N. R., Page, A. C., and Folkers, K. (1963). *Biochemistry* **2**, 200–203.

Gale, P. H., Erickson, R. E., Page, A. C., and Folkers, K. (1964). *Arch Biochem. Biophys.* **104**, 169–172.

Glover, J., and Threlfall, D. R. (1962). *Biochem. J.* **85**, 14P–15P.

Hollander, R., and Mannheim, W. (1975). *Int. J. Syst. Bacteriol.* **25**, 102–107.

Ishii, M., Kawasumi, T., Igarashi, T., Kodoma, T., and Minoda, Y. (1983). *Agric. Biol. Chem.* **47**, 167–169.

Isler, O., Ruegg, R., Chopart-dit-Jean, L. H., Winterstein, A., and Wiss, O. (1958). *Helv. Chim. Acta* **41**, 786–807.

Isler, O., Rüegg, R., Langemann, A., Schudel, P., Ryser, G., and Wursch, J. (1961). *In* "Ciba Foundation Symposium on Quinones in Electron Transport" (G. E. W. Wolstenholme and C. M. O'Connor, Eds.), pp. 79–96. Churchill, London.

Jeffries, L., Harris, M., and Price, S. A. (1967a). *Nature (London)* **216**, 808–809.

Jeffries, L., Cawthorne, M. A., Harris, M., Diplock, A. T., Green, J., and Price, S. A. (1967b). *Nature (London)* **215**, 257–259.

Jeffries, L., Cawthorne, M. A., Harris, M., Diplock, A. T., Green, J., and Price, S. A. (1967c). *Spisy Prirodoved. Fak. Univ. J. E. Pukyne Brne* **K40**, 230–235.

Karr, D. E., Bibb, W. F., and Moss, C. W. (1982). *J. Clin. Microbiol.* **15**, 1044–1048.

Katayama-Fujimura, Y., Tsuzaki, N., and Kuraishi, H. (1982). *J. Gen. Microbiol.* **128**, 1599–1611.

Kofler, M. (1946). *In* "Festschrift für Emil Christoff Barell," pp. 199–212. Hoffmann-LaRoche, Basel.

Kofler, M., Langemann, A., Rüegg, R., Gloor, U., Schweiter, U., Würsch, J., Wiss, O., and Isler, O. (1959). *Helv. Chim. Acta* **42**, 2252–2254.

Kröger, A. (1977). *In* "Microbial Energetics" (B. A. Haddock and W. A. Hamilton, Eds.), pp. 61–93. Cambridge Univ. Press, Cambridge.

Kroppenstedt, R. M. (1982a). *GIT Labor-Med.* **4**, 266–275.

Kroppenstedt, R. M. (1982b). *J. Liq. Chromatogr.* **5**, 2539–2569.

Kroppenstedt, R. M. (1984). *In* "The Use of Chemotaxonomic Methods for Bacteria," (M. Goodfellow and D. E. Minnikin, Eds.), Society for Applied Bacteriology Technical Series, No. 20. Academic Press, London.

Lester, R. L., and Crane, F. L. (1959). *J. Biol. Chem.* **234**, 2169–2175.

MacCorquodale, D. W., Binkley, S. B., McKee, R. W., Thayer, S. A., and Doisy, E. A. (1939a). *Proc. Soc. Exp. Biol. Med.* **40**, 482–483.

MacCorquodale, D. W., McKee, R. W., Binkley, S. B., Cheney, L. C., Holcomb, W. F., Thayer, S. A., and Doisy, E. A. (1939b). *J. Biol. Chem.* **131**, 357–370.

McKee, R. W., Binkley, S. B., Thayer, S. A., MacCorquodale, D. W., and Doisy, E. A. (1939). *J. Biol. Chem.* **131**, 327–344.

Mannheim, W., Stieler, W., Wolf, G., and Zabel, R. (1978). *Int. J. Syst. Bacteriol.* **28**, 7–13.

Moore, H. W., and Folkers, K. (1968). *J. Am. Chem. Soc.* **88**, 567–570.

Page, A. C., Gale, P., Wallick, H., Walton, R. B., McDaniel, L. E., Woodruff, H. B., and Folkers, K. (1960). *Arch. Biochem. Biophys.* **89**, 318–321.

Powls, R., Redfearn, E. R., and Trippett, S. (1968). *Biochem. Biophys. Res. Commun.* **33**, 408–411.

Redfearn, E. R., and Powls, R. (1968). *Biochem. J.* **106**, 50P.

Scholes, P. B., and King, H. K. (1964). *Biochem. J.* **91**, 9P.

Scholes, P. B., and King, H. K. (1965). *Biochem. J.* **97**, 766–768.

Shah, H. N., and Collins, M. D. (1980). *J. Appl. Bacteriol.* **48**, 75–87.

Sommer, P., and Kofler, M. (1966). *Vitam. Horm. (N. Y.)* **24**, 349–399.

Taber, H. (1980). *In* "Vitamin K and Vitamin K Dependent Proteins" (J. W. Suttie, ed.). University Park Press, Baltimore, Maryland.

Tamaoko, J., Katayama-Fujimura, Y., and Kuraishi, H. (1983). *J. Appl. Bacteriol.* **54,** 31–36.

Kuraishi, H. (1983). *J. Appl. Bacteriol.* **54,** 31–36.

Watanuki, M., and Aida, K. (1972). *J. Gen. Appl. Microbiol.* **18,** 469–472.

Yamada, Y., Aida, K., and Uemura, T. (1969). *J. Gen. Appl. Microbiol.* **15,** 181–196.

Yamada, Y., Inouye, G., Tahara, Y., and Kondo, K. (1976a). *J. Gen. Appl. Microbiol.* **22,** 203–214.

Yamada, Y., Inouye, G., Tahara, Y., and Kondo, K. (1976b). *J. Gen. Appl. Microbiol.* **22,** 227–236.

Yamada, Y., Ishikawa, T., Tahara, Y., and Kondo, K. (1977a). *J. Gen. Appl. Microbiol.* **23,** 207–216.

Yamada, Y., Yamashita, M., Tahara, Y., and Kondo, K. (1977b). *J. Gen. Appl. Microbiol.* **23,** 331–335.

Yamada, Y., Takinami, H., Tahara, Y., and Kondo, K. (1977c). *J. Gen. Appl. Microbiol.* **23,** 105–108.

Yamada, Y., Aoki, K., and Tahara, Y. (1982a). *J. Gen. Appl. Microbiol.* **28,** 321–329.

Yamada, Y., Hou, C. F., Sasaki, J., Tahara, Y., and Yoshioka, H. (1982b). *J. Gen. Appl. Microbiol.* **28,** 519–529.

Yamada, Y., Takinami-Nakamura, H., Tahara, Y., Oyaizu, H., and Komagata, K. (1982c). *J. Gen. Appl. Microbiol.* **28,** 7–12.

Yu, C.-A., and Yu, L. (1982). *J. Biol. Chem.* **257,** 6127–6131.

Index

in optical spectra, 300–301 (table)
amino acid sequencing, 313
function in respiration, 314
history, 285–286
oxidase test, 311–312
peroxidase test, 312–313
spectrophotometer design, 289–293,
    296–298, 304–305; *see*
    Spectrophotometers
spectrophotometry
  combination with other methods,
    309–310
  complex absorption spectra,
    numerical analysis, 306–309
  difference spectra of redox states,
    293–298
    at temperature 77 K, 295–298
  of extracted heme, 310–311
  ligand-bound form spectra, 298–303
  photodissociation spectra, 304–305
  at subzero temperatures, 305–306
  structure and nomenclature, 286–288
  taxonomic significance, 320–323
  X-ray diffraction, 313
Cytochrome oxidase a_1, function, 319
Cytochrome oxidase aa_3
  function, 318
  purification and structure, 317
Cytochrome oxidase d
  cyanide insensitivity, 318–319
  purification and function, 319
Cytochrome oxidase o, properties,
    318

D

Demethylmenaquinones, structure
    and distribution, 330–331
Desulfococcus multivorans,
    menaquinone MK-7, mass
    spectrum, 353
Diaminobenzoic acid, in DNA
    quantitation, 46
Dimethyl sulfoxide,
    lipopolysaccharide extraction
    with, 175
Diphenylamine, in DNA quantitation,
    45

DNA
  base composition
    biological and taxonomic
      significance, 29
    bromination, 19–21
    buoyant density centrifugation,
      5–6
    depurination, 21–23
    electrophoresis on
      polyacrylamide–agarose gels, 29
    high-performance liquid
      chromatography, 10–19
    thermal melting profile (T_m), 6–10
    thin-layer chromatography, 28
    UV absorbance
      after hydrolysis with perchloric
        acid, 24–27
      ratios at pH 3.0, 23–24
      ratios at pH 7.0, 27–28
  homology, biological and taxonomic
    significance, 71
  immobilized, sequence qualitative
    detection
    DNA-containing membrane
      preincubation, 65
    duplex thermal stability, 66
    labeled probes, 65
    reassociation, 65–66
  isolation
    for base composition study, 2–4
    hydroxylapatite method, 38–41
    Marmur method, 37–38
  reassociation rate, 54–70
    measurements
      optical, 57–59
      S_1 nuclease and hydroxylapatite
        procedures, 59–64
    melting temperature, effects of
      formamide, 55–56
      salt concentration, 54–55

E

Ectothiorhodospira halochloris
  absorption spectra of
    bacteriochlorophyll b, 266, 268
      (table)
    bacteriopheophytin b, 268 (table)

Sulfobulus acidocaldarius,
  caldariellaquinone, NMR
  spectrum, 361
Synechocystis, murein preparation
  from, 129

T

Temperature, subzero
 cytochrome spectrophotometry,
  295–298, 305–306
Thermodesulfotobacterium commune
  acid-stable ether lipids, 218–219,
  227
Thermoplasma, lipid composition,
  227, 228
Thermoplasma acidophilum,
  thermoplasmaquinone-7, NMR
  spectrum, 357
Thermoplasmaquinones
 NMR assay, 356–357, 359 (table)
 structure and distribution, 331, 332
Thin-layer chromatography
 carotenoids, 642
 DNA base composition, 28
 reverse phase partition, for
  isoprenoid quinones, 335–338
 on silicic acid, lipid separation,
  214–215
 silver ion, for isoprenoid quinones,
  344–345
 16 S rRNA oligonucleotide
  cataloguing, 78, 80–97
Thiobarbituric acid, in DNA
  quantitation, 45–46
Trichloroacetic acid
  lipopolysaccharide extraction
  with, 173

Tumoricidity, murein *in vivo,* 150–
151

U

Ubiquinones
 mass spectrometry, 350–351
 NMR assay, 360 (table)
 Q-10, UV spectrum, 349
 reverse-phase partition
  high-performance liquid
  chromatography, 338–340
 reverse-phase partition thin-layer
  chromatography, 336–337
 silver ion high-performance liquid
  chromatography, 346–347, 348
 structure and distribution, 332–333

V

Vitreoscilla stercoraria, ubiquinones
 silver ion high-performance liquid
 chromatography, 348

X

Xanthomonadins, distribution in
 Xanthomonads, 249–250

Y

Yersinia pestis, 16 S rRNA
 fingerprint, 76–77

Z

Zymomonas mobilis, hopanes, 223

Contents of Previous Volumes

T. L. Pitt and *Y. J. Erdman*. Serological typing of *Serratia marcescens*

R. Sakazaki. Serological typing of *Edwardsiella tarda*

M. B. Slade and *A. I. Tiffen*. Biochemical and serological characterization of *Erwinia*

G. Kapperud and *T. Bergan*. Biochemical and serological characterization of *Yersinia enterocolitica*

T. Bergan and *K. Sirheim*. Gas-liquid chromatography for the assay of fatty acid composition in gram-negative bacilli as an aid to classification

Volume 16

J. Stringer. Phage typing of *Streptococcus agalactiae*

M. J. Corbel. Phage typing of *Brucella*

T. Tsuchiya, M. Taguchi, Y. Fukazawa and *T. Shinoda*. Serological characterization of yeasts as an aid in identification and classification

M. Popoff and *R. Lallier*. Biochemical and serological characteristics of *Aeromonas*

E. I. Garvie. The separation of species of the Genus Leuconostoc and the differentiation of the Leuconostocs from other lactic acid bacteria

T. Bergan, R. Solberg and *O. Solberg*. Fatty acid and carbohydrate cell composition in Pediococci and Aerobocci identification of related species

A. Bauernfeind. Epidemiological typing of *Klebsiella* by Bacteriocins

U. Berger. Serology of Non-Gonococcal, Non-Meningococcal *Neisserieae*

G. R. Carter. Serotyping of *Pasteurella Multocida*

R. Sakazaki. Serology and epidemiology of *Plesiomonas shigelloides*

R. Sakazaki and *T. J. Donovan*. Serology and epidemiology of *Vibrio cholerae* and *Vibrio mimicus*

R. C. W. Berkeley, N. A. Logan, L. A. Shute and *A. C. Capey*. Identification of *Bacillus* species

H. Gyllenberg. Automated identification of bacteria. An overview and examples

K. Holmberg and *C. E. Nord*. Application of numerical taxonomy to the classification and identification of Microaerophilic Actinomycetes

K. B. Doving. Stimulus space in olfaction

Volume 17

P. M. Bennett and *J. Grinsted*. Introduction

V. A. Stanisich. Identification and analysis of plasmids at the genetic level

N. Willetts. Conjugation

J. R. Saunders, A. Docherty and *G. O. Humphreys*. Transformation of bacteria by plasmid DNA

L. Caro, G. Churchward, and *M. Chandler*. Study of plasmid replication *in vivo*

J. Grinsted and *P. M. Bennett*. Isolation and purification of plasmid DNA

H. J. Burkardt and *A. Pühler*. Electron microscopy of plasmid DNA

W. P. Diver and *J. Grinsted*. Use of restriction endonucleases

C. M. Thomas. Analysis of clones

T. J. Foster. Analysis of plasmids with transposons

P. M. Bennett. Detection of transposable elements on plasmids

G. Doughan and *M. Kehoe*. The minicell system as a method for studying expression from plasmid DNA

N. L. Brown. DNA sequencing